Creating Customer Focused Organizations

About the Author

Brian Dickinson is the founder and president of Logical Conclusions, Inc., a training and consulting corporation founded in 1981 and based in Lake Tahoe, California, with a branch office in Blackpool, England. Logical Conclusion's extensive client list includes many government agencies and major corporations throughout the world. Brian is also an active participant in the Citizen Ambassador Program and has represented the United States in China and Russia.

Brian has over 30 years of experience in teaching and consulting in Business Systems and all areas of Data Processing. He started his career in England with British Aircraft Corporation and then worked internationally as an independent consultant. In 1975 he moved to the United States and was one of the handful of people who originally helped spread the system engineering message across the United States. In this period Brian published papers and books, developed courses, and lead seminars on the emerging System Engineering techniques. Since then, he has consulted and taught globally on the subjects of Software and Information Engineering, Project Management, and Business Engineering. He is also a popular speaker at many national and international business and software conferences.

In 1979 Brian wrote his first book on the subject of System Development Methodologies which was many years ahead of its time. This work was titled ***Developing Structured Systems: A Methodology Using Structured Techniques***. It was translated into Japanese and it proved to be highly successful. Brian's next book was titled ***Developing Quality Systems***. This work was published by McGraw Hill in 1989 as part of their Software Engineering Series. After writing ***Developing Quality Systems***, Brian extended his focus into the area of Business Engineering and published ***Strategic Business Engineering: A Synergy of Software Engineering and Information Engineering*** in 1992.

To this newest publication, ***Creating Customer Focused Organizations*** (formely titled ***Risk-Free Business Re-Engineering***), he brings the perspectives and insights of almost twenty years of experience in working with and implementing methodologies and new methods for successful organizations.

Additional Titles by Brian Dickinson

- *Developing Structured Systems: A Methodology Using Structured Techniques*
- *Developing Quality Systems*

Books in progress by the same author in the Business Engineering Methodology Series:

- *Book II — The Business Engineering Methodology*
- *Book III — Logical Business Engineering Analysis*

The Logical Conclusions, Inc.
Business Engineering Methodology Series

Creating Customer Focused Organizations

by Brian Dickinson

A Practical Handbook for Business People & System Developers

LCI Press
Kings Beach, CA
A wholly owned subsidiary of Logical Conclusions, Inc.

Copyrights and Trademarks

Published by LCI Press
(a wholly owned subsidiary of Logical Conclusions, Inc.)
P.O. Box 699, Kings Beach, CA 96143.

Cover Design: Brian Dickinson & Jeff Sparksworthy

ISBN Number: 0-9629276-3-5

First Edition
Produced and Printed in the United States of America

The publisher offers a discount when you order this book in bulk quantities or for educational establishments.

For information on Logical Conclusions' Seminars and Consulting Services in the fields of:

- Business Engineering and Customer Focused Engineering,
- Business Process Analysis and Business Information Analysis,
- Computer Systems Design, and
- Project Management based on the Business Event Engineering Methodology

contact:

Logical Conclusions, Inc.
800 645-2226
530 546-2546
http://www.logical-inc.com

Acknowledgments

There are many events that shape a person's knowledge. For me these events included my world travels, a series of books, a select group of individuals, and, more recently, the physical relocation and restructuring of my own organization.

- My travels took me to many countries and into their cities and villages around the world and gave me the opportunity to see and compare their diverse cultures and political systems.
- Many influential books seemed to be presented to me at appropriate times along these travels (such as metaphysical books while in India and scientific and sociological books while in the United States).
- In addition, certain individuals had a profound influence such as school teachers who made learning a pleasure and brilliant business colleagues with whom I had the good fortune to work.
- A few years ago I decided to change the structure of my own organization from a formal office in San Francisco with desks, equipment, an 8-to-5 staff, etc. to a "virtual" office in a small, rural area high in the mountains above Lake Tahoe. I get to contemplate my navel a lot more here without the typical business environment getting in the way. This new structure and location allowed me to "stand six feet back" and see "systems" for what they are — something quite different from what I saw when I was immersed in the day-to-day operations under the old structure.

All of these contributed in some way to the contents of this book. However, perhaps the greatest factor influencing my business concepts was my being at the right place at the right time. In the late 1970s I joined a New York company headed by Ed Yourdon. He and many of the consultants there such as Tom DeMarco and Steve McMenamin had a great influence on my way of thinking about systems. It was Ed Yourdon who gave me the opportunity to write my first book in the late 1970s.

In the early 1980s I formed Logical Conclusions, Inc. and attracted some great people with whom I had long and fruitful debates. These people included III (yes, that's his name), Colt Rymer, Doug Brown, and Ed Barstad. To all of these folks I owe thanks.

I would also like to thank Jeff Sparksworthy of PubTechs Consulting Services for his wordsmithing skills while writing this book and for having helped me find the right words with the nuances I needed to put over some oftentimes difficult concepts.

I have been teaching the ideas in this book for over ten years and I have had the advantage of running them past thousands of students. Fortunately, they have been remarkably well received, and so, here I would like to thank all my students who helped me shape my ideas, with special thanks to the select few who believe in grasping these ideas and championing them in their workplace, invariably against all kinds of resistance to change.

Brian Dickinson
January, 1998
Lake Tahoe, California

Dedicated to

Dawana

Preface

In the formative years of an organization keeping the business alive is a major issue. Typically, you don't have the capital to invent a radical new design of the business. So, it's most likely you will form the same types of departments, jobs and tasks that every other organization has (e.g., Accounting, Stock Control, Marketing). Also, you are likely to purchase off-the-shelf software to support these departments (e.g., Word Processors, Spreadsheets, Accounts Receivable/Accounts Payable, General Ledger, Invoicing packages).

Once your organization is established and you know you're not going under, then you may find that your organization can expend some effort in cleaning up and improving your existing processes through methods such as Total Quality Management (TQM). Over the years a few engineered systems may get built and some integration of the now "legacy" systems may take place. You may even invest in a centralized database and supporting software to help organize all the disjoint and redundant data in file cabinets and isolated computer files that reside in separate departments and computer systems. However, there comes a time when the only improvements seem to come from upgrading to the next wave of systems technology. By "systems technology" I mean both human- and computer-based systems (e.g., human "empowered" teams and/or computer hardware).

This is the stage where you need to rethink the whole organization from the point of view of your customers, whomever or whatever they are. Depending on your organization's Mission Statement, a customer may be a person, an organization, a computer system, an external device, the environment, etc. The key thing to note is the customer is always outside the organization. This means there are no "internal customers." An organization's internal management structures and systems are always aspects of the organization's last design and its implementations.

In our organization we need to look at ("dis-cover") our true business and conduct what I like to call Customer Focused Engineering. This involves, first of all, looking at the organization's business processes and ignoring any past human and computer implementations.

Everything You See Is a Design

In every organization everything you "see" is a design and its implementation — not the actual business.

"Everything you see is a design" because every human- or computer-based system is an "implementation" of requirements. This implementation is derived from either human or nature's designs.

I guarantee that 100 years from now, everyone in your organization will be gone and that any software system in place today won't be there either, but the same business can still be running. What is it that's running when you remove the human beings and computer systems? The "business" is what's running.

The problem is, you never "see" the business, just the implementation of the business. You can't "see" the business because you can't see an analysis. You can see a design, but you can't see an analysis. However, to create a Customer Focused Organization, you have to be able to obtain a true business view that comes from conducting business analysis. As in any engineering profession, we need to use a model to "see" the business. When we utilize a model, we want to be careful not to fall into the trap of modeling the old or new design, but to model the underlying business itself.

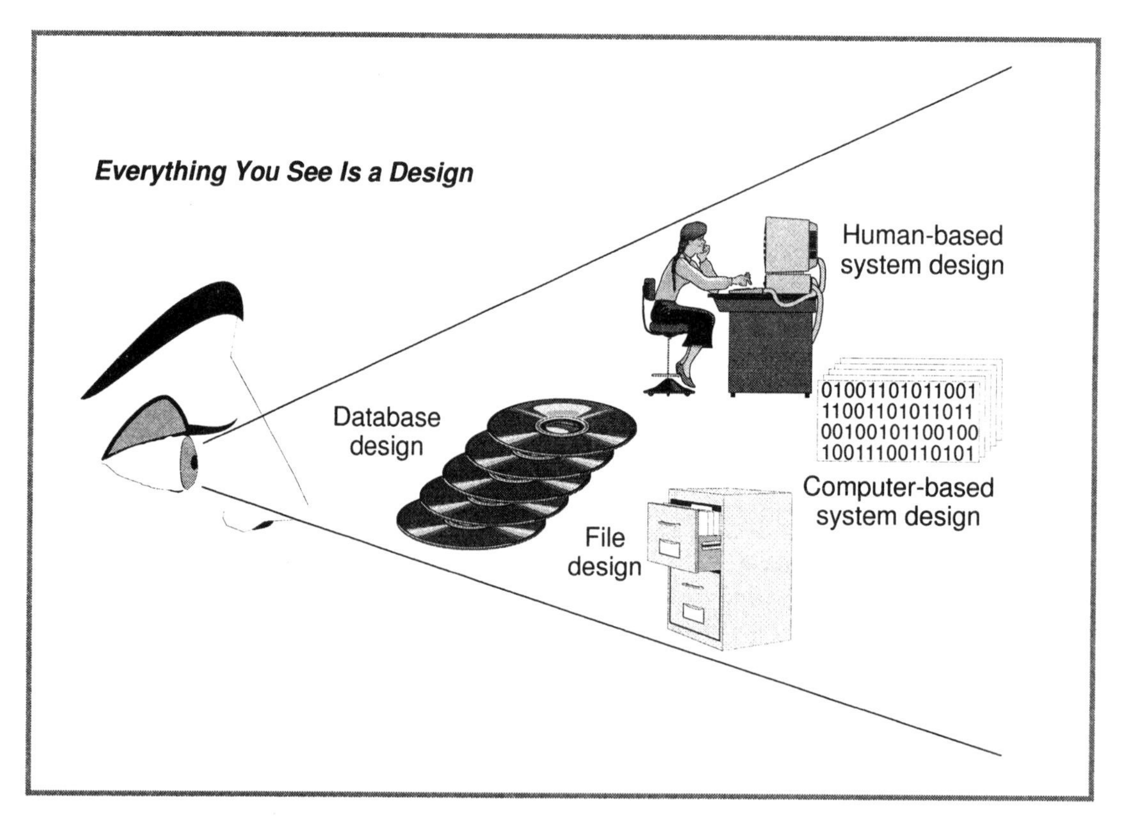

Based on over three decades of observing organizations I have found that the vast majority of them have simply evolved to where they are today. This evolution was typically without any conscious strategic "engineering" method applied to past growth and, as new systems were developed, it was typically without performing an analysis of the business issues required to satisfy the customer. So, the vast majority of organizations run with a disjointed collection of human and computer systems in place based on historical factors. Of course, it was necessary to go through this "do-what-everyone-else-did" evolution to get to where we are today. Any organization that finds itself in this position will have a whole closet full of historical/hysterical, systemic (design) limitations in place. Some organizations have a long history of designs to overcome (for example, the government of any country).

Using the concepts in this book we can analyze any historically/hysterically evolved organization to realize significant improvements. In the vast majority of cases, taking on the task of producing a Customer Focused System (or entire organization) gives us the opportunity to finally do it right the second (or Nth) time.

Removing Technical/Systemic Obstacles

Satisfying the Customer should be the central "reason for being" of any organization by using its human and automated systems to implement responses to customer requests and needs. We find, however, organizations where the focus seems to be on satisfying their internal hierarchies and giving the demands of their installed systems a priority over the needs of their customers.

This myopic focus on the needs of the "systems" is the result of a "warped" view that I call the "Clothes Have No Emperor" syndrome in which the old systems get perpetuated by things such as the old process of budgeting by department/divisions. As customers we sometimes wonder why organizations send us from department to department or have us interact with disjointed computer systems with no continuity of data between them. When working within an organization, we justify these customer inconveniences; however, when we are the customer, they seem silly.

> For example, as a "Trekkie", while I'm watching the TV and movie series Star Trek®, the crew oftentimes goes to the transporter room to "beam" themselves and their equipment from one place to another. At other times they show that it's unnecessary to go to the transporter room by "beaming" from anywhere on the ship to any other location. So, why do they need to go to the transporter room on the Enterprise anyway?" Obviously, it provides a way to cut the scene or to provide some suspenseful situation based on whether they will make it to the transporter room on time. Going to the transporter room is a leftover from the past — just like the fact that we don't need to go to our back yards to pump water from the well anymore. By the same token, the crew of the Enterprise doesn't really need to go to the transporter room anymore.

Familiarity Gets in Our Way

When one has lived with a human procedure or a computer system for a long time it's easy to believe that it is essential to the business without questioning whether it is an aspect of the old design. We all fall into this trap of believing the systems we "see" in an organization are the real business. However, they never are. Every structure and supporting system in the organization should be designed and operated so that it is part of a Customer Focused solution and not a high tech, reworked version of an old problem.

The workflow paths of the systems and processes and the data needed should be the shortest distance between the two points of the customer's initial request and the satisfaction of that request.[1]

When we have the opportunity to conduct a new project, we must avoid perpetuating a non-Customer Focused view within an organization by performing a genuine analysis to discover our business processes. Using the results of this analysis, we can then apply the available technology of today (i.e., carbon-based and silicon-based units) to produce new systems based on our analyzed customer-focused business processes. The new designs we create from this perspective will typically form design structures that are the opposite of what we see in place in a typical organization.

> For example, in our new designs we should see no such thing as an Accounting System (Accounts Payable, Accounts Receivable, General Ledger), Order Entry, Stock Control, or Invoicing Systems. Similarly, there should be no Edit, Update, Print or Input, Process, Output programs in our new computer systems. All these are leftovers from old manual/data processing environments.

Using Customer Focused business processes as the basis for your new designs gives you a logical vantage point from which you can create unbeatable, seamless customer-satisfaction compartments based on what I call Business Events. Orienting your organization's structures and systems around Business Events (rather than on departments or bureaus) provides a truly rational basis for creating a Customer Focused Organization.

Older and larger organizations face more obstacles in overcoming inertia to change. Some of the major obstacles to creating a Customer Focused Organization are the old human and computer system boundaries. These boundaries will also be the organizations' biggest obstacles to customer satisfaction. Re-vamping or eliminating these boundaries will be the major sources of improvements when you build a Customer Focused Organization. In this book I refer to these boundaries as "partitions." By "partition" I mean how an organization is divided. These include human boundaries such as jobs, departments, divisions, bureaus (with their associated "politics") and computer system partitions such as macros, programs, and systems. Many computer systems were built in a haphazard manner in response to the latest fad (hardware platform, computer language, system development methodology, etc.), or which departmental manager carried the most budgetary clout. This type of partitioning is what I call "dysfunctional partitioning" because it actually interferes with meeting a customer's needs.

1 "For an organization to function well, information must flow along the most efficient channel, whatever that may be." — David Packard

Significant and real business improvements will best be seen when an organization takes a complete look at its business and understands how it effectively responds to customers' requests coming from outside the organization.

I hope that the preceding "key point" sounds obvious. However, the older an organization is, the less responsive it will have become to its customers. What I hope you'll get out of this book is a clear understanding of what "functional partitioning" really is and the tools to fundamentally improve the organization itself by focusing on real business issues.

A Rose by Any Other Name

I don't care whether you call the process of organizational change "System Development," "Business Process Re-Engineering," "Organizational Restructuring," "Total Quality Management," "Workflow," "Team Building," or "Creating Customer Focused Organizations." Regardless of the name you give it, the goal should be to satisfy whomever or whatever you call the customer.

Since my first book in the late 1970's the basic idea I have taught has been to look at the old design and determine when the design gets in the way of satisfying the customer's needs. This book contains a methodology that advocates the best way to structure an organization without the design getting in the way.

In the draft copy of this book, I used the title ***Risk-Free Business Re-Engineering***. However, the word "Re-Engineering" implied that you already had an engineered system in place — something that I rarely found. However, the concept of Re-Engineering does have the goal to dramatically improve an organization's systems, something which is in common with the concepts of creating Customer Focused Organizations.

Unfortunately, the term Re-Engineering has been misused to describe downsizing, outsourcing, cutting middle management, or implementing the latest computer or human-based technologies, and not to improve customer service or to increase product quality. It doesn't matter what you call this methodology as long as your focus is on satisfying the customer's needs first and not the needs of the organization and its systems.

Growth through Customer Satisfaction

We've already seen that customers will migrate to the organization which gives them quality products and services, even if the organization is across the country or in another nation. We've also seen that an organization will outsource part of its business if it can be done more cost-effectively by another organization. I assert that by focusing on satisfying your customer's needs, your organization will automatically be more productive and hence will need to expand its human and technological resources to meet the expanding customer demand. So, the ultimate goal of the methodology in this book is organizational growth.

This does not mean jobs won't change as part of the move to become a Customer-Focused Organization. We may re-train staff or repartition management tasks and bring in new technology, but this is what keeps organizations alive. Trying to avoid this constant (albeit, at times uncomfortable) improvement can have a catastrophic cost. For example, does anyone want to buy stock in slide rules?

Creating Customer Focused systems in your organization also should ultimately improve the quality of life — not only from a customer's point of view, but also from both the employer's and employees' points of view while ensuring the longevity of the organization. Depending on the organizations size, age, and traditions, the creation of a Customer Focused Organization is a strategic project that can take months and even years to fully implement. These ideas are not the quick fix that will make you the hero in the next quarterly report or necessarily this political term. History has taught us that there will always be another organization (or country) that's willing to take the long-term view and do it right.

How this Book Is Organized

This book provides the conceptual tools you will need to create a Customer Focused Organization. It teaches how to understand **what** your organization does and how to conduct Customer Focused Engineering. Also, you will understand how to better implement the organization's engineered systems using an engineering methodology and discipline.

This book's chapters are organized into a series of building blocks to help you incrementally absorb its concepts.

- The initial chapter ***Evolving to a Customer Focused Organization*** discusses why we need to engineer the organization's systems and structures in the first place. This chapter lays out the goals for a Customer Focused Organization and some of the potential difficulties in obtaining them.
- The next chapter ***Creating a Customer Focused Organization via Strategic Planning*** briefly introduces some of the strategic planning issues to consider in creating a Customer Focused Organization. This discussion is important because if we begin a Customer Focused Engineering effort without a plan, it's almost certainly doomed to failure.
- The ***Systems Archaeology*** chapter focuses on how systems were designed and developed in the past, and how these designs are still haunting us today. This chapter exposes the mis-guided reasons for creating non-Customer Focused systems in the first place.
- The chapter titled ***Understanding the Nature of Systems*** describes the fundamental characteristics of every system. This understanding is needed before we try to analyze a system and convey the results using an abstract model.
- ***The Model IS the Business*** chapter discusses why and how we model systems. This chapter provides examples of using graphical tools to analyze and model the organization's business. This knowledge is critical when we select the set of models that are most appropriate for understanding our particular organization.
- The next chapter ***Our Event Horizon — The Boundary of Our Organization*** defines what a Business Event is and how we model it in our organization to be able to create successfully a Customer Focused Organization.

- The ***Recognizing the Five Types of Events that Stimulate Our Organization*** chapter shows how to differentiate between the five types of Events to which our organization responds. It also shows how the differences between Event types dictate which model they affect and how to model them correctly.
- The chapter on ***Partitioning by Business Events*** describes what we need to know to be able to functionally partition our business. We will use the concept of Business Events to partition our organization's model along Customer Focused lines (as opposed to design or implementation lines). This is important because our initial partitioning of our model will dictate all subsequent classifications, and will probably affect any future engineered systems' designs.
- ***Unfragmenting Events from Old Design Traps*** explains how we can find and integrate any Business Model Events into our new partitions. Once we break these fragmented Events free of their old partitioned contexts, we can conserve their data and processes and maximize our organization's effectiveness.
- In the chapter ***Achieving Organizational Process and Data Integrity*** we build on the knowledge from the previous chapters and talk about how we gain some incredible advantages by using engineered Business Event Partitioning. These advantages include such bonuses as Process Integrity and Data Conservation. These benefits are only available if we have an engineered Business Event Methodology to model and manage the organization.
- By this stage the hard part (the analysis) is over. The next part is inventing a new design, and in the chapter, ***Designing and Implementing Business Event Systems***, we will see how our new designs will be based on business (not technology) issues. This will lead to "seamless" business systems that give ultimate customer satisfaction, that are easily understood, that are easy to change, and that are ultimately more cost effective.
- After that is a discussion titled, ***Strategic Planning via the Business Event Methodology***. This chapter focuses on the organization's long term survival. We revisit the Strategic Planning issues given our new found knowledge of the Business Event Methodology.
- Then there's a short chapter called ***The Logical Conclusion — The Final Analysis*** that shows how we can now satisfy the goals of creating Customer Focused Organizations using the Business Engineering Methodology.
- At the end of the book are appendices that contain a ***Glossary***, a ***Bibliography***, and an ***Index***.

Summary

In this book I wish to pass on the knowledge and insights I have gained in over 30 years experience. It is my belief this book covers topics of vital concern to the long-term success of all organizations (in both public and private sectors). Albert Einstein said: "No problem can be solved by the consciousness that created it." I hope the ideas in this book will change your consciousness and that you will become one of the "champions" in your organization to improve the quality of life for all of us. Remember, we are all customers.

Table Of Contents

Preface i

Everything You See Is a Design iii
Removing Technical/Systemic Obstacles v
Familiarity Gets in Our Way v
A Rose by Any Other Name vii
Growth through Customer Satisfaction vii
How this Book Is Organized viii
Summary ix

Introduction 1

Conducting Pre-Engineering 1
Conducting Business Engineering 2
Conducting System Engineering 2
Quality through Engineering 3
What Is "Engineering?" 3
The Cost of Poor Quality 3
The Cost of Quality 4
Building In Quality 4
Quality Using Engineering Methods & Models 6
A Word about Words 7
Objectives vs. Requirements 8
A Word about Typographical Conventions 10

Evolving to a Customer Focused Organization 11

The Goals for Creating a Customer Focused Organization 12
Obstacles to Creating a Customer Focused Organization 13
Technologically Dependent Systems Obstacles 15
The "Clothes Have No Emperor" Obstacle 16
Outsourcing/Downsizing Obstacles 17
Not Acknowledging the People Issues Obstacle 18
The "Stage of Evolution" Obstacle 19
Old Management Practice Obstacles 21
Overcoming the Obstacles 22
Take Care of the "Three Ps" 22
Be Accountable for Each Deliverable 23
Summary 24

Creating a Customer Focused Organization via Strategic Planning 25
A Basis for Strategic Planning 27
Providing a Competitive Edge 27
A Strategic Business Planning Approach 29
The Key Organizational Questions 29
1. What Are Our Purpose and Mission? 29
2. What Is Our True Business Today? 30
3. Do Our Current Systems Support/Not Support Us? 30
4. Where Do We Want to Be in the Future? 31
5. How Do We Get to Where We Want to Be? 31
Summary 31

Systems Archaeology 33
Knowing Who's the Prisoner 34
What Obscures the Essential Business? 35
What Is Functional Partitioning? 37
"The Designer Made Me Do It" 41
Wrong Turn #1 — "People" Partitioning in Manual Systems 42
Wrong Turn #2 — Computer Partitioning of Automated Systems 44
Wrong Turn #3 — Program Partitioning in Automated Systems . 47
Wrong Turn #4 — Data File Partitioning in Automated Systems . 49
Wrong Turn #5 — File Partitioning in Manual Systems 50
Wrong Turn #6 — After-the-fact Quality Control 51
Summary 53

Understanding the Nature of Systems 55
Business Issues as Conceptualized by Humans 56
The Stimulus-Response Nature of Systems 57
The Process-Memory Nature of Systems 59
Summary 61

The Model ***IS*** the Business 63
Effective System Modeling 64
The Characteristics of an Effective Model 67
Models for Analysis 68

Process Oriented Models 70
Flow Charts 70
Functional Decomposition Diagrams 71
Process Hierarchy Diagrams 72
Information/Data Oriented Models 73
Entity Relationship Diagrams 73
Hierarchical/Network/Relational Data Oriented Models 74
Process and Information/Data Oriented Models 75
Data Flow Diagrams 75
Object Oriented Models 78
Control Oriented Models 79
State Transition Diagrams 79
Control Flow Diagrams 80
Choosing the Right Model 80
Example of Business/Analysis and System/Design Models 81
Summary 85

Our Event Horizon – The Boundary of Our Organization 87

The Foundation of the Business Event Methodology 88
A Little History of the Concept of Event Partitioning 89
The Business Event Methodology and Its Players 90
The Types of Organizational Events 92
Identifying Organizational Events via where They Occur 93
Summary 95

Recognizing the Five Types of Events that Stimulate Our Organization 97

Strategic Events 97
Definition of a Strategic Event 97
Recognizing Strategic Events 98
Don't Model the Modeler 99
System Customers and System Events 100
System Events 101
Definition of a System Event 101
Recognizing System Events 101
Business Events 103
Definition of a Business Event 103
Recognizing Business Events 103

Business Customers and Business Events 105
Business Event Naming 105
Regulatory Events 106
Definition of a Regulatory Event 106
Recognizing Regulatory Events 107
Dependent Events 108
Definition of a Dependent Event 108
Recognizing Dependent Events 108
The Business Model Event List 110
Unmasking Events 111
Summary 112

Partitioning by Business Events 115

Definition of a Business Event Partition 116
What Constitutes a Business Event Partition 118
The Business Event Source 119
Naming the Business Event Source 119
The Business Event Stimulus 120
Data-oriented Stimuli 121
Naming Data-oriented Stimuli 123
Material-oriented Stimuli 123
Naming Material-oriented Stimuli 124
Control-oriented Stimuli 124
Naming Control-oriented Stimuli 125
The Business Event Processing 126
Naming the Business Event Processing 127
The Business Event Memory 127
Beware of Design/Convenience Stores 128
Why We Need Stored Data 131
Logical Stores — the Concept of Business Event Memory ... 132
Naming the Business Event Memory 134
The Business Event Response 135
Naming the Business Event Response 135
The Business Event Recipient 135
Naming the Business Event Recipient 136
The Beginnings of the Business Library 137
Summary 138

Unfragmenting Events from Old Design Traps 139

Beware of Fragmented Events (Design Trap #1) 141
Transaction Types and Fragmented Events 142
Beware of Bundled Events (Design Trap #2) 144
Beware of Historical Events (Design Trap #3) 146
Beware of Fragmented & Bundled Data Stores (Design Trap #4) 147
Beware of Fragmented Stores 149
Beware of Bundled Data 149
Using Business Events to Drive Associated Regulatory and Dependent Events. 150
Dynamically Defined Business Events 150
Summary 153

The Detailed Business Event Specification. 155

Subpartitioning of Processing 156
The Reasons for Subpartitioning Processing 160
Subpartitioning Business Processes 163
Subpartitioning of Memory 166
The Reasons for Subpartitioning Memory 166
Repartitioning Business Event Memory 167
The Progression of Analysis Models and Their Levels. 172
The Business Library — Documenting the Analysis Business Event Specification 174
Managing the Library via the Business Library Conservator 175
Summary 175

Achieving Organizational Process and Data Integrity 177

Determining the Potential for Reusability 178
Reusability Incentives 180
Obtaining Process Integrity (Reusability) 181
The Business Model Event/Reusable Process Matrix 181
Process Reusability Is All in the Name 183
Obtaining Data Integrity (Data Conservation) 183
Maintaining "Meta Data" 185
Data Reusability Is All in the Name 186
A Note about Data Ownership 186
The Business Model Event/Data Element Matrix 187
The Business Model Event/Data Entity Matrix 191

The Business Model Event/Relationship Matrix 192
The Business Model Event/Object and Method Matrix 193
The Business Model Event/Engineered System Matrix 194
Saving System Development Time/Investment 195
Summary 196

Designing and Implementing Business Event Systems 197

System Support Issues 198
Business Event-Driven Systems Design 201
Customer Focused Engineering via Human Systems 201
Customer Focused Engineering via Computer Systems 203
The Benefits of Customer Focused Engineering Computer Systems 203
Business Event Driven Design 205
Three Design Scenarios for a Business Event 207
The Worst Case (or I'd Rather Walk) 207
The Current Case (or, Why's my luggage in Rangoon when I'm in LA?) 209
The Future Case (or "You Want It? You Get It!" Airline) 210
The Complete Business Event Specification — An Example 213
Sample Business Event Analysis Specification: Data on the Move 214
Sample Business Event Analysis Specification: Data at Rest 214
The Beginnings of Design: Data on the Move 215
The Beginnings of Design: Data at Rest 216
Summary 219

Strategic Planning via Business Event Partitioning . . . 221

The Key Organizational Questions 222
1. What Are Our Purpose and Mission? 222
2. What Is the Set of Business Events We Respond to Today? . . . 223
3. Do Our Current Systems Support/Not Support these Business Events Seamlessly? 223
4. What New Business Events Do We Want to Support? 225
Using the Business Model to Look to the Future 225
5. How Do We Become a Customer Focused Organization from where We Are Today? 228
Summary 231

A Logical Conclusion — The Final Analysis 233
Customer Focused Engineering's Objectives and their Conclusions 233
Championing Customer Focused Organizations. 236
Glossary A-1
Bibliography. B-1
Index C-1
The Author as a Stimulus-Response System

Introduction

One way to create a Customer Focused Organization is to take an old organization and try to make it customer focused using empowered teams or putting in place "customer friendly" technology. With dedicated individuals you will probably have some degree of success. However, my experience has been, that to realize the full potential of creating a Customer Focused Organization, you need to conduct a genuine engineering effort on the entire organization and its systems. This is what I call Customer Focused Engineering (CFE).

This organization-wide engineering effort is no different than engineering a custom home in that we must first define our requirements, then invent a design solution to these requirements, and finally implement/construct the house with built-in quality.

I would like to describe three flavors of organizational engineering effort:

- Pre-Engineering
- Business Engineering
- System Engineering

Let me briefly define these terms.

Conducting Pre-Engineering

If we started a business from scratch, and we didn't want to base it upon the same model used by existing organizations in our business arena, we would conduct what I call Pre-Engineering.[1] (This is like buying a vacant lot, defining our house living needs, drawing blueprint designs, and building the house.) This is where we can invent our business rules and then place a design around these rules that is seamless from the customer's point of view. This seamless Customer Focused concept runs throughout this book.

It's easier to Pre-Engineer a non-existing (new) organization than to re-focus/Re-Engineer an existing organization because there are no systems in place that obscure the essential business. There also are no political or institutional roadblocks to overcome such as historical/hysterical partitioning or archaeological wrong turns.

Another benefit of a Pre-Engineering project is we can perfect the entire organization at once so nothing is "outside" the scope of study. In this mode we have no existing design issues to accommodate in our business modeling. Also, we do not have to be concerned with taking pains to ensure the project does not impact the day-to-day operations of existing systems.

1 Thanks to my editor, Jeff Sparksworthy, for providing a name for this concept.

We will still need to be aware that if we are tasked with a Pre-Engineering project and have been taught "standard business practices" while in school (e.g., "every organization needs an Accounts Department, a Personnel Department"), then we would be pre-disposed to use these "standard practices" and take the same archaeological wrong turns (listed in Chapter Three). In this case, we wouldn't be conducting true Pre-Engineering at all.

Of course, the downside of Pre-Engineering is when starting to design a business from scratch, we have no existing systems from which to "dis-cover" the true business rules.

Pre-Engineering reminds me of the message I gave to my audiences on my trips to Russia and China with the People to People Citizen's Ambassador Program. This message was you don't have to go through the last twenty to thirty years' of business evolution to get to today's best engineered implementation of a business. You can leapfrog and start today without historical partitioning or pre-conceived notions of what constitutes a computer or human-based system.

Conducting Business Engineering

Business Engineering involves producing a Customer Focused Organization starting with an established organization. Established organizations have legacy systems in place and departments and other boundaries — all of which will get in our way when conducting Customer Focused Engineering.

> This is similar to buying a Victorian house and substantially remodeling it to fit our requirements today. In this example, we need to "dis-cover" the business rules that are in place and this is the most time-consuming part of creating a Customer Focused Organization.

In Business Engineering we invent our new design and create a strategic plan to take us from where we are today (with our legacy systems) to where we want to be (with Customer Focused Systems). Later on in this book I introduce two strategies to do this.

Business Engineering resembles Pre-Engineering in that the scope of change is the entire organization.

Conducting System Engineering

System Engineering is the process of taking each existing system within an established organization (one at a time) and replacing it (if necessary) with an engineered system. We must do this with the acknowledgment that each system (if it is still useful) will be merged with the next system to get engineered to create eventually a completely Customer Focused Organization. (This is like having an existing Victorian house and remodeling one room at a time with the understanding that the connecting walls will be where we have to pay the most attention and incur some extra work in the process.)

The major difference between Pre-Engineering or Business Engineering and System Engineering is one of scope.

I would prefer that you try to conduct Pre-Engineering or Business Engineering in that overall, they are less costly and less time-consuming. However, System Engineering is what I've seen attempted the vast majority of the time when transforming an organization. This has led to many costly failures under the guise of such techniques as Re-Engineering, TQM, restructuring, or streamlining.

Quality through Engineering

Having said all this, let me state two major concepts I believe are essential for us to share if you are to fully understand and successfully use the ideas I'm trying to put across in this book:

- the need to use an "engineering discipline"
- the need for quality in manual and automated systems

What Is "Engineering?"

The word "engineering" is extremely important. The word must have substance when we use it in a title or activity. From a process point of view it means that we're going to apply a discipline (not necessarily rigid) to ensure quality throughout a development effort to produce a final quality product.

Notice that the terms "Pre-Engineering" and "Business Engineering" (which are techniques I advocate for building Customer Focused Organizations) both contain the word "Engineering." When we start creating a Customer Focused Organization, we must first have a good mission statement that contains measurable objectives. We use these objectives to maintain and measure the quality throughout the entire development life cycle of any new system.

The Cost of Poor Quality

As stated previously, a high degree of quality is a cornerstone in any engineering effort. Quality has to be built-in if the system is intended to last. Attempts to shortcut quality may lead to contempt (both internally and externally) for the system, the project team, and ultimately, for the organization that produced the poor-quality product. This reminds me of a sign I once saw on the wall of a Colorado store:

> ***"The Bitterness of Poor Quality Remains Long after the Sweetness of a Low Price Is Forgotten."***

Of course, we can throw a system together for a lower price. But, if we do, we should also factor in the cost of maintenance, dissatisfaction on the part of business people, and the long-term business loss from the customers' dissatisfaction.

As we shall clearly see by the end of this book, an "engineering quality" approach will lengthen a system's life span, improve the return on investment, and shorten the system development life-cycle (especially when testing time and time for "getting the bugs out" are included as part of development time). It will also deliver a system that reflects a commitment to quality.

The Cost of Quality

In my early days of teaching I pondered the question: "What is quality?" In my own work it seemed to come down to "attention to detail," but that didn't ring true when I built something in quick-and-dirty mode, such as a set of shelves made from concrete blocks and press-board planks, and knew that it satisfied my needs perfectly. Then I came across *Quality is Free* by Philip Crosby and the quote:

"Quality Is Conformance to Requirements."[2]

A set of requirements provides something against which we can measure our work, its quality, and our progress. When we develop a new system, there are requirements for analysis, design, and implementation of that system. We can apply this definition of quality to the development of a product, a service, or software.

Crosby's title, *Quality Is Free,* means quality costs nothing if it's built into the product at the time of manufacture. When we decide to build a system, we can choose either to build in quality as we go, or just to throw something together and worry about quality later with inspections and testing phases. We can take the latter approach if that's what our organization wants us to do, but we should frankly acknowledge that that's what we're doing and make sure management and business people document this approach as a Project Objective in a Project Charter. [3] We should also make it clear the cost of adding quality into a finished system/product is always much higher than building it in as we go. Unfortunately, we can't go to a store and buy a pound or two of "quality" to be added in after product completion.

Many studies have shown correcting an omission in the final stages of the development life-cycle (typically where I see it performed) will cost significantly more than in an earlier stage. Therefore, building in quality from the very first stage of the development life-cycle, rather than trying to test and remove defects at the end, will actually save time and money, and deliver an overall superior product. So, I'd like to amplify Philip Crosby's statement and say:

"Quality Can Save You Money in Development and Make You Money in Production."

Building In Quality

Cost overruns and poor project statistics are also the result of not having a useful set of development and project management techniques.

Throughout the book I use some real-life examples taken from my experiences in building my own homes as extended analogies. Since writing my first book on methodologies[4] I have built two complete homes. On my first house I did the majority of the work myself — carpentry, plumbing, electrical, and so on (I even developed muscles). On the second

---

2 *Quality is Free,* by Philip Crosby — see Bibliography

3 Many of the largest software companies in the U.S. put out products they know contain numerous bugs. These companies even use their customers to find bugs in "beta" (pre-release) versions of their products. I buy these products myself. However, when I buy software from these companies for $99.00 and use it in my company, I know the consequences and accept them (even though the "bitterness of poor quality" sometimes comes out of me during times of critical use). The point being that I acknowledge along with everyone else that these companies put out "buggy" software (nobody can hide that fact).

4 *Developing Quality Systems,* by Brian Dickinson — See Bibliography

house I contracted out some tasks and took on the role of project manager. Throughout this book, you'll find that I make much use of house building analogies because, having built many systems and then having built two houses, I clearly saw the parallels between the two processes.

Let me ask you some questions relating to the concept of building quality in:

Have you ever seen a house being built?

A new house typically seems to go up very quickly; what we don't see are all of the planning, up-front analysis, and design work that went on before the first load of concrete was poured for the foundation. Let me ask another question:

When the house was completed, did you ever see the entire construction crew pushing against a wall to see if it would fall over? Or, did you see them set up huge fans to see if it would withstand strong winds? Or, have you passed by when the fire department had hoses trained on the roof to see if it leaked?

You haven't? Neither have I! We just don't expect to have to "test" a new house after it's built to compensate for poor building techniques. Problems which show up after the house is completed show the poor construction methods and materials used during development and/or the lack of quality in the design. I'm not saying: "Don't test systems," rather that after-the-fact testing is not a quality assurance activity and it should not be used as a substitute for thorough business analysis and system design. My house building experience has taught me "quality" is required and has to be demonstrated at all stages of the project.

For example, land recording/surveying, ground surveys, architectural drawings, blueprints, structural calculations, heat loss calculations, and so on, all have to be externally approved prior to any construction. Moreover, local government staff (building inspectors) also inspect every stage of the construction itself. Any aspect of the construction failing to meet the national and local building code standards must be brought up to code by law before construction can continue. And, of course, as each stage is completed, the cost of making a change (such as modifying the plumbing or raising the ceilings) goes up dramatically.

Quality Using Engineering Methods & Models

Established engineering disciplines are supported by business, system, and project models (produced throughout the development life cycle) that allow potential errors and conflicts to be identified early in the development process. Here are some simple definitions of these terms:

Business Model: An implementation-independent model of business rules and policies. This is a business tool for declaring what has to happen to satisfy customer needs.

System Model: An implementation of the Business Model. This is a design tool for declaring how the organization runs its operations in the real world.

Project Model: A methodology or life cycle for building systems. This is a management tool for declaring an organization's systems development methods and standards.

These models provide a framework for defining issues, identifying the real conflicts, quantifying problems, and evaluating solutions in relation to the overall goals of a development effort. Also, they serve as the first point of reference in a future modification to a system.

Obviously, the house building analogy applies here in that each stage of the process is well documented. We need to make sure that the developers of a system have a good set of goals, and a Project Charter for any project that they undertake, as well as the skills and tools needed to do their job. The design of the actual human environment should be created with the idea of satisfying the organization's business people and customers. In the past (as we'll see in the chapter on ***Systems Archaeology***), human-based and computer-based systems have not been designed with the main intention of satisfying the customer, but more to satisfy the hierarchical implementation of the organization or the needs of the technology. This is what I call the "Clothes Have No Emperor Syndrome" which I'll discuss in Chapter One. The house building analogy I use may break down in places when compared with System Engineering. However, in general it works very well because house building and System Engineering share the following characteristics:

- Specific start and end points
- Significant deliverables
- A strong client/developer relationship
- Periodic, intermediate quality assurance checkpoints
- Risk (financial, project, and finished product)
- Dramatic escalating cost-of-change as the project proceeds

Moreover, both house building and systems development:

- Involve many people and require complex coordination of many different skills and professions
- Require specialized building tools, including "power tools"
- Use models for conceptualization, communication, and overcoming complexity
- Use off-the-shelf, modular, and reusable components whenever possible

I lived in the San Francisco Bay Area of California where I built my first house. I was pleased to see the building inspector come in periodically and tell me where I had not adhered to standards, and why; I looked upon the inspector's approval as a real benefit and I had respect for his knowledge. The inspector showed up before the cost of change had escalated. The inspector ensured quality in the development cycle as opposed to being an exercise in after-the-fact testing. I was building the house for myself and I wanted to have a safe and long-lasting home. I knew my life might depend on it *(California is beautiful, but it has its faults!).*

In October, 1989, a significant earthquake (7.1 on the Richter scale) hit Northern California during the evening rush hour. Property damage in California was major, but limited to older buildings, and approximately 60 people were killed. In October, 1994 a 6.0 earthquake hit a region 140 miles from the city of Hyderabad, India. The Indian quake (although smaller by a factor of over ten times), by contrast, killed 25,000 people even though the epicenter wasn't near a major city, it wasn't during a peak commute hour, and there were no 52-story buildings nearby.

Why the difference? Of course, the emergency services in California responded magnificently, but the degree of damage in each place depended simply on the prevailing civil engineering disciplines.

In house building, the soil engineers, structural engineers, electricians, carpenters, plumbers, and all the other professions rely on each other to do their jobs correctly (e.g., you can have a wonderful home with beautiful carpentry and efficient plumbing, but if the wiring behind the walls is sub-standard the house may burn down). In a Business or System Engineering project, everyone must also rely on the others to do their jobs correctly.

> In the software world for example: the analyst must rely on the designer; the designer must rely on the application programmer; the application programmer must rely on the database technician; the programmer must rely on the compiler programmer; and the compiler programmer must rely on the operating system programmer, etc. Similarly, each manager and technician must employ professional practices when creating a Customer Focused Organization.

I lived in my first house for ten years and had virtually no maintenance costs even though I added on to the original house significantly over the years. Creating manual or automated system(s) that have virtually no maintenance costs is exactly what we should see as the result of a Business or System Engineering project.

A Word about Words

This book introduces and uses a set of words and phrases to describe various key Business and System Engineering concepts. Some of these words or terms may be familiar to many readers while others will be more obscure. Please note that I have defined some words to have very specific and explicit meanings within the context of Customer Focused Engineering. Wherever any of these words or phrases are used in this context, they are initially capitalized such as "Business Event." I have taken care to provide my definitions for these terms where it seemed appropriate. In addition, any capitalized words or phrases are defined in the glossary.

An example of my word use is the word "organization." I use this to cover both public and private sector enterprises. I reserve the word "business" to mean the implementation-independent activities organizations perform in the course of satisfying their customers needs. In this sense, a governmental postal service is in the "business" of delivering packages. Both the postal service and a private sector package service are "organizations" and each is in roughly the same "business."

Another important tenet of this book is that the concepts and methods I advocate can be used to create or modify any type of system, manual or automated. I take pains throughout the book to make it clear you can create human-based systems that rely on individuals following engineered procedures just as easily as you can use the latest technology-based tools to implement automated systems. For the purposes of Customer Focused Engineering, human beings and computers systems are both means of *implementing* business needs. I want to show that what we're trying to improve are systems, regardless of whether the implementations' Central Processing Units are carbon-based gray matter or silicon-based wafers.

We must be aware of the difference between business issues and system issues when we perform Customer Focused Engineering. Business issues are those things pertaining to **what** the organization does regardless of how they are implemented. System issues are those things pertaining to **how** the organization is designed to run in its day-to-day operations. System issues involve technology aspects (e.g., human beings, computers, robots).

Objectives vs. Requirements

Objectives can be looked upon as achievable goals for the project. I would like to distinguish between Requirements and Objectives here.

I use the term "goal" to apply to an Organizational Goal of which there is only one. This Goal manifests itself in one line or paragraph documented in the organization's Mission Statement. This one Organizational Goal is broken down into individual Objectives which may be a collective set for the whole organization or partitioned by project. These Objectives become Requirements.

There is a definite hierarchy of Goals, Objectives, and Requirements. There is one Organizational Goal from which derive Business Objectives. We in turn break down these Business Objectives into Business Model Events which in turn break down into Business Requirements.

If it doesn't sound too esoteric:

> ***Objectives are a description of measurable reality as of a future instant in time.***
>
> ***Requirements are data flows, data stores, data relationships, and processes, that must be in place for a system to work. These requirements, operating through time, satisfy the objectives.***

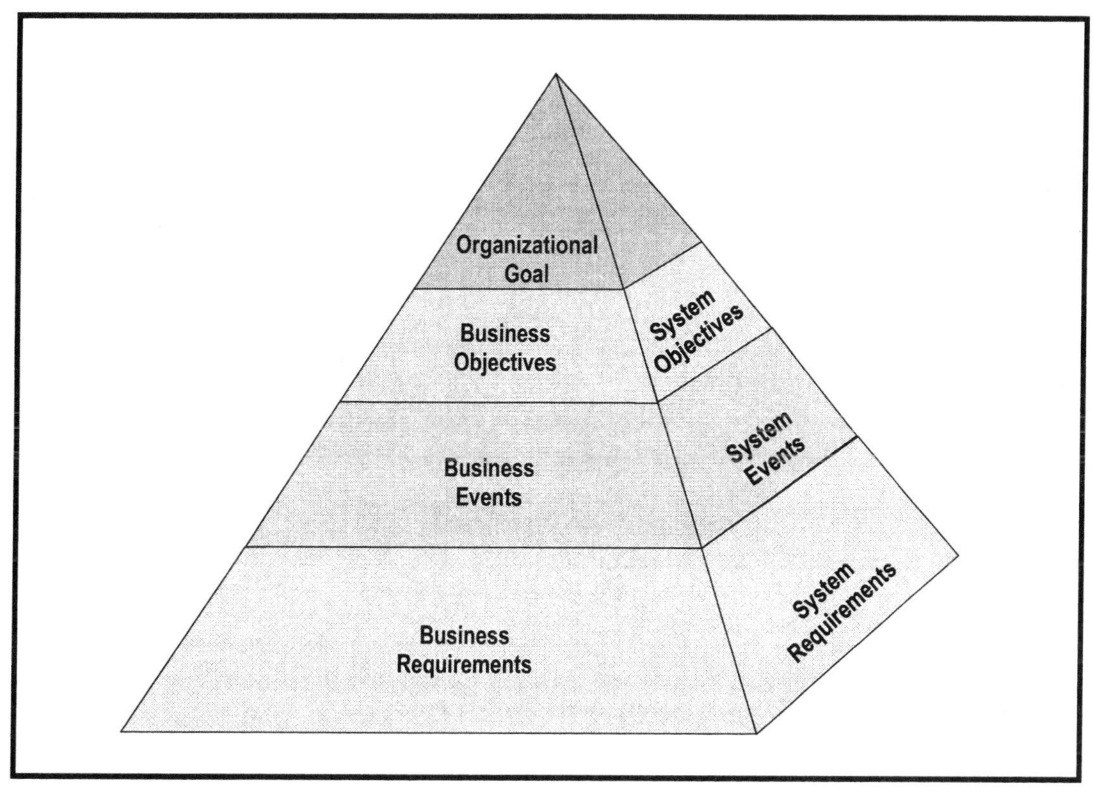

Fig: 1-1 The Hierarchy of Goals & Objectives

Let me clarify this by using another one of my analogies. When we build a system, it's kind of like building a radio. We have requirements for what the radio must do. We design the radio around those requirements and implement/build the radio. Now, just like systems, the radio isn't exactly what the customer is looking for; they're looking for music or news, etc. to come from the radio. When we put a system in place, procedures and the code are the radio. By running data through our procedures and code, we produce the necessary output(s) that the customer really wants.

I like to separate Objectives into at least three categories: Business Objectives, Project Objectives, and System Objectives. In Strategic Planning we want to focus on Business Objectives. We recognize Business Objectives as being applicable to what the resulting system should accomplish for the business, in business terms. They are measured from day one of production onwards. They are typically measured against current system statistics.

> Here is an example of a Business Objective: Reduce by at least 50% the turn-around time between customer order and delivered product and accomplish this with no more than a 5% increase in existing operating costs.

We can measure from "day one" of production onward whether we accomplished this objective.

We recognize Project Objectives by the fact that they apply to the project issues such as planning, scheduling, budgeting, etc. and can be fully measured by the last day of the project (i.e., the day before production).

Here is an example of a Project Objective: The project must not exceed a cost of $1,000,000.00, and must follow an engineering development discipline.

We recognize System Objectives as being design related, usually technical, and are further recognized by the fact that they are accomplished and fully measured during the project, usually in the design and implementation stages.

Here is an example of a System Objective: The system must use the in-house data-base management system and be implemented with client server architecture.

The organization conducts projects in order to produce systems and systems accomplish the day-to-day operations of the business.

Several key roles must be recognized and assigned to the appropriate individuals and/or groups in the course of creating a Customer Focused Organization. I will define these roles as I go through the book. However, I want you to think about the roles introduced throughout this book as "hats." A single person of the Customer Focused Engineering team may wear one or more hats.

I understand you may have to back off from some of the ideas in this book due to politics, budgetary constraints, limited resources, etc., but it's my task to put forth one comprehensive way of doing it right. I hope you enjoy and profit from the contents of this book.

A Word about Typographical Conventions

This book uses an extensive set of models, graphics, and other elements to convey important ideas. To help tie together these elements to the matching text, this book uses a set typographical conventions related to words and phrases in the models and elements. An example is the use of the convention shown below to set apart examples to show that they are for optional reading.

> Extended examples appear in this gray bar paragraph format. If you understand the idea preceding an example paragraph such as this, you can save time by skipping ahead in the text.

Business Model Event Names

Customer Wants to Order Our Materials, File End-of-Year Tax Report, Supplier Sends Invoice, etc.

Model Element Labels

When a word in a paragraph or example refers directly to an item in a model the paragraph is about, that word is **Bold** and capitalized. For example, on a Data Flow Diagram, the name of an External Interface might be **Customer**. By the same token, any data flow names such as **Purchase Order** or any process names such as **Verify Customer Address** are shown using the same convention. Note that there may be paragraphs in which you see a word such as "customer" shown as both **Customer** and customer. In this case only the bolded **Customer** would refer to a model or other diagram.

Control Flows

On a Information/Data Model (such as a Data Flow Diagram), a line that indicates something that controls a process is called a Control Flow and its name is shown in **'Single Quotes and Bold.'**

Chapter

1

Evolving to a Customer Focused Organization

No problem can be solved by the consciousness that created it — we must work to see the world anew.

Albert Einstein

I hope this book changes your consciousness, or, by the time you've read this book, I hope it will provide the ammunition you need to sway the old consciousnesses of any skeptics in your organization.

The legacy of everyone in an organization today is to keep the founder's vision alive. With the exception of those organizations that were intended to be temporary, the ultimate tragedy of not working towards being the best structure for an organization would be to see the loss of all the effort and struggle your predecessors went through to build the organization. Even though creating a Customer Focused Organization will be challenging, costly, and time consuming, the ultimate benefit is your organization keeps its doors open and hopefully provides a pleasant environment for its workers and benefits to its community. Note that because realizing the full potential of Customer Focused Engineering could take many years, the successors of the initiators may be the ones who reap the final credit.

When we use an engineering discipline to create a Customer Focused Organization, what we're really engineering are the new systems that make up the organization, not the business itself. In other words, after we've created a Customer Focused Organization, we'll still be in the same business. So, we really engineer an organization's manual and automated systems through new systems development efforts (typically known as projects).

Projects produce systems which are then implemented, so let's agree on two key terms before we state our goals for Creating a Customer Focused Organization:

Business Issues These are those things pertaining to **what** the organization does regardless of how they are implemented.

System Issues These are those things pertaining to **how** the organization is designed to run in its day-to-day operations. System issues involve technology aspects (e.g., human beings, computers, robots).

These definitions help us delineate between analysis (the "**what's**" of the business) and design and implementation — the "**how's**" of the systems (see Figure 1–1).

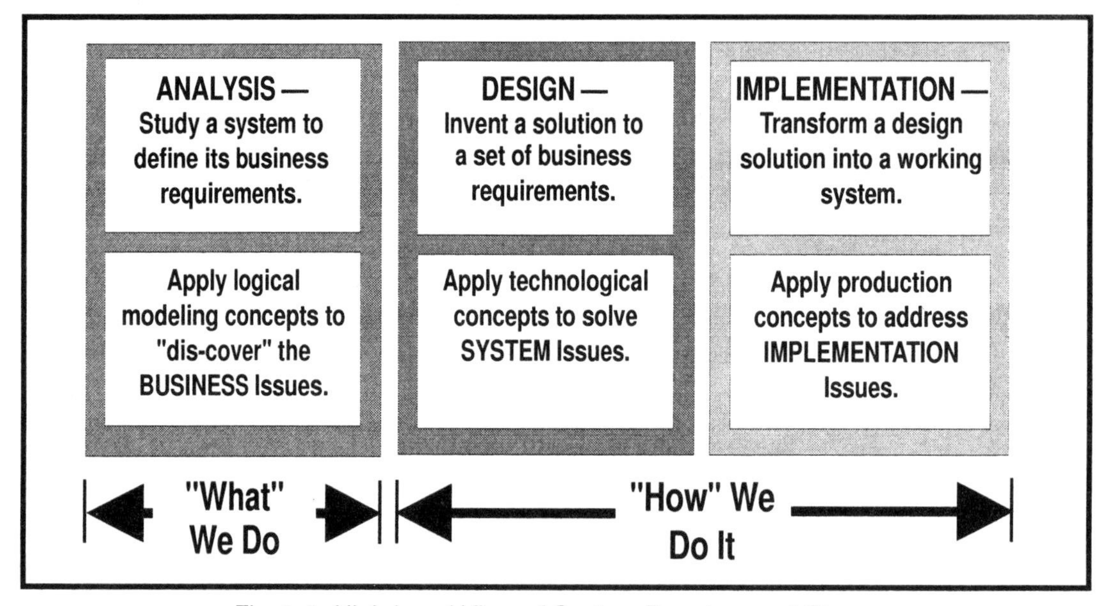

Fig. 1-1: High Level View of System Development Stages

The Goals for Creating a Customer Focused Organization

In this section we'll discuss the goals for creating a Customer Focused Organization and how to attain them with the understanding they apply to both our future implemented human-based systems and computer-based systems. These goals are:

- To put the customer first — satisfy customers' needs and expectations by structuring our organization to respond "seamlessly" to these needs.
- To get back to business basics — focus on **why** we're in business and on **what** we do. These **why's** and **what's** should be derived from the organization's Mission Statement.

- To cut red tape — achieve dramatic and measurable improvements in the performance of the organization's old implemented systems by creating the most effective processes for delivering products and services to our customers.
- To engineer the organization — replace old systems that may be hurting the organization today with quality engineered systems through an engineering development discipline. The new engineered systems (be they manual or automated) should be faster and easier to install and maintain than any legacy systems.
- To create systems that promote data conservation — create systems with no dead data or inconsistent data and take advantage of the opportunities for re-using data and processing whenever possible.
- To attain the highest quality in the development and delivery of our products and services — empower employees and use the best available technology in implementing systems.
- To satisfy the organization's strategic mission and allow for constant improvement in the organization by producing a flexible environment for future change. This means that we create an environment in which we can make business changes that will not affect technology issues, and where we can make technology changes without affecting business issues.
- To remove those aspects of your organization that do not directly satisfy the organization's strategic mission — remove old design/implementation issues in place today that no longer add value to our business.

Having stated our goals for creating a Customer Focused Organization (and hence the goals for what I want to teach in this book), let me go through some of the factors that will be in your way when trying to accomplish these goals.

Obstacles to Creating a Customer Focused Organization

There are many texts published today that seem to imply that using the latest computer technology and support systems is the way to improve customer service. However, we may accomplish the important goals above without new computer systems.

> For example, we can go from an existing completely human-based system to a new human-based system, and can even replace an existing computer-based system with a human-based system and still accomplish the goals.

We can always overcome any old technical limitations simply by replacing old technologies with those available today. You should see improvements in at least timing, capacity, or access to data, even if they're not dramatic. However, these are not necessarily Customer Focused improvements — after all, do your customers really care if you're using the fastest equipment with state-of-the-art technology? Your customers care about good products and services. You can obtain far more benefits by improving the processes and data you use to respond to your customer's needs.

The gains achieved by switching technologies are often incidental improvements based upon this increased speed and capacity, but the organization gains no competitive edge because its competitors can use the same technology, or even leapfrog it by waiting and implementing the next generation of technology.

Be cautious when turning to technological solutions first to produce a Customer Focused Organization. By the time you've read this book, you will realize that any computer technology, or even manual implementations, are not the main subject of Customer Focused Engineering.

To produce a Customer Focused Organization, the design at which you arrive may indeed use the latest technology, but don't let this obscure your view of what it is your organization actually does. The true nature of your business will almost certainly be unrelated to its implementation technology.

If your project approach focuses on new technological solutions before analyzing the organization's business issues, I recommend you schedule a lot of beta testing and allow for "unscheduled user field testing."

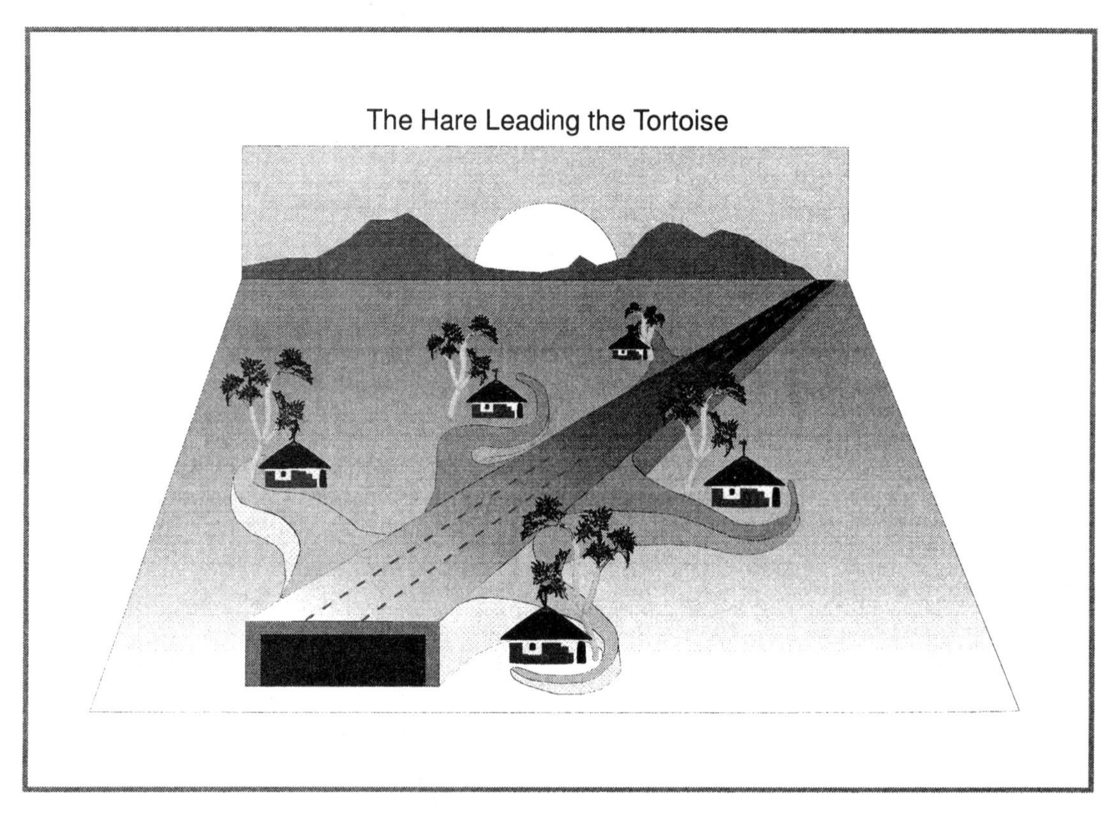

The world-wide Information Super Highway would be a tragic waste if all it did was connect every organizations' bad (non-engineered) systems and their inaccurate data. This is analogous to having a modern highway connecting a bunch of mud huts and dirt roads. It is our responsibility within our organization to engineer its internal systems.

Others have taken on the responsibility to engineer the Information Super Highway that will link our systems with those of other organizations. However, it would be another tragedy to have our new, wonderfully engineered systems connected by "dirt roads." So, we can only hope that the Information Super Highway's builders also apply an engineering discipline in their efforts.

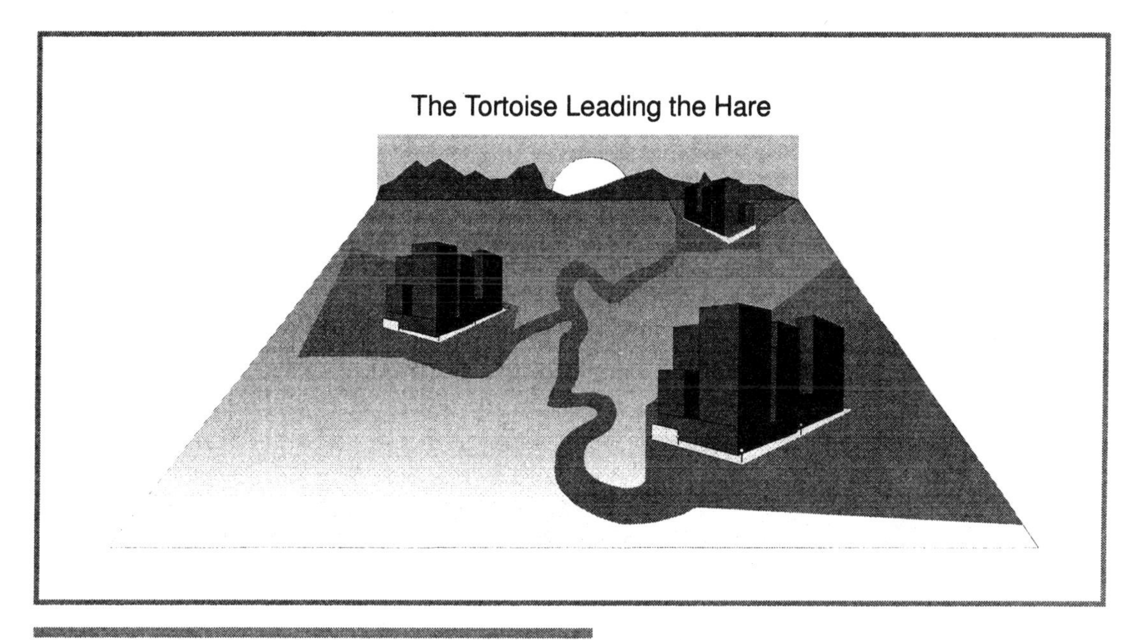

Technologically Dependent Systems Obstacles

One major reason for systems becoming obsolete early in their lifetimes is that many of them are built around a particular implementation technology. That is, they are designed to run only within an existing departmental context or on a specific computer, use a specific database management system, or rely on specific system support software. Therefore, the advances in hardware and support systems that are constantly being announced lead to premature obsolescence, short lifetimes, and "patches" for these technologically-dependent systems. Technology changes rapidly (have you ever tried to sell an old PC?). Oftentimes the project that implemented the "last" system technology is usually classed as a failure to justify the cost of that system's early replacement. Sometimes the failure of these projects gets blamed on the preceding generation of technology used. At other times the blame goes on the old methodology used to build the existing system. We want to be cautious not to fall into this trap when creating Customer Focused Organizations.

Any system built around a particular technology will last only as long as that technology before it requires rebuilding in response to the next new wave of technology and support systems.

In my past data processing career I've seen organizations adopt fads and new methods, models, and technology. They constantly change the designs of their systems, but the nature of their true business rarely changes (or changes quite slowly).

The "Clothes Have No Emperor" Obstacle

Perhaps the worst obstacle to the satisfaction of our goals is "thinking in the box."

When an organization embarks on a project, someone (usually a human being in a management position) has to declare the objectives that define "success" in a Project Charter. Unfortunately, that "someone" in many cases has to put themselves in the hot seat and be willing to admit that they may be part of the problem. He or she may be associated with the "Clothes that have no Emperor" obstacle. Let me explain.

Much of what we see in place at an organization is there to support the "design structure." In other words, most systems were put in place to make the business run as designed for the real world at some point in the past. This structure is the biggest target for improvement in creating a Customer Focused Organization. It is there, of course, to accommodate human systems and computer systems, as these are design aspects of an organization. These structures are the Clothes (the systems) that cover the Emperor (the business).

I have seen many senior managers set limitations on a project. These limitations are invariably based on system boundaries such as their department, division, bureau, etc. It can get worse where the use of new computer technology is the major goal of a project (e.g., on-line automation of tasks, client-server computer systems, relational databases, use of the Internet, and such). Again, with this technology limited to an old system boundary, this is technology "driving" the business and it's incredibly common in many organizations. In many cases these people have already mistakenly decided on a sexy new hardware solution before they've really analyzed the problem. They've lost sight of the fact that an old human or computer system boundary (and the hardware and/or software systems contained in that boundary) are really only one solution to the organization's business requirements.

If we do not conduct an accurate analysis as the primary part of a Customer Focused Engineering effort, we will probably fail to remove our blind acceptance of preconceived design notions of "normal business practices." These "normal" practices include human structures such as divisions, departments, agencies and bureaus and their "turf wars." They also include their computer equivalents such as Order Entry systems, Accounting systems, Edit/Update/Print programs, Input/Process/Output programs, 4 Kilobyte subroutines, and the programmer/D.P. mentality that perpetuates these structures.

Missing this point may lead us to engineer a system or a significant part of a system that should have been tossed out long ago. In fact, failure to see through the old designs leads us to create faster bad systems and hence, brands the project as a failure.

I do realize that one of the problems is Strategic Planners must often make decisions regarding long term equipment acquisition, site management, facilities expansion, and so forth, based on the needs of the proposed design. Even so, please don't let this lead to the myopic view that it's the new equipment that matters most.

Outsourcing/Downsizing Obstacles

If Customer Focused Engineering is being done at all, it is because the existing implementations of an organization's systems are perceived to be (or actually are) inefficient, ineffective, or more costly than a system we can engineer to take their place. Another problem that occurs when this is perceived is that of leaping to one design-oriented goal such as downsizing, outsourcing, or flattening the management hierarchy.

The potential tragedy of not seeking to focus on the customers' needs is that you may have your job (and maybe your whole department) outsourced or downsized. Outsourcing is the practice of using an external agency or vendor to perform tasks that were previously performed by an organization's own staff.

Be cautious of jumping at random outsourcing. We may, in fact, do a disservice to our customers by fragmenting our response to their needs. If a department's tasks are outsourced, the organization loses some control over satisfying the customer's needs, and we may realize operational savings, but with a loss of customer service.

By the way, this book is intended for your potential outsourcing vendors too. They must impress on your organization's management it is more cost-effective for the organization to use them instead of you. What I'm describing here is the nature of competition, which I find very healthy; also, it is inevitable in a market-driven environment. Any organization that sits back on its laurels will eventually suffer for its complacency (especially in hard times). These trends do not sneak up on us as we've witnessed in the global competitiveness of German and Japanese companies.

When I speak of outsourcing, I'm not talking about forming strategic alliances with our organization's existing vendors. This is a good Business Engineering strategy. Rather, I'm talking about outsourcing to a vendor the tasks that your department currently performs. If an organization found an aspect of its business that it could outsource and better satisfy the needs of its customers, it would be foolish not to outsource it (especially if it cost less). However, the organization would be more foolish if it didn't give its staff at least one chance to engineer their operation. After all, the knowledge and skills of an organization's staff are one of its greatest assets. In the face of downsizing, some of your best people are the first to find new jobs rather than waiting for the inevitable axe.

You may have realized from the above comments that I'm not a believer in downsizing. I am, however, a believer in right-sizing (which could just as well mean the need for more staff or less staff in certain areas). I find the people who set downsizing as a goal only see their organization as numbers of people. If you're measuring people as the cost of your business, you will cut people in lean times. If you measure in terms of the quality and quantity of responses to the needs of your outside customers, you can then cut unnecessary and non-critical responses. Rather than cutting a percentage of people, you may actually find a better solution is to relocate resources to more profitable and critical areas of the business. We can best save costs by removing unnecessary waste caused by historical systems and their maintenance. In this respect, I believe that management should budget **what** the organization does as opposed to **how** it's implemented (i.e., don't budget departments, agencies, systems, etc.). Later in this book, I'll recommend a more business-oriented view of budgeting.

You have to create an environment where you are the best possible implementation for your customers' needs and where you are more efficient and cost-effective than any vendor can be, so that downsizing or outsourcing are ridiculous options.

Not Acknowledging the People Issues Obstacle

I want to emphasize here (as I do in my seminars) that engineering computer systems is a lot easier than changing human values and management practices. The challenging part of creating Customer Focused Organizations involves everyone overcoming obstacles such as:

- Resistance to changes of management style and reporting structures (especially at the middle management level).[1]
- Politics, i.e., empire building, turf protecting, us versus them, etc. (across departments, divisions, bureaus, etc.).
- Fear of job loss (throughout the organization).
- The need for instant results as reflected in the quarterly profits obsession and the need to immediately appease the stockholders/taxpayers.
- The lack of a clear vision of purpose from the leaders of an organization as expressed (or not expressed) in an organization's Mission Statement.

Employees who are not informed of what's happening, and their resultant resistance to change are probably the biggest factors in the failure of any project. Therefore, be honest and openly discuss your organization's "politics." Put your staff back into their comfort zones by stating that you want to ultimately grow the organization and keep the doors open.

---

1 In fact, I believe that the word "management" should be abolished in the business world. We should instead have the concepts of leaders and facilitators. These are not new concepts. Theodore Roosevelt said: "People ask the difference between a leader and a boss.... The leader works in the open, and the boss in covert. The leader leads and the boss drives." I'd like to emphasize that the titles that we put on people and use are important. We don't want to manage people; we want to help and lead people. "Management" to me implies a control or top-down view. Whereas "Leadership" implies "at the same level" and "I will guide, help, and facilitate you in getting the job done."

There are other benefits for the people in a Customer Focused Organization:

- We can create an organization that truly fulfills its mission.
- People benefit by producing results in an engineered environment because people should ultimately be rewarded for meeting customer needs.
- Ultimately, your organization should be able to deliver services and produce products faster, making the employees and the organization appreciated.
- As an employee you should obtain more job satisfaction by being empowered and staying up-to-date by using new technology.
- Becoming proactive instead of reactive to produce a less stressful environment (no more putting out internal fires).
- Less frustration from re-work and re-inventing the wheel because of the engineered environment.

Having the right and the resources to overcome the above challenges and being able to realize the benefits is predicated upon having the leadership of an organization directly involved in championing the Customer Focused Engineering effort. Everyone must be totally enrolled in the process. There needs to be a watershed defined when the Customer Focused Engineering effort begins, and to affirm that the leaders of the organization are fully behind the effort.

The start of a Customer Focused Engineering effort requires that everyone is accurately informed of the Organizational Goal and Objectives and the direction and the progress points in the effort.

The "Stage of Evolution" Obstacle

The attainment of our Customer Focused Organization will be easy or hard, depending on how the organization currently runs its operations.

Figure 1–2 depicts this level of difficulty.[2] In general, the larger and more steeped in its own history an organization is, the more challenges a Customer Focused Engineering effort will face.

- If an organization's standard operating procedure is to expect and respond to internal emergencies (i.e., "putting out fires" and where planning is non-existent), then the systems in this organization (human- and computer-based) will have been built in the same haphazard manner. This reactive rather than proactive method of operation poses the most difficulties for Customer Focused Engineering because of the large differential between "constant crises mode" and an engineered environment.

2 A similar view to this is given in Watts Humphrey's book, *Managing the Software Process* — See Bibliography

- If an organization is using standardized development methods and models to run its day-to-day operations (manual and automated), then it still faces challenges, but not as hard as those in the first scenario. This is because this organization at least acknowledges the potential benefits of using standardized tools even though it may not use the same methods and models in the engineered environment.

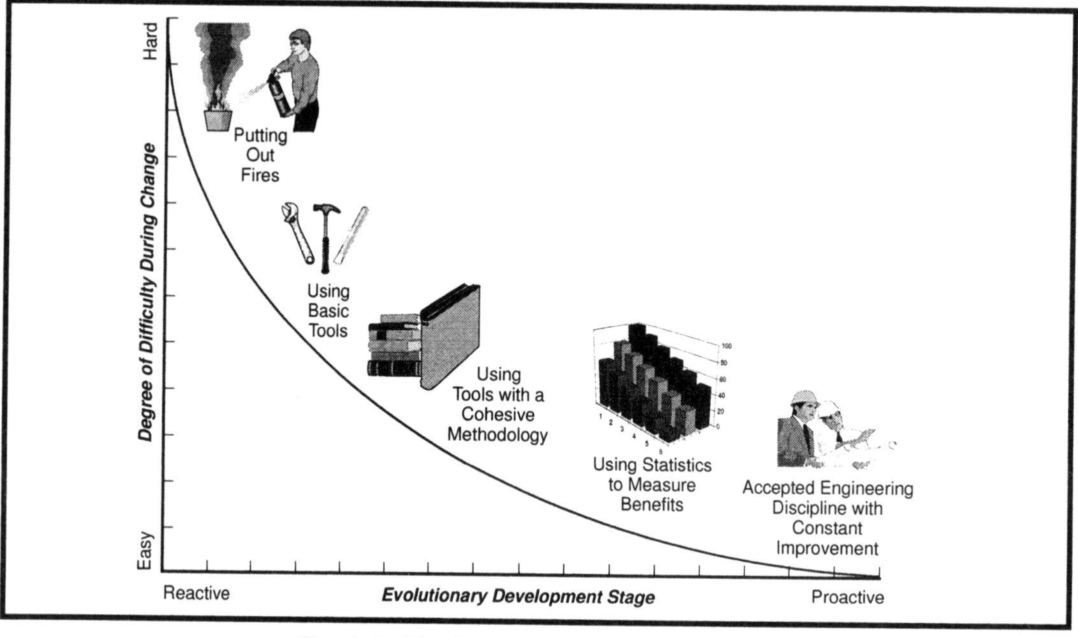

Fig. 1-2: The Evolution of an Organization

- If the organization defines an all-encompassing method within which to use standardized tools (i.e., it adopts a practical methodology), the task will be even less daunting because this organization is used to having structure in its operating procedures.
- The next easiest environment to engineer is one in which standardized tools and a methodology are firmly in place, and the organization gathers statistics and measures results to see what works, and what doesn't; i.e., it uses metrics (statistics) to plan and measure its successes.
- The easiest type of organization to engineer is the one in which an engineering discipline is already accepted as the standard method of operation, and the environment accepts constant improvements (even though the current methods may change after Business Engineering takes place). With my definition of "Business Engineering," this last example is the only one to which the term really applies, because we are engineering an existing engineered organization.

If you are low on the above organizational evolutionary scale, don't let this deter you because you will also realize the most benefits from Business Engineering. However, be aware that your investment will be the greatest as other training will be needed to support Business Engineering (for example, Total Quality Management, people/team skills, and training to facilitate change). I find that most organizations fall between the tools and methodology stage.

Old Management Practice Obstacles

Even the way we conduct projects is an aspect of the old design in which we artificially separate the technical issues from the managerial issues. This division is an archaeological wrong turn because, in reality, a project's technical issues are intrinsically bound to its managerial issues. Managerial obstacles stem from someone placing arbitrary and uneducated limits on the project's Resources (time, people, money, support). Managers of Business Engineering projects have a delicate balancing act on their hands. It's like trying to maintain an equal balance on a four-ended seesaw (see the following illustration).

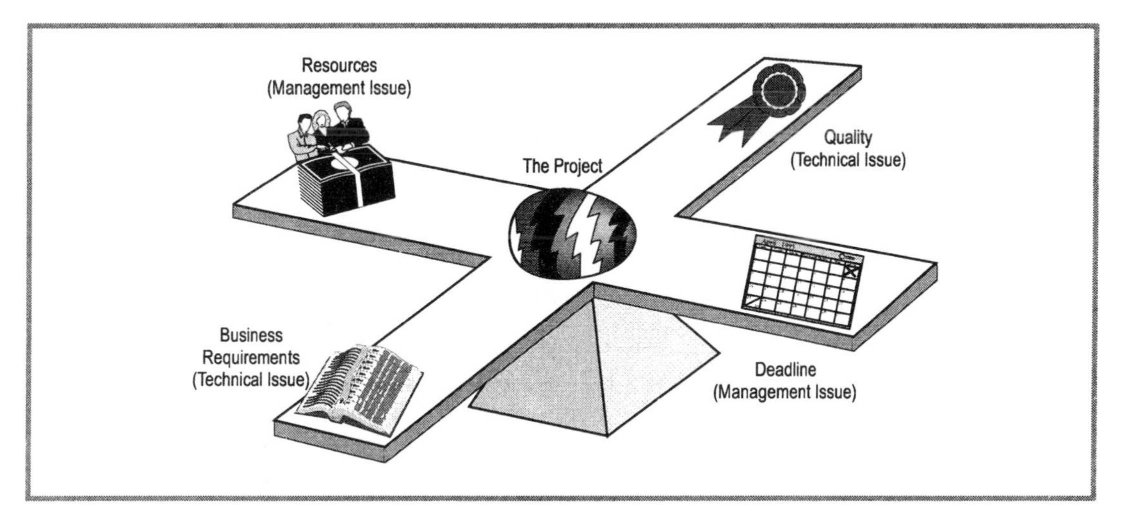

At one of the four ends of the seesaw sits the Project's Scope (identified by the total set of Business Requirements) and on another sits the allotted Resources. On yet another end of the seesaw sits the effort Deadline. Sitting on the remaining seat is the resultant system's Quality.

Knowing the Project's Scope (via the affected set of Business Events) is vital to its success. The ultimate scope is, of course, to apply Customer Focused Engineering to the entire organization to create a completely Customer Focused Organization. However, it is perfectly valid to engineer a portion of an organization as long as you understand functional partitioning and don't simply re-implement an existing, dysfunctional area of the organization's business. A small pilot project that takes on just one Business Event can be extremely beneficial by providing a success story to convince the leaders that it can be repeated organization-wide.

If someone sets a restrictive project Deadline and the date set is not based on empirical data (from other, similar Business Engineering projects), then one of the other factors on the seesaw has to "give" to maintain a realistic balance. With a preset Deadline you can achieve a balance if you can prune the number of Requirements (i.e., don't try to engineer the entire organization at once) or if you can put into the project more Resources (money, people, support tools, experts, etc.). Obviously, the same goes for any of the other factors; e.g., trying to increase the Requirements without increasing the necessary Resources upsets the balance as does trying to shorten the Deadline without upsetting Quality or one of the other factors. It's up to the project's resource providers to maintain this balance.

In my view, the most important rider on the seesaw is Quality. If you lower Quality by shortening Deadlines (or skew Quality through a mismatch of Requirements and Resources on the other beams), you may as well throw in the towel. This is where you may end up being chronicled as "another methodology project failure" when what has actually failed is not the concept of Customer Focused Engineering itself, but rather the management of the project.

Overcoming the Obstacles

The goals for creating a Customer Focused Organization are no different than those for developing any successful system.

Take Care of the "Three Ps"

Our goals can most effectively be be accomplished if we take care of what I call the "Three Ps."

- Take care of *project issues* with such things as having a clear and measurable Mission Statement.
- Take care of *product issues* (the deliverable) with such things as building Quality into the product by building quality into each step of the systems development process.
- Take care of *people issues* with such things as open communications, empowered employees, and effective leadership. I like to view the "Three Ps" as three legs of a stool. It's most stable when all three legs are in place (see Figure 1–3).

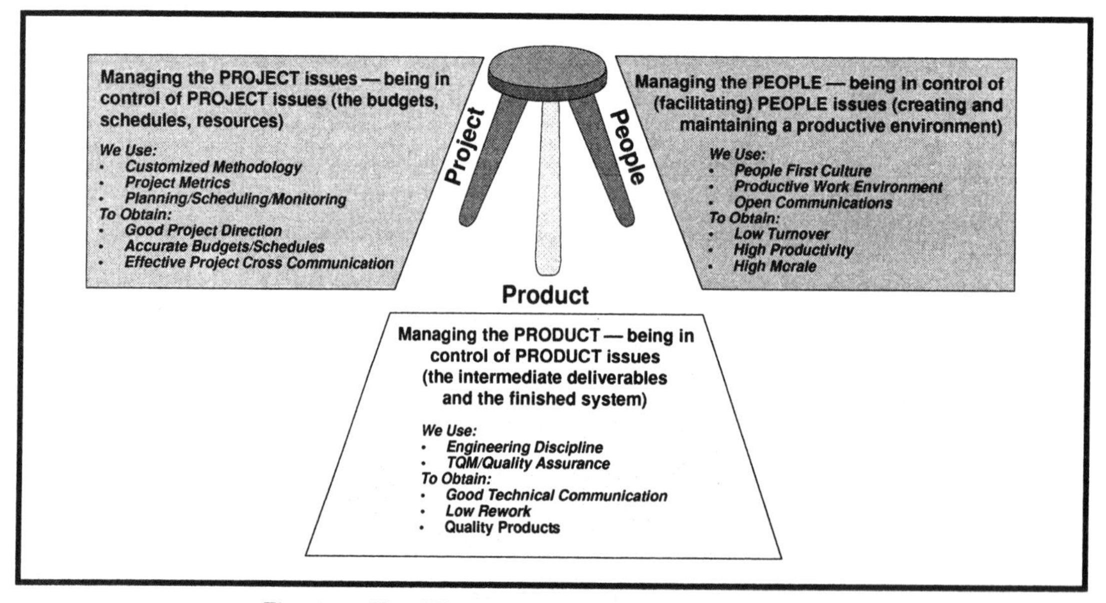

Fig. 1-3: The "Three Ps" of any Development Effort

Be Accountable for Each Deliverable

When we embark on a Customer Focused Engineering project, we have an opportunity to change the existing implementation of an organization to a better, Customer Focused implementation. To arrive at a quality finished product, each participant/player in the project must be accountable for his or her part of the project — both technical and managerial (see Figure 1–4).

Although I will address some issues having to do with Strategic Planning and project management, the majority of this book will concentrate on the project team and business policy user issues (i.e., the technical and production issues of creating a Customer Focused Organization). I will address the managerial/leadership issues in Book II of this series.

Managers/Leaders have the responsibility of declaring standards, realistic deadlines and quality levels, and for monitoring compliance (just as technicians have the responsibility of producing a quality product).

PLAYER	DELIVERABLE	GOAL ORIENTATION	METHODS/MODELS
Strategic Planners/ Upper Management	Mission Statement	Strategic/Organizational	Long Range Forecasting/ High-level Business Models
Project Manager	Project Charter/ Customized Methodology/ Project Plan	Project	Metrics/Communication/ Organizational Methodology
Project Team	Manual & Automated System(s) (via Analysis, Design, & Implementation)	Technical	Development Models/ Implement Technology
Business Policy User/ Operations	Completed Work	Production	Manual & Automated Systems/ Schedules

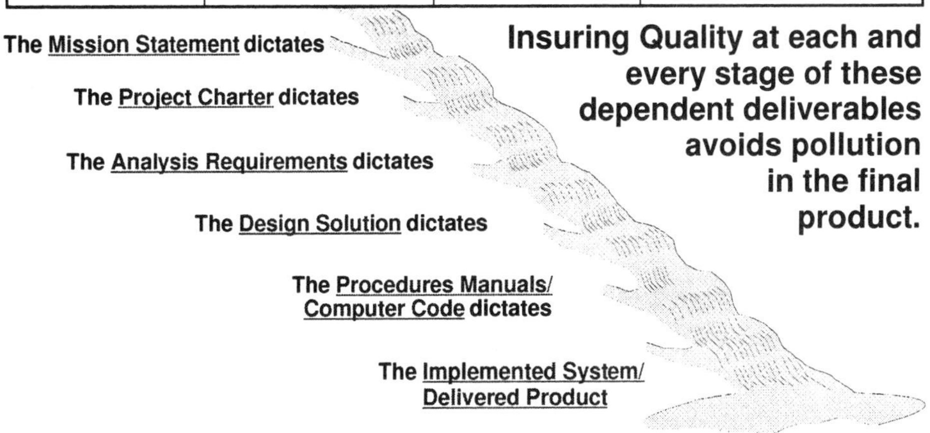

The Mission Statement dictates
The Project Charter dictates
The Analysis Requirements dictates
The Design Solution dictates
The Procedures Manuals/ Computer Code dictates
The Implemented System/ Delivered Product

Insuring Quality at each and every stage of these dependent deliverables avoids pollution in the final product.

Fig. 1-4: The Waterfall of Accountability

We must ensure that the various Product, Project, and People goals of any major effort do not conflict. If the goals do conflict, they usually result in a project failure and loss of jobs. In the worst cases these conflicts lead to loss of life such as in the cases of the near-meltdown at Three Mile Island in the Eastern United States and the explosion of the space shuttle Challenger.

Summary

Creating a Customer Focused Organization is an excellent opportunity to introduce an engineering methodology for the development and production of manual and automated systems, and to launch a continuous quality improvement program. The underlying focus is on achieving as perfect a match as possible between the needs of the customer and the organization's responses.

A successful Customer Focused Engineering effort may preempt any kind of reactive, short-term-profits driven downsizing, or outsourcing, because we will have the option of proposing a better design based on a rational study of the organization. I'm sure that everyone would rather grow their organization than shrink it.

The hallmarks of creating a Customer Focused Organization are:

- Bold, measurable business goals (e.g., "Obtain *the* best rating within our industry within three years," "Win the Deming or Malcom Baldrige Award within five years," "Immediately introduce a continuous improvement program," "Set up a Reusable Library of zero-defect data and processes immediately and reward its use by every project").
- A potentially total, top-to-bottom overhaul. A continuous and possibly fast-paced effort to minimize the period of disruption when only some parts of an organization's response to the customer's needs may be engineered. Please note that the parts of the organization being engineered must interface with the organization's old systems during the entire effort.
- A charter from the highest organizational level, along with a well-thought out strategic direction/plan available to everyone in the organization.
- An entirely new way of looking at the organization's structure as consisting of a set of partitioned, cohesive processes that satisfy the specific needs of a customer.

I believe that the Business Event Methodology, as described in this book, reveals a true business view, and as such, it can be the basis for engineering organizations that will grow and last as long as the businesses they were created to support.

The benefits of Creating a Customer Focused Organization are:

- Staying in business.
- Increased business revenue, greater market share and therefore increased satisfaction of shareholders or taxpayers, etc., due to improved customer service and competitiveness
- Reduced operating costs through efficient systems and improved quality.

Now that you know the goals for creating a Customer Focused Organization and the level of difficulty facing your organization, the first step is to consider some Strategic Planning issues as outlined in the next chapter.

Chapter

2

Creating a Customer Focused Organization via Strategic Planning

The world will not evolve past its current state of crisis by using the same thinking that created the situation.

Albert Einstein

Let me talk here a little about Strategic Planning and return to this subject later in the book after we know more about the concepts of creating a Customer Focused Organization.

We can achieve small successes within an organization by producing individual Customer Focused Systems. However, to achieve the ultimate efficiency and savings for the whole organization, we should look upon creating a Customer Focused Organization as a strategic issue.

> ***Strategic Planning is a long term planning approach aimed at addressing where the organization is today, where it expects to be in the next decade and beyond, and producing a plan of how to get there.***

If you talk to the manager of a specific department or division they will obviously see the organization from the viewpoint of their particular specialty.

> For example, if an organization decides to offer a new product or service, the accountant may look at the change in terms of new line items in the general ledger; the production manager in terms of new production schedules; the procurement manager in terms of new parts to be ordered; the sales manager in terms of new promotional programs to be launched; and the Data Processing (D.P.) manager in terms of new transactions and programs.

Under this traditional regime, several systems in different locations may play a part in responding to a customer's need. And, because they are implemented separately, these systems fragment the overall response to new business issues or to changes in the existing business. The leaders of an organization, however, need a more strategic view for overseeing the introduction of a new product or service and for restructuring the existing organization to create a Customer Focused Organization that is not fragmented across departments, divisions, etc.

This strategic view should enable leaders to see the organization as a unified system that responds to outside customer needs, which is, after all, the main reason why the organization exists.

Of course, this relies on the organization having a specification of its essential business needs available to take advantage of new business opportunities, new designs, and new technology.

For swift response to a new or changing environment, an organization should ask some strategic questions:

- How quickly can we introduce a new line of business, or an extension/modification to an existing line of business?
- How much of our existing systems can we re-use?

And, to maximize profitability on existing products and services, the organization should ask:

- What are our key lines of business?
- Which ones are the most profitable?
- How can we streamline these areas of business to reduce response cost and time?

These strategic business questions are answered more easily in a Customer Focused Organization than in one with a traditional, hierarchical/fragmented design.

It's imperative for Strategic Planners to have available to them a model of the true business showing the whole organization, unencumbered by any design and implementation characteristics.

Many techniques for modeling an organization's operations focus on an organization's internal functions rather than on its interaction with the outside world. They concentrate on mapping processes and data onto *existing* departments, and on the proper configuration of *existing* procedures, including human and material resources. Even techniques that focus in on modeling the data (like data normalization) and ones that focus on modeling the processing (such as functional decomposition) are too often performed on the implementation data and processing of old environments. In this, they fell short of being true *strategic business* tools.

We will see in this book that many systems in place today were based on a faulty understanding of the business they were intended to support. These systems were built for reasons of old design and internal politics. Unfortunately, new systems are built simply to replace old systems that were themselves based on the partitionings of earlier manual and automated systems, so that the effects of old partitioning still haunt us today. Many of today's systems reflect how an organization partitioned its activities and information several years, if not decades, ago.

The fact that an organization's systems may support its response to *today's* business, however, is not enough to protect it from the devastating effects of not being able to support its response to *tomorrow's* business. Having invested huge sums in manual and computer systems, an organization may encounter the disaster of inflexible systems that stifle the fast reaction time needed to respond to changes in the market place from both a business and a technological view. Trying to change a crucial system to offer a new product or service can become a quagmire of *ad hoc* quick fixes that buries an organization in ongoing production and maintenance problems and even greater inflexibility. The Business Event Methodology discussed in this book will greatly aid in the Strategic Planning of an organization.

A Basis for Strategic Planning

To support effective Strategic Planning, we need a method for determining the essential business to which the organization must respond now and in the future, and for evaluating how well its internal configuration enables it to make those responses. The heart of such a method, I believe, is in the methodology stated in this book. With it, we can:

- Produce a Business Model in which changes to departmental boundaries, job descriptions, or implementation technology would not affect business issues.
- Clearly show where changes in business policy will require modifications in the Business Model (and hence its implementation), making it easier to identify and manage changes.
- Enable both business and technical staffs to provide realistic resource statistics for how the organization responds to its business, and provide statistics for the cost of new lines of business and potential changes to existing business.
- Easily add or remove entire business products or services with minimal confusion.
- Easily assess the costs and impacts of any technology changes.
- Support the re-use of functions and data.
- Support existing system integration to create a Customer Focused Organization.

Providing a Competitive Edge

Organizations in today's competitive environment are trying to bring their products and services to market faster than their competitors, or to respond more quickly to customers' demands. Speed often spells the difference between success and failure, or growth and stagnation of an organization.

> In banking, for example, the first bank to modify its systems to support a new financial service, or to take advantage of new legislation, can corner the market. Or, a delivery service might save a significant amount of time by seeing through the old design and implementing the same essential business needs with a new design to deliver items more quickly and/or for a lower price.

Technologies such as telecommunications and computers are already dissolving the old boundaries around organizations. Today, the technology is available to allow a company to install terminals at its customers' sites for direct order entry, or to order materials directly by hooking into its suppliers' computer systems. But removing the physical barriers between systems requires an organization to understand the *essential boundaries* between systems versus the *non-essential boundaries*. Essential boundaries are those over which we have no control — they must exist for the business to run. Non-essential boundaries are those based on some implementation or historical reasons.

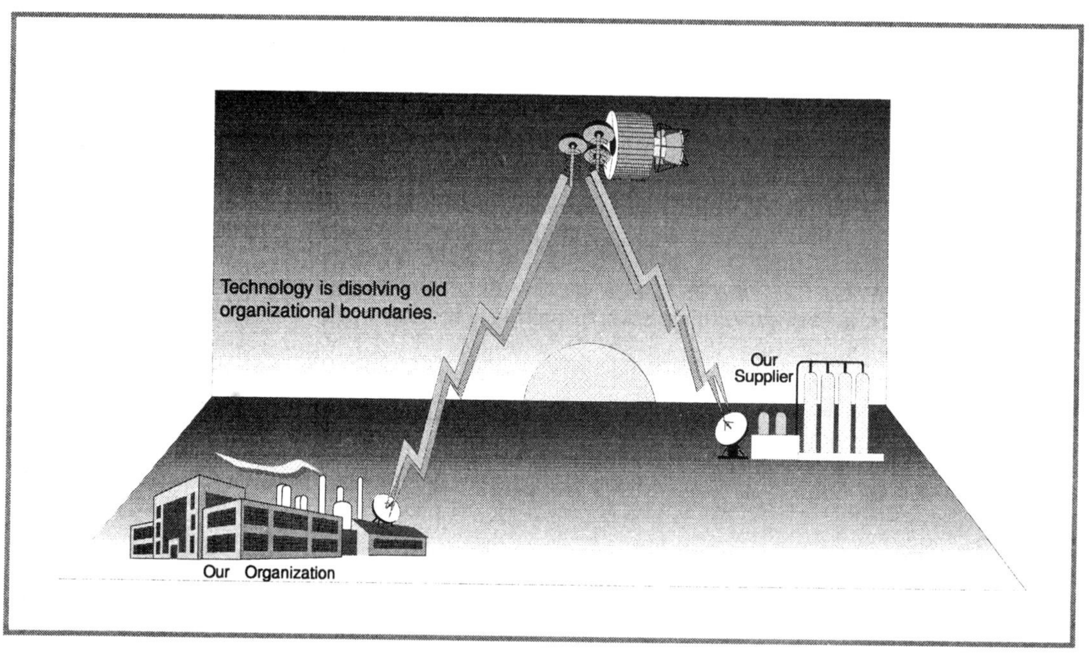

Needing to rewrite whole systems to benefit from new technology, such as the Internet, erasable optical disks, or voice recognition, is ridiculous and unnecessary. Similarly, not being able to absorb a complete new business product or service is equally ridiculous. An engineered system should accommodate both new business and new technology readily and it should give an organization a significant head start over any competitor with a non-engineered environment.

It's typical for people in a medium-to-large organization to be overwhelmed by the volume and complexity of its current implementation, and to not know the essentials of its true business functions. Organizations spend much effort on untangling disorganized existing systems when the effort could be applied to building new, engineered ones.

Until an organization introduces well partitioned, engineered systems, and keeps comparative statistics of the old and new designs, the degree of wastage will not be known, and senior managers will continue to be deprived of the tools they need for effective Strategic Planning.

A Strategic Business Planning Approach

The focal point of any Strategic Planning approach is the set of Organizational Objectives that defines where the organization is today, and where it expects to be in the future. Without this set of objectives the organization cannot define the processing and information it needs to support this transition.

The Key Organizational Questions

Figure 2–1 shows the strategic questions that an organization must answer. They are discussed briefly in this chapter and returned to in the ***Strategic Planning via Business Events*** Chapter.

Since Strategic Planning is more of a managerial issue than a technical issue, so I will only touch on it briefly here. ***Book II:*** of ***The Business Engineering Methodology*** series will go into more details on Strategic Planning leadership issues.

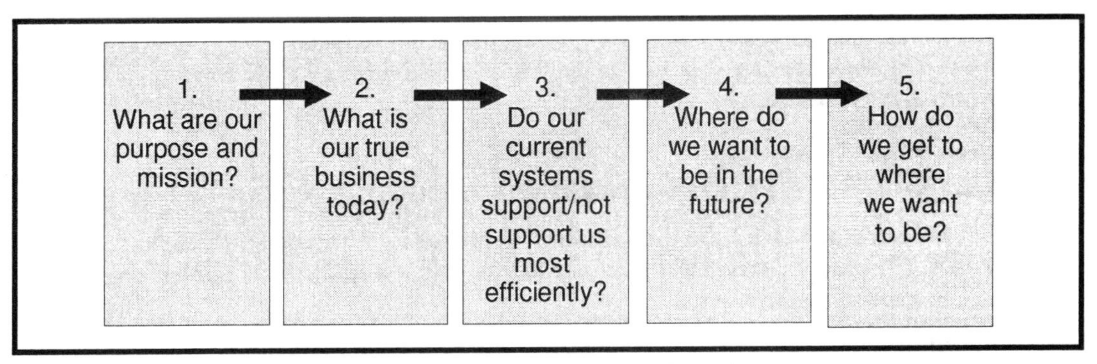

Fig. 2-1: Key Organizational Questions

1. What Are Our Purpose and Mission?

This is a description of the business our organization is in and what we want to be or do in current terms. This may be stated in a single line or paragraph Organizational Goal or it can be broken down into a set of Organizational Objectives. To be meaningful and useful these objectives should be quantifiable and measurable. They contain the "vision" of the Strategic Planners. All projects use them to develop individual objectives. The organization's Goal and its subsequent Objectives, as stated in the Mission Statement, should be displayed throughout the organization.

> ***A Mission Statement is a brief, clearly stated expression of an organization's reason for being in existence. It is ultimately what each and every employee is working to support.***

2. What Is Our True Business Today?

We have already mentioned that the managers in each department of an organization will typically have their own view on what the actual business is. Since the heads of an organization cannot take any particular department's view, I recommend answering this question with:

- A list of all customer requests or Events at the boundaries of the whole organization and the outside world to which it has decided to respond, with definitions of who or what is the customer and the Event. [1]
- Identification of critical line(s) of business (i.e., which of these Events are our reason for being in business, or our main revenue generators). (In a resource crunch, these would take priority.)
- A high-level map of our existing implementation (e.g., departments, divisions, computer systems, and computer platforms).
- A Business Objective or set of objectives for each customer Event to which the organization responds.

These Business Objectives are quantifiable, measurable statements of required performance. The objectives must be stated such that it will be quite clear when they are, or are not, being met.

3. Do Our Current Systems Support/Not Support Us?

We must evaluate how well the organization achieves its purpose and mission. We do so by analyzing how the currently implemented systems respond and contribute to the organization's purpose and mission.

One way of assessing how we are doing is to see how much any specific customer request is fragmented across existing systems (manual and automated). Also, given each business objective, we should be able to assess how we are meeting these objectives by projecting the optimum response/performance statistics and compare them with the current statistics. *(I find that organizations have reasonable statistics or metrics of their operating costs for their manual systems, but not for their computer systems. So, this step may be difficult, but nonetheless essential, for the computer environment.)* Thus, the organization's purpose and Mission Statement (Goal) decomposes into specific Business Objectives, which in turn decompose into specific Business Policy for each customer request.

1 I say "whom or what" is the customer because the customer isn't always a human being. For example, a computer system may be the customer of an automatic system or of a fax machine from an outside organization. In one project that I worked on with the Bureau of Land Management, the land was the customer and the Events were such things as fire, earthquake, flood, etc. to which the organization had to respond to meet its mission.

4. Where Do We Want to Be in the Future?

In addition to evaluating the organization's current position we must also identify what changes, additions, or deletions are required for future business opportunities. These can also be stated in high-level terms, but again, they must be measurable. Whenever I think of a future Goal with measurable Objectives, I remember the one declared by President John F. Kennedy on May 25, 1961:

> ***"I believe that this nation should commit itself to achieving the goal, before this decade is out, of landing a man on the moon and returning him safely to the earth."***

This was a succinct and measurable organizational goal.

The Strategic Planner is not just responding to the competition's actions. An organization's leaders need to set new directions, but this is not easy when they themselves are typically blinded by the existing design of their organization. One easy way to avoid this is again to focus on the customer and their Events (e.g., to which new Events can we respond?).

When an objective, business view of the organization is developed, it provides a platform for the introduction of many new products and services. Seeing the business essentials of the organization, as opposed to how it is currently implemented, opens up a world of opportunities for expansion and improvement.

5. How Do We Get to Where We Want to Be?

Unfortunately, we can't do a "big bang" restructuring of an existing organization into a Customer Focused Organization by completely shutting it down and starting it over from scratch. We have to engineer the organization's activities *in situ* via a Strategic Plan.

Preparing a Strategic Plan presents us with an opportunity to examine the relevance of an existing organization's partitioning with objectivity, to select what works, and to replace what doesn't, all with the overriding goal of optimizing the organization's response to its customers for its critical areas of business.

Summary

In many organizations, systems have grown historically over the decades. The departments, divisions, computer systems that are in place were typically built with an internal focus. Strategic Planning should take an external focus from the point of view of the customer.

To do this effectively, the leaders of the organization need to create a Strategic Plan. The best basis for this plan is from the point of view of the customer and the Events that they initiate from outside the organization. Even though individual, isolated gains can be made by using the concepts in this book on a system-by-system basis. To gain the optimum benefits, I advise applying this Strategic Plan organization wide. The core team responsible for this organization-wide task should not be aligned with any particular organizational unit.

An organization *must* be prepared to change at every level. The hardest part to change will be the business culture. A Strategic Plan based on Business Events will help facilitate human change. It will also assist the staff in identifying with the big picture of improving customer service, meeting the challenges of global trends, and increasing the quality of life.

Chapter

3

Systems Archaeology

What perception sees and hears appears to be real because it permits into awareness only what conforms to the wishes of the perceiver. This leads to a world of illusions, a world which needs constant defense precisely because it is not real.

When you have been caught in the world of perception you are caught in a dream. You cannot escape without help, because everything your senses show merely witnesses to the reality of the dream...

We look inside first, decide the kind of world we want to see and then project that world outside, making it the truth as we see it.

A Course in Miracles
The Foundation for Inner Peace[1]

I believe that the vast majority of systems in place today are badly designed. This statement applies equally to manual and computer systems. These badly designed systems create many problems for the overall business and a serious impediment to satisfying customers.

If we look carefully at these old systems (and especially how they are partitioned), we see that they cause business problems such as:

- *Slow Response to Customer Needs* — because the separate systems wait on each other.
- *Inflexibility in the Face of External and Internal Change* — because processing and data are fragmented across systems.

1 The quotation on this page is from the Preface to *A Course on Miracles* ®, © Copyright 1975, and is reprinted by permission of the copyright owner, Foundation for Inner Peace, Inc., P.O. Box 1104, Glen Ellen, CA 95442.

- *Unnecessarily High Costs of Doing Business* — because the organization expends effort and capital in supporting the systems instead of the systems supporting the organization.
- *Duplicated Processing* — because one system receiving data from another may not "trust" the source system's accuracy and repeat some of its processing (e.g., validation) out of self-defense.
- *Duplicated Data* — because individuals and computer systems tend to have their own redundant sets of data accessible for their processing (coupled with expensive or no consistency checks to support this duplication), individual data elements up to whole files are duplicated.

Even though many existing systems have the above problems, we can't just throw them away and start from scratch. First, because they typically represent a huge financial investment and can't be replaced overnight, and second, because poor documentation often leaves these dinosaurs as the only existing repository of the underlying essential business logic and data. This especially applies to old computer systems. Many organizations evolved in an environment of "quick fixes" and deadline-driven, non-engineered systems projects. This environment presents an enormous obstacle to organizational flexibility and competitiveness. However, there is a benefit to studying these old systems, as George Santayana said:

Those who cannot remember the past are condemned to repeat it.[2]

Knowing Who's the Prisoner

In my younger years I traveled the world and, on one of my sojourns, I ended up in India. Many of the ways of thinking I encountered there have stayed in my memory. A person by the name of Ram Dass (a former Harvard University professor) also traveled to India. His experiences — like mine — changed his life. In his book, *Be Here Now*, he quotes George Gurdjieff who once said:

"You don't seem to understand. You are in prison. If you are to get out of prison, the first thing you must realize is that you are in prison. If you think you're free, you can't escape."[3]

I believe that many of the ways we have implemented a business imposed prison-like restrictions on us. Knowing we're restricted is the first step to change. I hope this book demonstrates that the business world is trapped by old, outdated designs, and that the book offers a means of escape.

---

2 *The Life of Reason*, George Santayana — See Bibliography

3 *Be Here Now*, Ram Dass — see Bibliography

What Obscures the Essential Business?

When we analyze an existing system, we take off the cover (dis-cover) that was put in place throughout the life of the system. However, we also go back further into the system development effort that originally put the design in place.

When I did real work for a living, I worked in data processing. My career path took me through computer programming, then into computer systems design, and then into system analysis. In systems analysis I realized that it didn't matter whether the system I was analyzing was currently implemented with human beings or computer programs. Both of these "implementation" views had to be removed to get a non-corrupted view of the essential business.

Deriving a view of the business that removes any implementation details is the most important step in creating a Customer Focused Organization.

Figure 3–1 shows the phases in the "old" (and for that matter, "new") system development life cycle that deliberately obscure the essential business. When not seen as old design issues, the existing "cover" tends to lead designers of new systems into the trap of "cloning" outdated, bad designs.

Fig. 3-1: Discovering the Essential Business

Many existing systems, be they manual or automated, are still partitioned for historical reasons. *(Remember, I prefer to call them "hysterical" reasons.)* Often this historical partitioning obscures the true nature of the business. Everyone can relate to the term "government bureau" as well as to "corporate department" because they both obscure the mission of the organization. I believe this partitioning is the biggest obstacle to becoming a Customer Focused Organization.

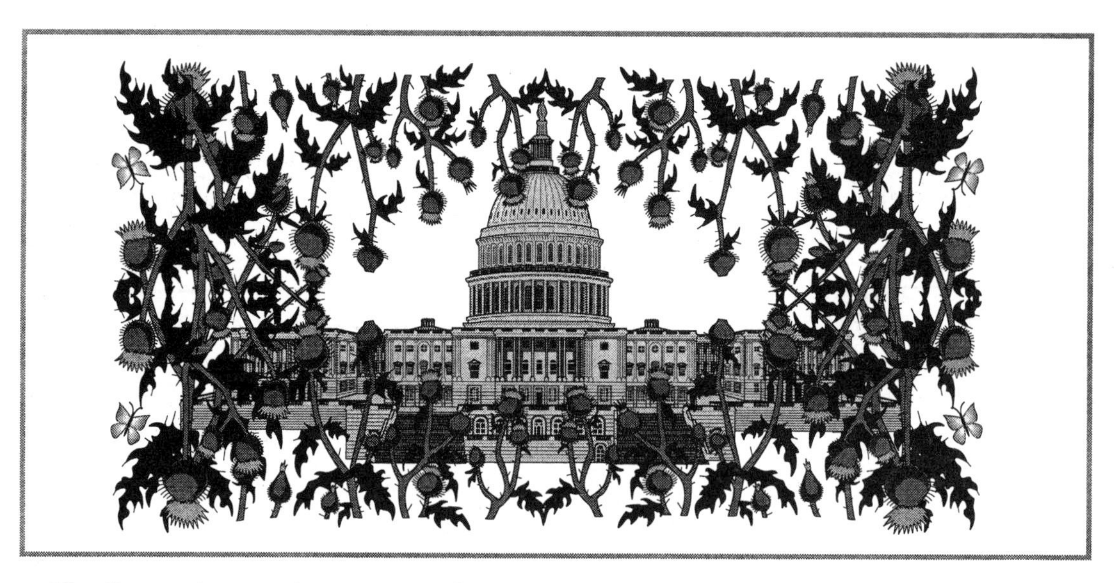

The first major step in creating a Customer Focused Organization is to "dis–cover" the underlying business purpose. Then we can engineer a new design that will support this essential business view and not just use new technology that re-implements the old system faster. The original needs of an established organization are invariably buried in the over-growth (the "cover") of the organization 's processes and data (as shown in Figures 3–2 and 3–3).

OVERGROWTH (THE COVER)	EXAMPLES OF OVERGROWTH		
PROCESSES: Computer System Boundaries	Accounts Receivable	Accounts Payable	General Ledger
Computer Program Boundaries	EDIT	UPDATE	PRINT
Computer Processing Limitations	PC 8086	64 KB Memory	8 Character Filenames
Computer System Controls	Job Control Language (JCL)	Batch Tape System	Online System
Human System Boundaries	Bureau of the Exterior	Accounts Dept.	Shipping Division
Human Job Titles & Boundaries	Accountant	Sales Clerk	Dispatcher
Human Physical/ Logical Procedures	1. Unlock cabinent.	2. Copy invoice.	3. Lock cabinet and store keys in safe.
Human Processing Limitations	to Work Day	Speed	QWERTY Keyboard

Fig. 3-2: Process Overgrowth Examples

All the examples in Figure 3–2 are process-oriented designs (i.e., what we "see" associated with the processing in place at an organization). We must see though these when analyzing our business. We "dis-cover" the organization further when we remove the overgrowth associated with its data (see Figure 3–3).

OVERGROWTH (THE COVER)	EXAMPLES OF OVERGROWTH		
DATA: Computer Screen & Report Layouts	Name Address City State Zip 80 Column Screen	Name Address City State Zip 132 Column Report	AUTOEXEC.BAT 8 Character Names
Computer File Access Methods	Relational DB	Sequential Tape	Parent Child Child Search Parameters
Computer File & Database Capacity	10 Gigabytes	1.44 Megabyte	Record Layout
Human Preprinted Forms	8.5 x 11	3-part Forms	Index Card
Human Data Access Methods	Product Catalog	Random Storage Brain	Alphabetical Supplier List
Human Data Storage Capacity & Location	Fred's Customer List	7 ± 2 Things 4-digit PIN	Long/Short-term Memory

Fig. 3-3: Data Overgrowth Examples

All of the examples in Figure 3–3 are data-oriented designs (i.e., what we "see" associated with the data in place at an organization). Again, we must remove these aspects when analyzing our organization. Both the process- and data-oriented "overgrowth" prevent us from seeing an *essential* or *functional* view of the business. Before we delve deeper into systems archeology, let's briefly examine a little of the history of functional partitioning. Then later in the ***Partitioning by Business Events*** Chapter, we'll arrive at a definition of *Essential Business Functionality*.

What Is Functional Partitioning?

In the late 1960s two people in the Data Processing industry, Ed Yourdon and Larry Constantine defined in their book, *Structured Design*, seven levels of "cohesion" of a computer system module or process.[4] They defined the best level of cohesion as "functional." A piece of software at this level of cohesion does only one thing and has minimal coupling (interfaces) to other pieces of software. Figure 3–4 illustrates the difference between a non-functional "do everything" software module and a functional, "single-minded" module. The main point is we can evaluate the functionality of those modules by looking at the data that couples modules together.

4 *Structured Design* by Ed Yourdon and Larry Constantine — see Bibliography

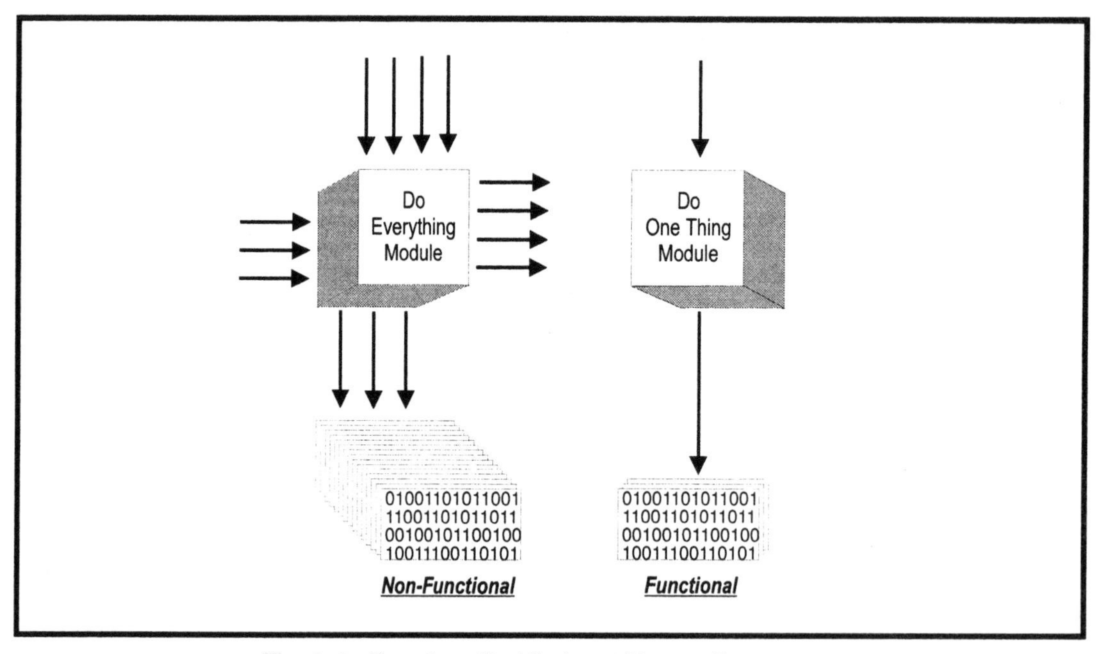

Fig. 3-4: Functionality Via Input/Output Parameters

In the late 1970s another D.P. professional, Tom DeMarco, wrote the book, *Structured Analysis and System Specification*[5], and in it defined the most functional business process as the one with fewest interfaces (Data Flows). The book advocated constantly re-partitioning the model of a system (via a Data Flow Diagram or DFD) to obtain minimal Data Flows for one complete task. Figure 3–5 illustrates the difference between a non-functional and a functional process. Again, the main point is that too many inputs and outputs around a task or process indicates poor functionality.

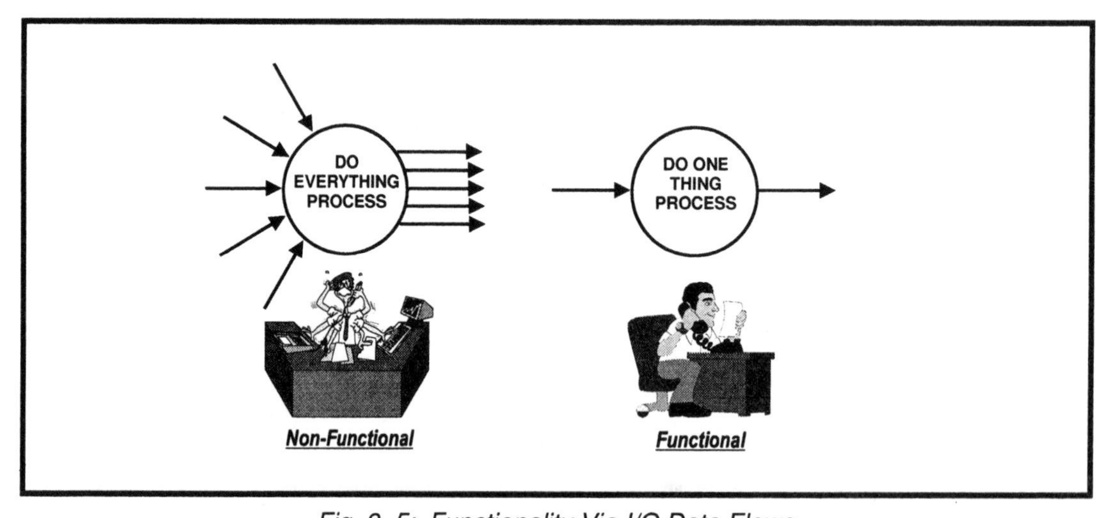

Fig. 3–5: Functionality Via I/O Data Flows

5 *Structured Analysis and System Specification*, Tom DeMarco — see Bibliography

Based on these definitions we can see that one functional view, or *partition,* would be based on grouping together only the processing and its essential data that our organization needs in order to satisfy a single request from outside the organization. Given one interface from outside our organization, we bring together into a single partition all necessary business logic and only the data needed to support this logic.

In contrast to this view, suppose we took an incoming Customer Order and followed it through a **typical**, traditionally partitioned system (a non-Customer Focused system) we would probably find:

- A ridiculous "trail" left over from poorly-fragmented designs.
- A series of poorly implemented, quick fixes.
- A history of poorly planned system development efforts.
- Massive duplication of processing and mass redundancy of data capture and storage.

We could draw pictures of this incoming order, charting it as it bounces through the various programs and systems (see Figures 3–6 and 3–7). Figure 3–6 shows a simplified representation (model) of a typical organization's order processing. In this type of environment each order has to find its way through a variety of totally distinct departments and systems, before the customer receives the requested product or service.

If we took the same order going through the same set of departments and systems, and we modeled the processing boundaries that the order has to traverse, it would look like the tire tracks left by a demolition derby (see Figure 3–7)!

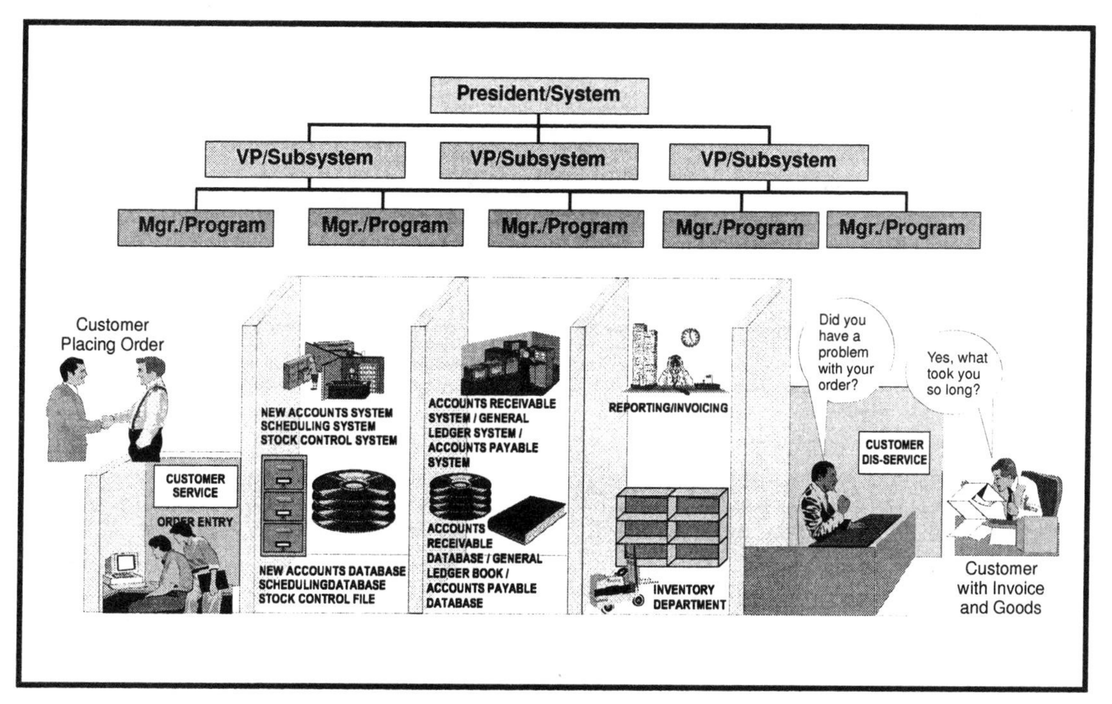

Fig. 3-6: Old Partitioning to Fill a Customer Order

Note that Figures 3–6 and 3–7 do not show a functional partitioning of systems, even though this common partitioning into department and system boundaries is called "functional" in many organizations. Although we might expect this chaotic partitioning in a human-based system, this poor design is also often perpetuated in automated systems.

Our first task in creating a Customer Focused Organization is to remove historical system barriers — both technical and managerial — that stand in the way of providing ultimate customer satisfaction.

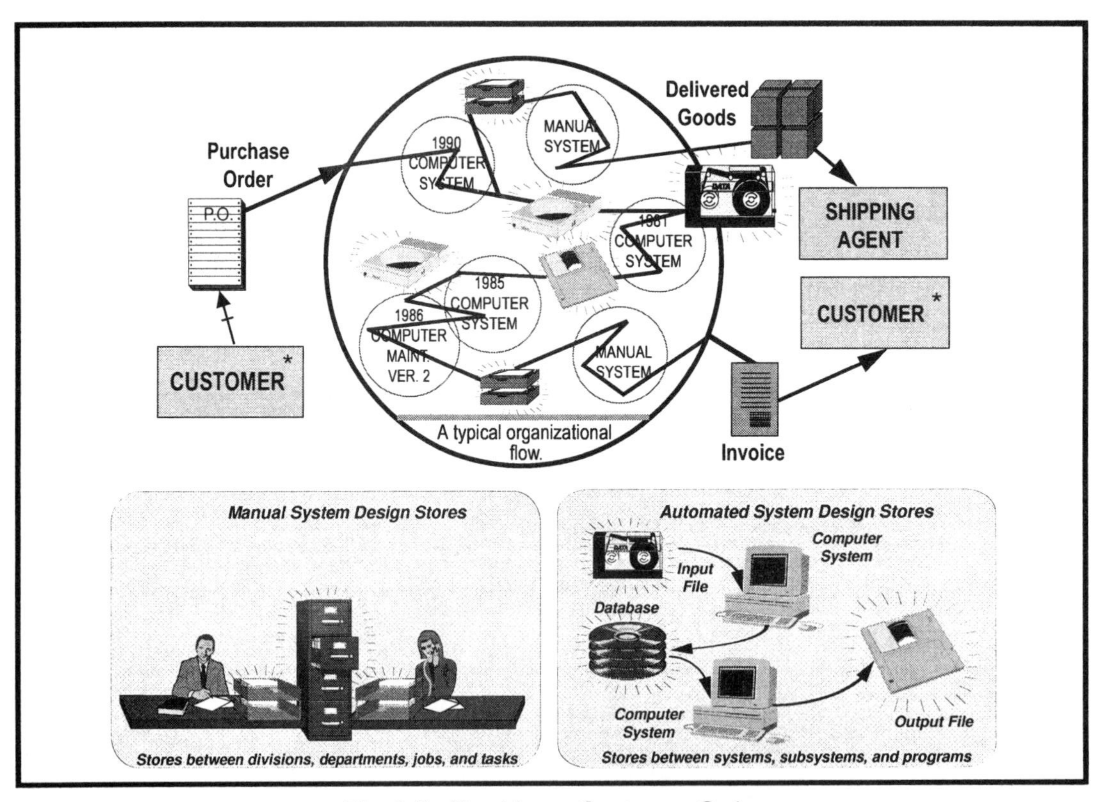

Fig. 3-7: Tracking a Customer Order

Poor and non-functional partitioning leads to unnecessary complexity and "bugs." It seems everyone I meet understands the term "system bug." It's not uncommon to see front page headlines or a TV news commentary on an organization's silly procedures that sometimes have devastating results in the outside world.

For example, a utility's customers would be irate if they were double-billed or had their service cut off due to a computer system bug. Or an employee following poorly-designed procedures "to the letter" could cause discontented customers. In these examples somehow the implementation of the procedures corrupted the intent of the Business Policy Creators.

This brings home the fact that we need good design in computer and human systems. Despite the fact that we now have the tools to **engineer** manual and automated systems, we still find today that we're rewriting computer systems from the last decade, and even some that were built less than five years ago. In addition, we see organizations revamp their procedures and management structures constantly. The useful life expectancy of some systems is often not worth the development investment.

We should be building systems that last as long as the business they support.

"The Designer Made Me Do It"

So, how did we get in this situation and how is it affecting our organizations? Designers of systems (myself included) have made a number of mistakes in the past. These mistakes (the "overgrowth" I listed in Figures 3–2 and 3–3 can be summed up as a series of what I call: "Archaeological Wrong Turns." They include:

- **Wrong Turn #1 — "People" Partitioning in Manual Systems**
- **Wrong Turn #2 — Computer Partitioning in Automated Systems**
- **Wrong Turn #3 — Program Partitioning in Automated Systems**
- **Wrong Turn #4 — Data File Partitioning in Automated Systems**
- **Wrong Turn #5 — File Partitioning in Manual Systems**
- **Wrong Turn #6 — After-the-fact Quality Control**

Imagine that we are system archaeologists digging about in the remains of a typical existing organization. What might we find? How did it get to be designed like that? The archeological dig we're about to undertake can be costly depending on the degree of "digging" you want to perform on your existing systems; however, you will reap more benefits as you dig deeper (see Figure 3–8).

Figure 3-8: An Archeological Dig Through Systems

Wrong Turn #1 — "People" Partitioning in Manual Systems

An "archaeological dig" of an organization partitioned with human beings in mind might uncover the reason for certain partitions. These include reasons such as being divided into an Accounts Department, a Stock Control Office, a Manufacturing Division, and so on.

Figure 3-9 shows the departmental partitions and lines of communication we might find in our archaeological dig of a typical manual environment.

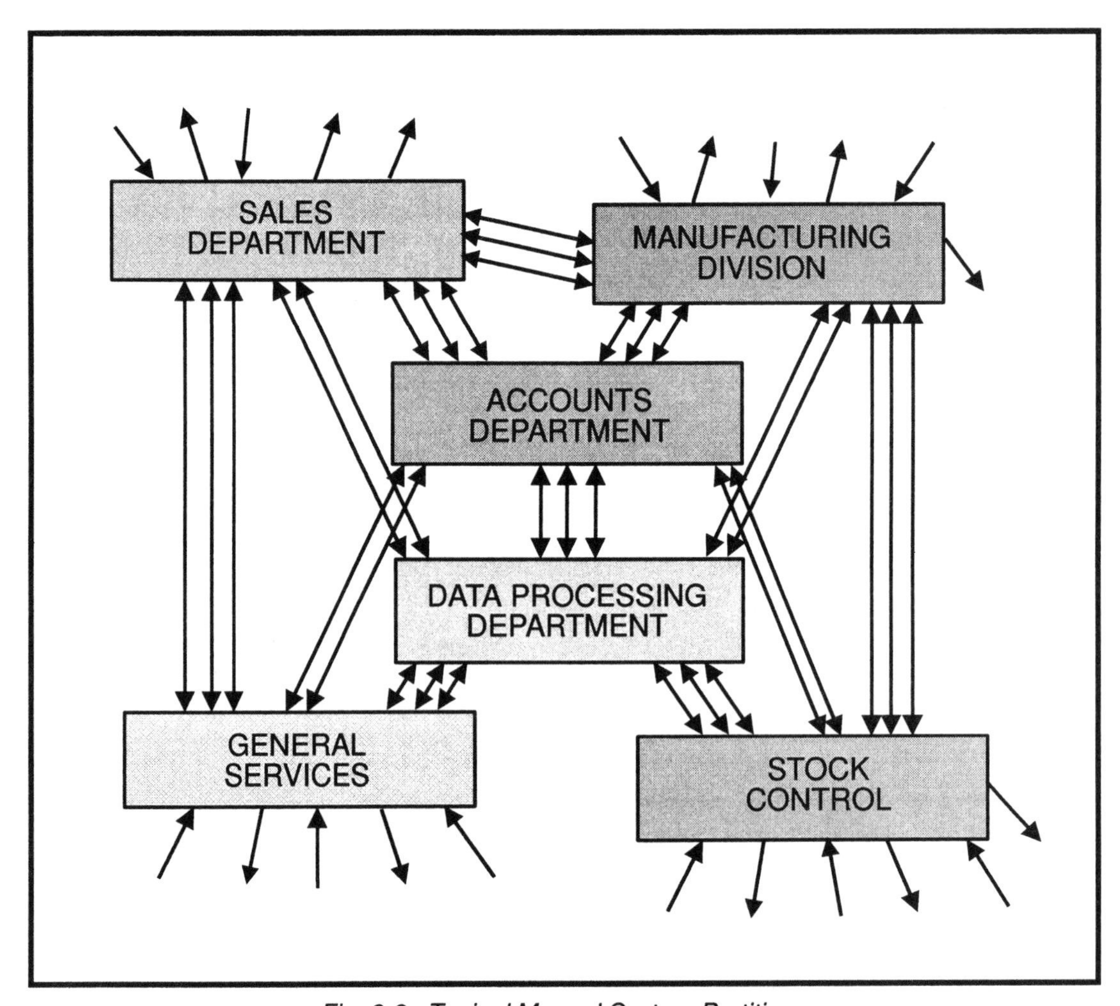

Fig. 3-9: Typical Manual System Partitions

The partitions of Figure 3–9 are very common in manual systems for a number of reasons:

- People with the same kinds of skills get grouped together.

For example, people who are good at selling get hired into the "Sales Department," people who are good at bookkeeping are assigned to "Accounting," and people who can write computer programs have glamorous job titles and get to work late in their cubicles in the "Programming Department." (In the past putting together in one manual environment all the people with the same certain aptitude and skill made a lot of sense from the point of view of economy of scale.)

- People or facilities may be concentrated in one area, building, or branch location based on the location of raw materials, ease of transportation, or customer base distribution.
- Physical space availability may have split a group of people up into different buildings or cities, or forced different groups into being situated together regardless of having different functions.
- Security considerations often isolate one group of workers from others.

For example, Payroll is often a separate group, physically as well as administratively

- Historical reasons such as "It's always been that way." *(I call this a "hysterical" reason because it's the irrational response usually given when someone asks why things are grouped unreasonably.)*
- Political reasons, such as when a manager of a division who has the political power to create deparíments under him or her captures work that really belongs to another division. This gives the manager more power in the organization.

The most common partitioning basis in the above examples is *similarity of task.* We will come to see that this is not a functional partitioning.

I hope to convince you that similarity of task and the other rationales I've listed are very poor reasons for creating departments. None of the above reasons for creating a department reflect an organization that is Customer Focused.

Similarity of task is not the basis of a truly functional partitioning.

These implementation issues are often the sole basis for many *so-called* "business" partitions. Although these all seem to be valid *manual* partitions, and the structure of Figure 3–9 is an extremely common partitioning, it's not a *functional* partitioning.

Wrong Turn #2 — Computer Partitioning of Automated Systems

If we extend our archaeological dig to the world of computer systems, we find a remarkable similarity between the partitioning of computer systems and that of manual areas.

Figure 3–10 shows the structure and lines of communication we might find in an automated environment. Notice how this is a similar partitioning to Figure 3–9. The exception is the absence of a General Services System. The General Services Department tended not to get computer system support, possibly because it had no budget for automation or had insufficient political clout.

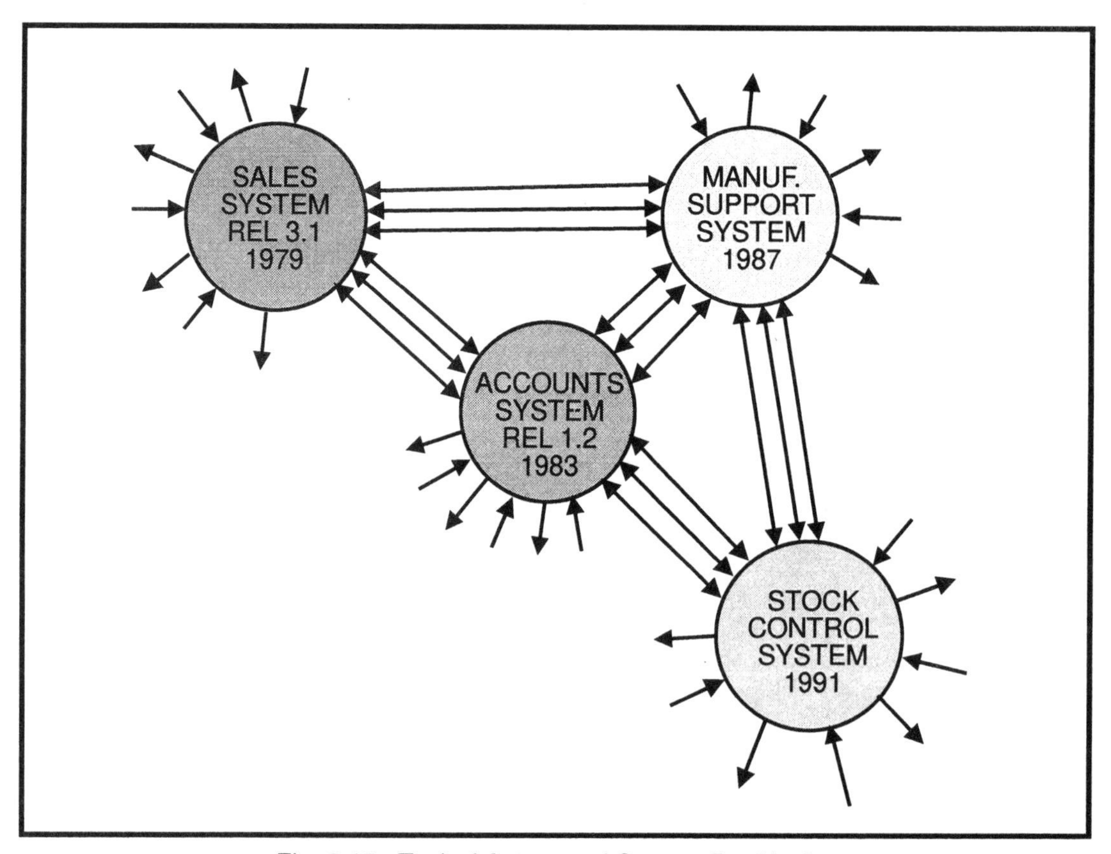

Fig. 3-10: Typical Automated System Partitioning

What this obviously points out is the reasons for creating manual system boundaries were blindly used as reasons for computer system or program boundaries. This led to computer systems that were partitioned for silly reasons when looked at from equivalent computer system partitioning viewpoint.

For example, we don't put all mathematical logic together in one program because of the math skills of that program — all programs can do mathematics (i.e., they have the same skill). It would be equally silly to put all the "IF" statements together in one program because that program was good at making decisions?! So, it's ridiculous to use this old manual partitioning for computer system or program partitioning; however, every company I've seen has these types of systems.

Another example of how the past can influence us into repeating an archaeological wrong turn is the use of double entry bookkeeping in automated accounting systems. Double entry bookkeeping was invented around the turn of the 14th century by Italian accountants to reduce the chance for human error when maintaining records for banks and merchants. Because computers are not prone to making errors while adding columns of numbers, there's no logical reason why we should design this time-consuming, double checking redundancy into a modern automated system (other than the fact that it's now considered part of "Standard Accounting Practices"). If a human being outside the computer system really needs to see a balanced set of debits and credits, we can print the same figure(s) from the computer system twice in two separate columns.

The partitions of Figure 3–10 are very common in computer systems for a number of reasons:

- *Automated systems have traditionally been built along manual system lines — accounting systems, sales support systems, stock control systems, and so on.*

  It seems only natural to assign project teams to build systems along the lines of internal boundaries, such as Accounting, Sales, Stock Control, and Manufacturing. After all, the money for systems development is usually partitioned and acquired (by the internal departmental boundaries and their management). *(This can be salvaged by a "neutral" D.P. Department budget, but if the department has a budget, it is very often used to buy computer hardware and support software, or to rewrite bad batch systems into faster, bad on-line systems.)*

- *The availability of machine time and capacity often led us into batching system functions together.*

  Even though a function might logically belong on its own, the lack of machine capacity may have forced the designer to squeeze together functional units that did not really fit. Functions which could be separated on-line, such as individual transactions, found themselves batched with other transactions into a reporting system to keep the number of separate computer jobs down and save machine time. In this "batched mentality" category there are many old design partitioning reasons, and all of them are very strange from a Customer Focused point of view.

For example, how about the practice of putting together a system just to enter all orders (i.e., an Order Entry System)? Grouping different orders together is as bad as grouping different reports together.

- Budget/Political boundaries are as influential here as they are in manual systems.

  In my consulting work, I've often encountered systems that would have been far more business-efficient if their boundaries could have been stretched a little to complete some functional processing that was outside the project scope. The complexity of budget administration is often so great that it forces system boundaries to align with the budgets of the project requester's departmental boundaries. Also, a manager of one department may not want to give away any power or have their budget used to study a competing department. So, computer system boundaries matched political power boundaries.

- *External constraints often affect system boundaries.*

  Government deadlines, for example, may mean an organization must focus all its efforts on the immediate scope of the mandated change or the new requirement boundary. It may also mean that we can't afford the time or resources to look beyond it. Given the time to study further, we may fold the new requirement into existing functional areas instead of separating it out with duplicate extracted data files and processing. (You should be starting to see a vicious cycle — with a poorly partitioned system, changes are difficult to make so separate systems/programs get built out of place to satisfy immediate demands.)

- *The packaged software industry and PC hardware often imposes its own partitioning.*

  Software packages obviously have to target specific, already existing departmental partitions in order to make sales. So again, they follow traditional departmental boundaries for their packages to sell to department management, and hence reinforce those boundaries.

For example, payroll systems are sold to the Payroll Department, stock control systems to the Warehouse Manager, etc. Also, PC hardware and PC-based software packages are often "kludged" onto an existing system (as opposed to being engineered in). This often results in fragmented and duplicated data processing efforts.

Our data processing systems are rarely based on true functional partitioning, but rather on old manual design boundaries or new technological boundaries.

Wrong Turn #3 — Program Partitioning in Automated Systems

Let's now go one layer deeper to look at program boundaries inside a typical computer system. In Figure 3–11, we see typical computer program boundaries. These programs were based on similarity of tasks, such as the "Edit program, Update program, Print program" or "Input program, Process program, Output program" structures.

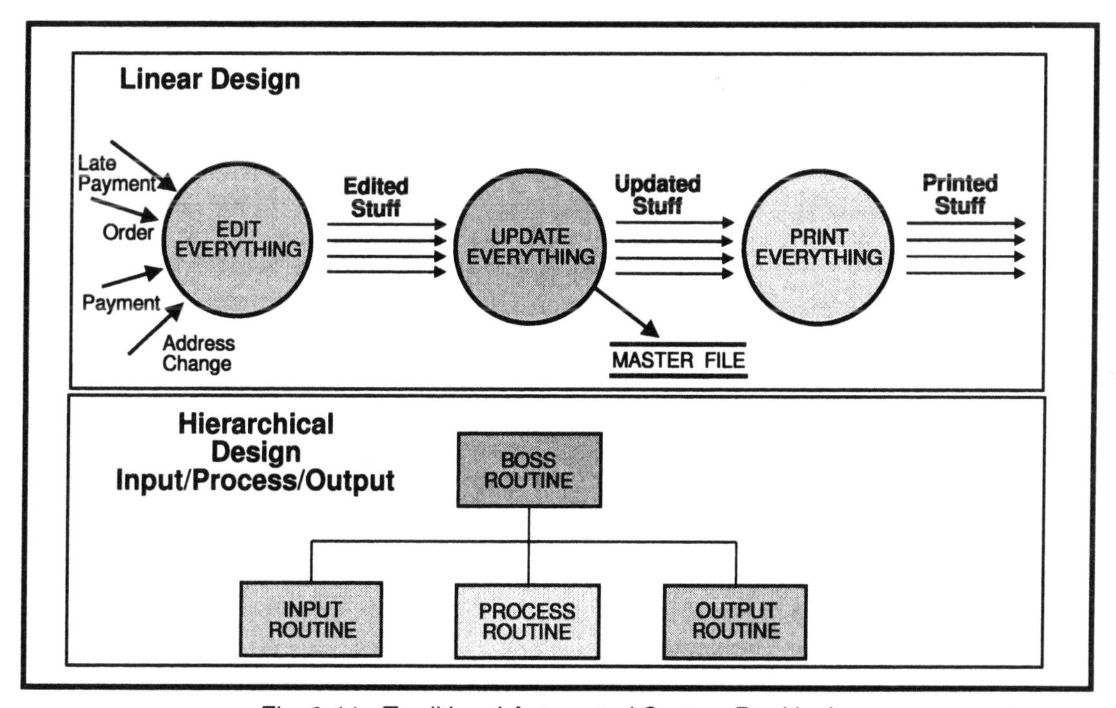

Fig. 3-11: Traditional Automated System Partitioning

Obviously, these kinds of programs were large. They typically formed many thousands of lines of code because all transactions traveled through them. With this type of design, multiple inputs are passed through common system partitions, such as one huge, hard-to-maintain, monolithic edit program or an "everything-but-the-kitchen-sink" program.

Many computer programs are also poorly partitioned because:

- *Technology constrained design.*

  The D.P. Industry historically began by focusing primarily on the *processing* aspect of automated solutions to business problems, and the data aspect was seen as secondary. The data aspect was something to be addressed by the programmers. The low speed of early computers and peripherals meant that individual processes had to be batched into monolithic "lumps" and a programmer would strive for "efficiency" by processing as much as possible with one program or job. The data was lumped together into large transactions and master files to be fed to the monster, nightmarish, multi-thousands of line programs each night. Some of these unmaintainable programs still cause organizations problems today.

Another excellent example of the constraint of technology is the anachronism of the QWERTY keyboard layout. This keyboard was originally arranged to slow down typists on old, manual typewriters, otherwise the strikers would jam together. This layout is still around and still limiting keying rates, even though there are no strikers to jam on today's electronic keyboards. We're therefore stuck with a keyboard layout purposely designed to slow down data input and hence, it is the worst possible design for today's needs.

- *Programmer skills often constrained early systems.*

  It's obvious that if you're skilled in only one technique, such as how to build batch systems, you'll keep building using that technique. I still see many on-line systems that have internal logic from batch processing days. Another example would be to have somebody who is good at a particular computer programming language (e.g., COBOL or Assembler) write every process they are asked to automate in that language, even though it may not be the most appropriate language for the type of processing required.

- *Highly-procedural programming languages also constrained processing structure.*

  Languages constrained the way in which business systems were viewed by program developers. This is why Object Oriented Design and Programming views can be difficult to pick up by programmers who started with a Process Oriented language view. Also, some old computer languages dictated the design of your program (e.g., RPG and COBOL with its internal sort feature).

- *Programs were often built around a master file or a particular file structure (e.g., sequential or hierarchical).*

  It is not uncommon to find a "monster" program into which every function that wanted access to master file data was incorporated. (This is analogous to a manual system that revolves around, say, "Fred's Ledger," and the only way to access the data in this ledger is to work your way into Fred's busy routine.)

- *Resource availability often influenced the structure of a system.*

  Some program designs may be based on the availability of file data, machine speed, storage capacity, or machine availability. These are ridiculous reasons, when you think about them, for structuring new business systems. You know this hardware platform will be obsolete in a few years and then the programs will have to be rewritten for the next platform.

  Using the house building analogy, can you answer the question, "Can you tell how many carpenters worked on building a house by counting the number of rooms?" Of course not. A house's structure is not based on such an arbitrary physical partitioning as how many people were available to build it. But, if you look at Figure 3–11 and ask, "How many programmers worked on this system?", the answer is likely to be four or some divisor of four because there are four programs. If the project team had consisted of five programmers, we would probably see five programs. Creating separate programs for each programmer working on the project is based solely on implementation resources and not on functional partitioning.

Wrong Turn #4 — Data File Partitioning in Automated Systems

So far, we've been concentrating on the processing side of our archaeological dig. We see exactly the same problems, however, when we start digging through old data partitioning. We see old automated file boundaries based on:

- *Everything anyone could ever want to know in one record.*

  This is the old "kitchen-sink" master file containing a record structure of several hundred or even several thousand characters (bytes). On one of my consulting assignments, I encountered a 22,000 byte-record customer file, built initially with the best intentions of satisfying one system's needs, but which just grew and grew over the years as each new application wanted to store more data. This file is typical in that it contained masses of "dead" data no one had ever removed because they assumed someone somewhere might be using it.

- *Limited storage devices which limited the file structure.*

  This reminds me of when I was a young programmer, working at a company where the D.P. Department owned only two tape drives and no disk storage. We were *forced* into batch processing (reading from one tape drive and writing to the other). But when direct (random) access storage devices became available we just put the batch file structure onto disk with no structural changes! The resulting structure looked something like Figure 3–12. (It's unfortunately quite common to see a blind evolution from old file designs to new file designs, without taking advantage of the opportunities for improvement — especially where a conversion deadline drives the task.)

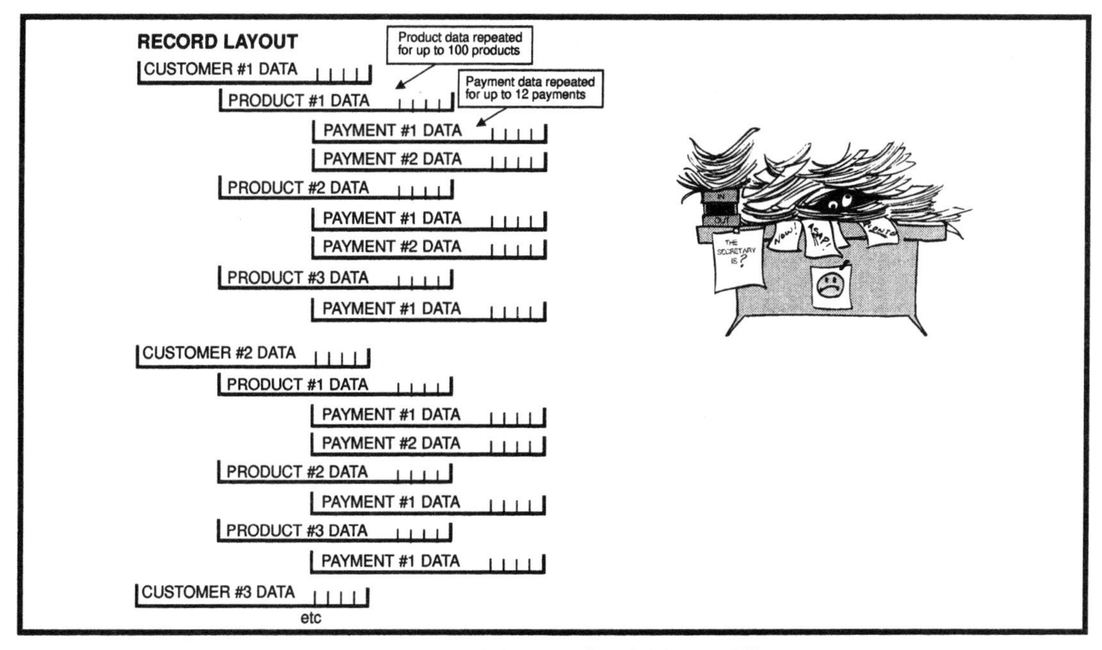

Fig. 3-12: A "Kitchen Sink" Master File

These files contained bundled information and arose when the previous designer grouped separate sets of data together for the sake of efficiency. This was very common of the files used in "edit-update-print" computer systems because this data partitioning was based on process bundling for technology and efficiency. In Figure 3–11 the data between the processes Edit-Update-Print would typically be bundled into the Edited data file and the Updated data file.

- *Misguided access efficiency decisions.*

  These decisions produced file structures based on the least number of READ instructions in a program, so that the data we needed was all "right there" in one record. *(When combined with the "kitchen sink" master file wrong turn listed above, the resulting huge record structure dimmed the computer room lights whenever a program READ statement was executed. This is like receiving the total **Encyclopedia Britannica** when you asked for the answer to a simple question.)*

- *PC/Local Workstation Partitioning.*

  With PCs becoming so inexpensive it's easy for the average organization's employees to requisition computers to store and process data at their desks. Of course each PC user then needs to have data that is really *global* in nature stored *locally* on their machine (and in their software packages' formats). This leads to mass data redundancy, stored dead data, and data synchronization problems (e.g., they have a customer's new address on their *local* PC, but the company's *global* data doesn't get updated with this new address and besides, it's in the wrong format). Client/Server/Networking technology is helping with this, but again only when analysis is performed first, not when the flaws of old designs are brought forward in new designs.

Wrong Turn #5 — File Partitioning in Manual Systems

We can find the same problems with old file boundaries in a manual environment (e.g., monster tub files or a doctor's office manual files), but it is more common to see the opposite (fragmented data files resulting in mass redundancy), e.g., my name probably appears in at least five files at my bank) and non-synchronized data (e.g., my address is potentially different across all five files). In Figure 3–13, we see Accounts Receivable, Manufacturing, Customers, Collection Offices, etc. all have their own "local" files. The old system's designers had to create these files so that one department could accumulate documents while waiting until the scheduled time for another department to pick them up.

With these local files in Figure 3–13, it's common to find the same customer information in all of the files. Customer details may appear in three places in the name of easy manual access, because to deal with a customer inquiry in Sales, the Sales staff must have access to Customer Orders without running over to Accounts or Manufacturing. Similarly, the Accounts Department staff must have access to customer data without going to the Sales Office. If a customer sends in a change of address, the chances are that only the file in the department that happened to receive the notification will be updated.

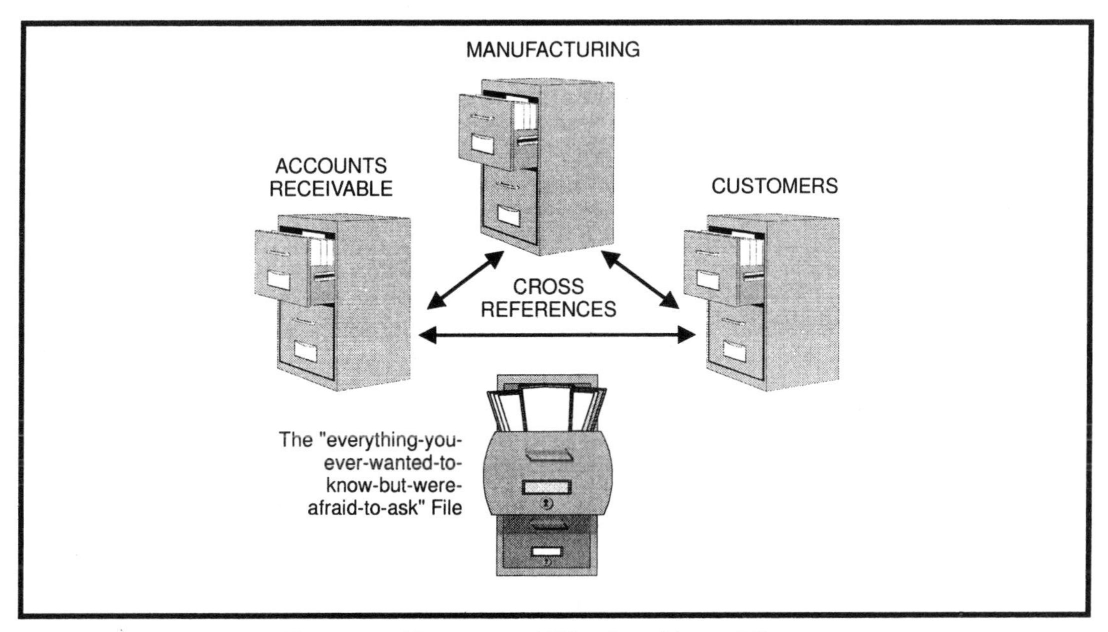

Fig. 3-13: Fragmented Files in a Manual System

There are many other reasons for the fragmentation and lumping together of manual data. Some you may hear are:

- "We've always kept this file updated — you never know when someone will ask me for this data one day."
- "It's my private file."
- "We lock our files up overnight in separate locations for safety reasons."
- "I look entries up by name and they look them up by account number."
- "We keep multi-part paper copies for backup and security, but of course, no-one uses all the data on each form."

Notice between each and every old design boundary (human or computer) there will be some kind of store. This store is usually unnecessary in the engineered business world. If the analyst does not recognize this old design characteristic, the old store is very likely to become part of the requirements for a new file/database. This will perpetuate unnecessary stored data and the housekeeping processes associated with it, such as Creating, Deleting, and Updating data.

Wrong Turn #6 — After-the-fact Quality Control

We introduce another wrong turn when we believe that inspections and after-the-fact processes are needed to produce a quality product. This introduces a double-edged sword. Workers may not put their ultimate effort into doing the job right the first time since they know that somebody else will be looking to catch their errors in the products they are creating. This leads to less quality up front. The inspectors have the inverse view. They assume the folks who were putting together the product knew what they were doing, so that they, as inspectors, don't have to scrutinize every detail. We find whole departments assigned to this "after-the-fact" checking task (typically known as Quality Control or the Inspection Department).

In the computer environment, this after-the-fact quality control is typically seen during systems development where we see multiple testing phases which outnumber the other development phases (i.e., unit testing, system testing, performance testing, acceptance testing). On the other hand, we typically have only one analysis phase and one design phase.

The very reliance on after-the-fact inspections simply indicates that something was wrong with the process that produced that product.

What were the reasons for putting inspections after the fact, or insisting on management approvals and sign-offs? Obviously, the reason was because under the old paradigm, we thought that human beings needed to be watched and were not empowered. Management was always "higher" than the technical tasks. A lack of trust was also an issue.

> We didn't, for example, trust another department, so we inspected their result when it came into our department.

A lack of skilled workers or training also led to the feeling that inspections after the fact were needed. Possibly the biggest *hysterical* "reason" of all is that it's always been that way. All the textbooks told us that inspections were part of how we implemented a manufacturing organization or managed a development process.

When people who came from the shop floor (or from the technical ranks of the Programming Department) were made managers, they were already indoctrinated into the old methods of testing after the fact. Even worse, we expected defective returns from the customer and set up a special department to handle returns, to track numbers of errors per thousand, etc. Again, like Quality Control, this department was self supporting. It was expected to have statistics on defects and to be a growing department. In some cases the idea was to put more people into processing returns than into putting more effort into the development of the product itself.

This way of thinking can be self-perpetuating, such as in the case of finding people monitoring the phone calls to the people in the Complaints Department to see if these people are doing their jobs correctly. In this case we have quality control of the Complaints Department. What this leads to is poor customer satisfaction. We may never even hear from those who don't return the product, or don't like our service, but they will never do business with us again.

Another factor to consider in after-the-fact inspections is the wastage. We came to accept that a certain amount of the product is going to be thrown away. Therefore, the cost of producing that product and the amount of materials going into that product is already factored into purchasing and ends up being scrap.

We also see organizations overproducing a product anticipating returns, or to have them on the shelf as replacement parts. In the computer software world we have duplicate processing in things such as re-editing data and cross checking the last system's output (which is our input). Obviously, this is also a problem from the duplicate data point of view (i.e., backups, tapes, many generations of backups, histories, etc.), because we don't trust the software or hardware.

All these inspections and audits are not business issues, they are system and implementation issues.

Summary

Based on all these archeological wrong turns, it's obvious we need to repartition the typical organization to have a Customer Focus.

Of course, none of these wrong turns are unique to any particular profession. House building started with mud huts and no architectural plans. It was only when houses fell down and "outside" government mandates were introduced, that the building industry started to apply an engineering discipline. Only then did the industry adopt the mandating of detailed requirement specifications prior to building and the documenting of codified regulations for the profession.

Customer Focused Engineering professionals would be doing a disservice to the business people who use the resulting systems if the system design procedures make systems any more complex than the business functions they are supporting. We have to make sure that the system building profession adopts an engineering discipline.

The sooner we start building manual and computer systems around business issues instead of technological, or old design issues, the sooner we will be able to end the short payback periods of today's systems. Also, business people will be able to relate to automated and manual system designs, thus enabling modification and system building to be no more complex than the business at hand.

When conducting system archeology, the deeper you dig (the more you invest), the greater will be the return.

Chapter

4

Understanding the Nature of Systems

The first step towards language was to link acoustically or otherwise commutable signs to sense impressions.... A higher development is reached when further signs are introduced and understood which establish relations between those other signs designating sense-impressions.... When man becomes conscious of the rules governing relations between signs[,] the so called grammar of language is established.... When language becomes partially independent from the background of impressions, a greater inner coherence is gained. Only at this further development where frequent use is made of so called abstract concepts, [does] language become an instrument of reasoning in the true sense of the word.

Albert Einstein
Out of My Later Years
The Common Language of Science

Most organizations have become too complex to be comprehended by one person without using an abstract model of the organization. As stated in the previous chapter, one major complexity is the design that covers the essential business functions. When we are creating a Customer Focused Organization, it's important we first study (analyze) the nature of the business (the **what**) before we introduce a new design (a new **how**) for that business. We need a clear understanding of **what** the business rules are before transitioning from **how** we implement them today to **how** we will implement them in the future. This Business Systems Analysis is critical for Customer Focused Engineering. In fact, once we understand what our organization does we may not use any new technology and still create an effective Customer Focused Organization.

You can't do Customer Focused Engineering without Business Analysis.

It always seems ludicrous to me when I encounter a major organization that doesn't know its essential business details. By *details*, I don't mean manual procedures, rule books, program flow charts and such, but a design- and implementation-independent Business Model of **what** the organization does. Such a view declares which data are used where what processing is performed where and so on.

We need an engineered business view of the organization to be able to conduct such basic tasks as pinpointing where modifications are necessary in our response to a business change, and how to make those changes with minimal disruption to other parts of the business.

Business Issues as Conceptualized by Humans

The question is: "How do humans conceptualize the organization's essential business issues?" We have a problem as human beings when we try to analyze business issues because to some degree we're all designers and that gets in our way. For those of us who easily fall into "solution mode" — trying to solve problems as soon as we find them — it's very difficult to understand what analysis is and what its deliverable is. It's actually quite difficult to see an analysis.

> For example, a typical response to the request, "Show me an analysis of a chair" might be, "Do you want me to show you a specification that results from analysis?" A specification is usually a drawing or representation of a chair, but that's actually a response to a request to "Show me a design of a chair." Coming up with a design is an easy thing to do, but what was the analysis of the chair?

If you try to identify the requirements for a chair as a mental exercise, it's actually quite difficult. What I've arrived at in my discussions with students is a chair is a product that will support a human body in a seated state comfortably. It can be movable; it should support the back, etc. Given those requirements, we can have a chair that we kneel on; we can have a bean bag chair; and most importantly, we can invent something we haven't seen before. There are many designs that satisfy the requirements of a chair once we know the requirements.

Because we can't "see" an analysis we have to produce some kind of model. This model specifies the important aspects of the item we are analyzing.

It's even difficult to rely on a dictionary during analysis because the dictionary itself is design oriented (i.e., the dictionary uses design words and defines design issues). That doesn't mean there aren't any logical analysis terms in the dictionary; it just means that many dictionary entries are related to design because those are the kinds of things we see as human beings out there in the world. Many times we are forced into thinking about analysis using design terms. In my seminars, the easiest way I've been able to state this in simple terms is to talk about **what** we want versus **how** we implement the **what**.

The chair you see is **how** you've implemented the **whats** — the chair's requirements. It's implied in the design of a chair that we have implemented the requirements. Now, there are many ways we can view (model) a chair. If you take a photograph or draw the actual chair, then you're capturing a design model and not a model of the requirements of the chair. We can fall back on the modeling tool our schooling system gave us (i.e., text), but this is no where near being the best tool in many cases. *As an example, take a map of the United States (a mostly graphical tool) and redo it in text. (You're on your own with that one!)* It's hard enough with a tangible chair or map, but it gets even more difficult to develop the model of something that's intangible. For example, how do you model a system or a service? This is our problem when we analyze an organization and its systems.

The Stimulus-Response Nature of Systems

All systems we encounter have some fundamental characteristics in common — they're all ***stimulus-response*** mechanisms with ***process*** and ***memory***. The ultimate stimulus-response aspect occurs on the outside edges of the business and defines the organization as viewed from the outside. The process-memory aspect occurs within the organization and represents the organization's internal tasks and procedures and their associated data.

Webster's Dictionary's definition of "stimulus" is:

Something that incites to action or exertion.

An organization's stimuli and responses can be implemented with voice, paper, electronics, touch, sound, natural occurrences, non-verbal communications, etc.

The definition of "response" is:

An answer or reply, as in words or some action.

The stimulus-response nature of systems applies to all types of systems — physical, biological, and informational — and gives us a "black-box" view of a system.[1]

A rock is an example of a physical system. An example of a biological system is that of a person accidentally hitting his or her hand. In this case pain is the stimulus to the brain and rapid withdrawal of the hand is the usual response (often accompanied with a verbal response).

This stimulus-response view also applies to an organization. An example would be a mail order company that receives a stimulus (a **Order**) from a **Customer** in the outside world and responds by shipping **Ordered Materials** and an **Invoice**.

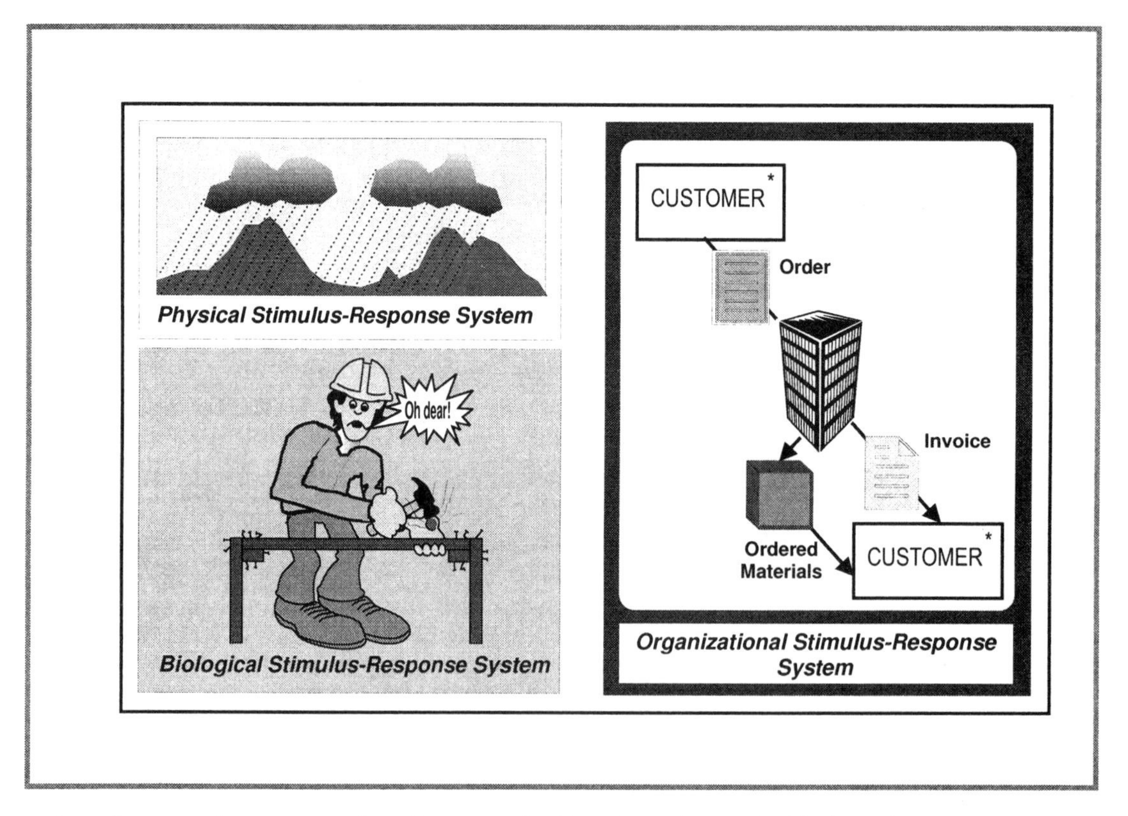

In other words, we can view the organization from just **what** it does in response to its inputs and outputs without getting involved with **how** it is implemented. Figure 4–1 shows this stimulus-response nature of a system with the system shown as a "black box."

1 I used to state that stimulus-response applies to systems and not things that are inanimate (such as a rock). Then I did a significant amount of teaching and consulting with folks who take care of the environment of the United States and they quickly made me realize that such physical things as rocks are stimulus-response mechanisms. (If you don't believe that, take a look at the Grand Canyon in Arizona.) Rocks respond to flowing water, acid rain, etc. So now I find it very difficult to find anything that isn't a stimulus-response system.

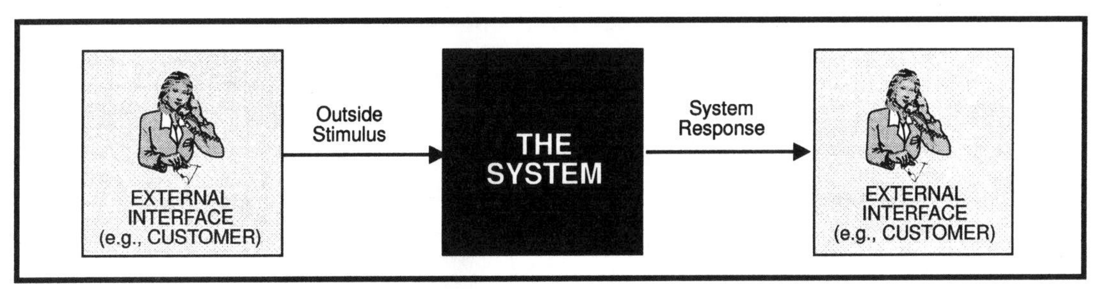

Fig. 4-1: The Stimulus-Response Nature of Systems

The Process-Memory Nature of Systems

Here are my definitions for process and memory:

A process is the set of forces, actions, laws, rules, changes, or operations that react to a stimulus. The process typically generates the response, and usually alters the state, memory, or material within the system.

Memory maintains a record of the system's internal state and the relevant data that the system needs to Create, Retrieve, Update, or Delete in response to a stimulus.

In an organization its system's processes are implemented by procedures, and/or computer code. Its memories are implemented by records, human memory, files, tables, and databases (see Figure 4–2). For too many years I looked upon systems only from an internal process-memory point of view, without acknowledging the stimulus-response aspect that provided the reason for being for the process and memory.

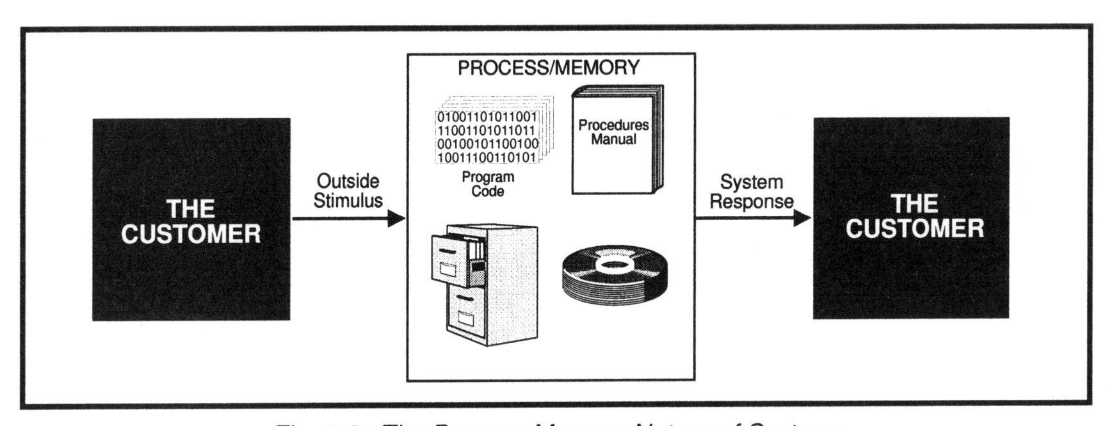

Fig. 4-2: The Process-Memory Nature of Systems

"Process typically needs memory" and "memory is processed," so they are not necessarily exclusive of each other. Be careful not to isolate process and memory from each other (especially by forming two groups of people to deal with them in the organization).

In some organizations, I have seen the situation where one group may believe that it can understand a business by looking only at what the business needs to retain in its memory, that is, a *data-only* view. Another group may believe that it's only necessary to look at things from a *process-only* view. Each view believes that the other is important only in so far as it supports the primary view. The truth is the two should not be looked at in isolation, especially when the goal is to support the business.

The response to an external event may alter the system, so that a system may not respond in exactly the same way to two identical stimuli.

For example, if I have four hundred dollars in my bank account and stimulate a banking system with a three hundred dollar withdrawal request, the response will be a successful transaction and I will receive three hundred dollars. But, if I immediately try to withdraw another three hundred dollars the response will be different.

As part of the first transaction, the withdrawal process updated the system's memory and the second withdrawal transaction retrieved the updated memory as part of its process and created a different response.

From an analysis point of view, the same system was invoked, but with different values for the memory. The business rules didn't change, just the values, and hence, a different part of the business logic was executed.

Notice that I started this discussion by talking about stimulus-response but as soon as we begin to look inside a system we must acknowledge process and memory. Unfortunately, the classical analysis techniques have tended to focus almost exclusively on *process-memory* aspects of a system.

Table 4–1 shows the past emphasis on an organization's internal processes and data in analysis and design without any focus on what happens on the outside.

	TYPE OF VIEW	ASPECT VIEWED
BUSINESS VIEW	Business Process View	Business Policy
	Business Memory View	Business Stored Data
DESIGN VIEW	Design Process View	Process Architecture/Structure
	Design Memory View	Data Architecture/Structure

Table 4-1: Internal Business and Design Process and Memory Issues

Figure 4–3 shows a design view indicating the stimulus-response and process-memory aspects of an organization that fulfills customer orders

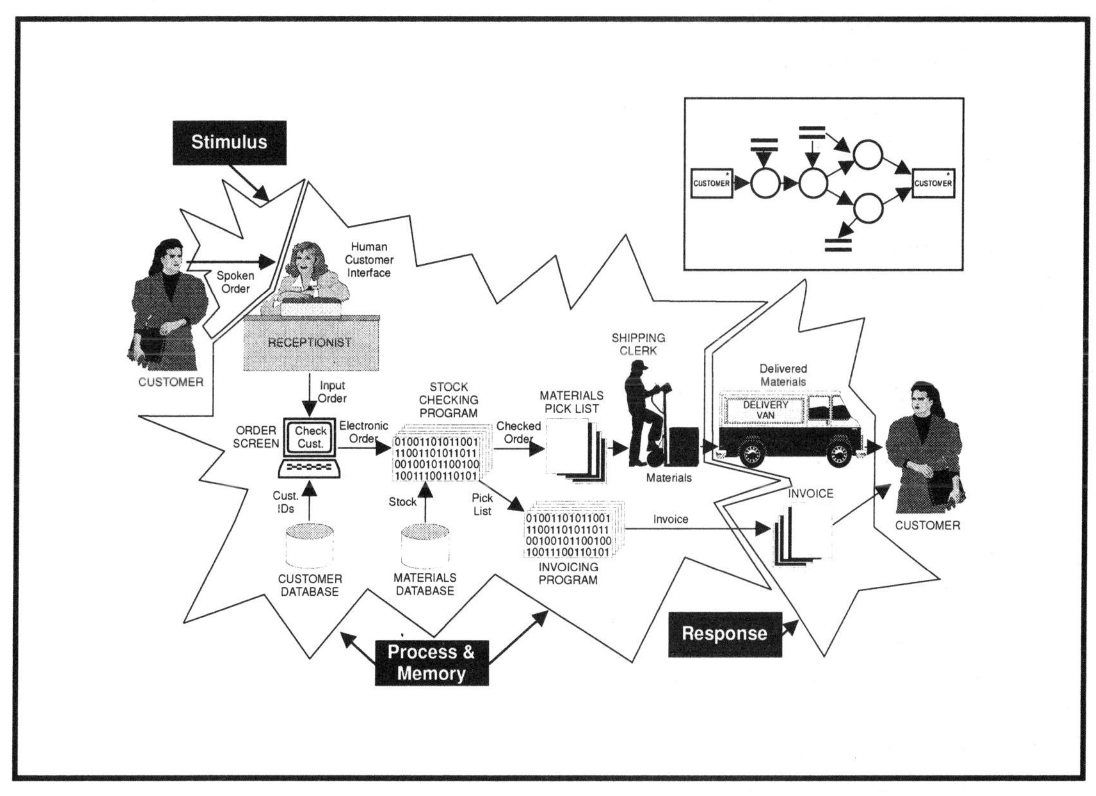

Figure 4-3: Sample Stimulus/Response & Process/Memory

In the next chapter I will show you the equivalent example from a business point of view.

Summary

Systems are merely stimulus-response mechanisms, regardless of how they are implemented (human systems, computer systems, biological ecosystems, etc.). We need to know our stimulus and response and then acknowledge our processes and memory associated with these.

To get a meaningful, seamless view of an organization, the stimulus-response, process and memory aspects need to be looked at in total and specified as the result of the analysis. Our problem is how do we represent these aspects of the business. If you realize from our previous discussion that everything you see is a design, we need some way of removing this design and seeing only the analysis requirements.

The next chapter covers one way to represent this stimulus-response, process-memory view of an organization using models.

Chapter

5

The Model IS the Business

If your only tool is a hammer, the [whole] world looks like a nail.

Abraham Maslow

As discussed in the previous chapter, we need some way to view the essential business without seeing the implementation of that view. This is where a model becomes very helpful, to conceptualize something that might not be easy to visualize. Now, to deliberately contradict this chapter's title, any model we produce isn't really the business; however, models are the closest we're ever going to get to seeing the business until we can do Spock-style mind melds by touching the side of an office building (*for all you Star Trek® fans*).

In many situations, graphical models are superior to text as a communications tool. Most engineering disciplines emphasize the importance of drawing graphical models before building actual products. This is because it's too expensive to build the real system before we are sure that we correctly understand the business requirements. Especially in the case of some particular physical product, it's vital to test the model of the product before manufacturing it in volume. This is analogous to drawing the blueprint and getting it approved by the customer before building the house.

When conducting an analysis of an organization, we need a model to communicate between the business people and customers (the users of the product or system), the analysts and designers (the people who will build the product or system), and those who perform system maintenance to modify the product or system as the needs of the business change.

When I go out to a consulting assignment and someone tries to explain their business to me the first thing I do is to try to "picture" it using a graphical model. There have been many models introduced over the past few years that can represent the basic set of business facts of an organization and its systems. I have found it useful to select from a variety of different models, depending on who is going to review and verify them and what I am trying to model.

Let me give and overview of some of the more common, tried-and-tested models. The basic families of models described in this chapter are: Process Oriented Models, Information/Data Oriented Models, Process and Information/Data Oriented Models, and Control Oriented Models. If you are already quite familiar with these models you may want to skip this chapter.

If the people I talk to are trying to describe their organization (or part of it) from the point of view of what data are stored and the Relationships between that data, an Information/Data Model is a good choice. If they're trying to describe something that's based on complex control issues, I'll use a State Transition Diagram or Control Flow Diagram. If they're trying to describe something that involves data flowing through a system and the transformation of that data (processing), I'll use a Process and Information/Data Model such as a Data Flow Diagram. If they're trying to describe their organization by relating to the things (objects) in it and all of the processing (methods) that can act on these things, I'll use an Object/Class Model.

I'll describe these model types in detail, but before I do, let me talk about the effectiveness of a model.

Effective System Modeling

When we represent the aspects of an organization in a model, we must take human issues such as readability into account.

> Using one model to identify all the details (facts) in a house, for example, can produce a complex, unreadable model because a single diagram including structures, wiring, plumbing, heat loss calculations, structural calculations, drainage, topography, etc. would not be very readable. This is also true when we model an organization.

To be able to see aspects of our essential business, we must use a model. Our success depends on the suitability of the model for the job at hand.

It's worth noting here that although graphical models provide a clear, understandable specification, they don't stand alone. Text is still usually required, but it no longer appears in huge, monolithic blocks. It is used, instead, as backup documentation for parts of the graphical model and is organized along the same lines as the model.

We must acknowledge that we human beings can't effectively assimilate and comprehend large, monolithic specifications. Presenting these for review allows errors to pass undetected and leads to reduced quality because people can easily get overloaded with too much information. Unfortunately, the most commonly taught modeling tool has been text.[1] Text is perfectly fine as a tool to describe something that flows in a linear manner.

1 Text is a model of sorts; the funny shapes you are looking at, letters, numbers, and punctuation mean things in your brain. This tool was given to us by our country's educational system.

For example, text is adequate to describe the flow of linear control, i.e., what do I do next?

In the English language we are automatically given the idea of going to the next line to carry on reading. Text is also perfectly fine if we are telling a story. However, text is not very good when we try to use it to "picture" something to do with a system that isn't linear in nature. As a modeling tool, text has a number of undesirable characteristics. The following table contrasts the weaknesses of text as a modeling tool versus the strengths of graphical models.

TEXTUAL MODEL SHORTCOMINGS	GRAPHICAL MODEL STRENGTHS
Linear — It must be presented linearly and is usually read from start to finish. Systems, on the other hand, can be both synchronous and asynchronous networks of processing and data does not flow linearly through them. It's only when we get down to a fine enough level of detail that some systems may appear to be linear.	Depict non-linear systems — Graphical models make it possible to represent non-linear structures and such processing features as modeling concurrent tasks.
Monolithic — It is not naturally partitioned although we do have the concept of major and minor breaks: chapters, sections, paragraphs, and sentences.	Support partitioning — They allow easy representation of the natural partitions of systems and their interfaces, leading to tangible, measurable deliverables and task allocation along the natural lines of business systems.
Difficult to change — correcting errors and omissions is extremely laborious, time-consuming, and error-prone because text is linear. Also, it's difficult to spot a global or referential change (e.g., page 63 disagrees with page 328).	Flexible — Graphical models are far easier to change than narrative text. Also, associations and referential integrity can be depicted easily.
Verbose — A word is one thousandth of a picture.	Succinct — A picture's worth a thousand words.
Not measurable — Lines of text in a requirements specification are difficult to use to quantify completion.	Verifiable — Graphical models support the validation and review activities during system development via easily-changeable representations of the system.
Ambiguous — The English Language is deliberately rich and therefore has the potential to be vague and misinterpreted.	Unambiguous — Graphical models break the language/semantic communication barrier by using the universal language — specific symbols.
Subjective — Almost all text is subject to individual interpretation.	Non-subjective — No interpretation is necessary given the rules of the model's symbology.
Not easily leveled — Textual material consists entirely of two-dimensional details.	Can easily show levels of detail — Graphics can support a diverse audience and depict high-level, abstract views and low-level, detail views.

Table 5–1: Text vs. Graphics for Models

We obviously use models to:

- Aid our own thinking and understanding.
- Be able to analyze something because we can picture it using the model.
- Defer detail in order to get a high-level view before we get drawn into the lower-level details.
- Communicate our ideas to others.

For example, we use models (blueprints) to communicate the design from the architect to the person who will actually build the house. Also, the architect uses models to communicate the design into the mind of the customer to make sure that the architect understands the desires of those who will live in the house.

It's important to keep in mind that any models we produce need to be specified to the necessary level of detail for the readers (whoever they may be) to be able to verify their requirements for analysis, design, or implementation.

The audience for the model may be ourselves (or our peers) who at some point in the future may need to refer to the model to remind ourselves (or those who follow us) why we did something in a particular way.

Using a graphical model helps to make an intangible problem tangible. Even though the final product is still intangible, we can conceptualize the problem more easily with these modeling tools. This is no different than a line on a blueprint to indicate a wall or a meter on the side of a house or apartment showing how much electricity has been used. The dials or gauges are not showing us the electricity; they are simply representations (models) of the electricity. You're probably already familiar with using models to depict real-world issues.

An example would be your family tree as represented in a hierarchy diagram or a map of your town. When we look at a map, we understand that the lines represent roads. The thicker lines (or, in some cases, the red lines) may indicate freeways. The size of the text may also indicate size of towns. Once we understand the model's symbology, we can read that particular model.

The Characteristics of an Effective Model

What if we are not depicting a tangible item?

An example of this is when we are trying to specify **what** an organization does versus **how** it is implemented)?

We need a model that can effectively represent the complexity of the organization's business and it should be:

Comprehensive	We need to use our model to represent both the process-memory (inside view) and stimulus-response (external view) aspects of our business completely.
Partitionable	We want to avoid complex models where we have to understand everything in order to understand anything.
Supportive	The model should assist us in dealing with complex and voluminous information that exceeds the capacity of human memory.
Intelligible	Graphical symbols, especially, help understanding because a picture can be worth a thousand words. The few words we do use should be concise, precise, and specific.
Non-redundant	If we have more than one model of the business, such as a process view and a data view, they should not duplicate each other. They should complement and correlate with each other.
General and/or Specific	The model should be able to show levels of detail from the highest to the lowest; for example, from an organization-wide enterprise model down to individual fields or Data Elements (attributes).
Abstract	We should be able to represent and describe an idea, or the qualities of a thing apart from the thing itself. For example, the memory model should allow us to describe the types of data we hold about our customers, quite separate from the actual data values describing each customer occurrence.
Economical	We should be able to simulate a business, or those aspects in which we are interested, without going to the expense of creating the business itself. Obviously, a wind-tunnel model or a computer simulation of an airliner is much cheaper, safer, and more convenient to work with than flying the real thing into a hurricane.

Table 5–2: The Characteristics of an Effective Model

Models for Analysis

As I stated earlier in this chapter, there have been a number of models introduced over the years for modeling systems. In fact, the profession that I'm most used to (Data Processing) has gone through an evolution of its modeling techniques. A number of proven models emerged from the Data Processing Industry's evolution that we can use in capturing the essential business of an organization. As analysis is a business issue rather than a system issue, we can quite effectively use the same analysis methods and models that were developed as part of D.P.'s evolution.

The vast majority of business systems deal with both data and processing (and some control issues) that are part of the analysis requirements. So, we can put on our analysis hat and take a look at data (the information that flows and that we store regarding a system) and model that information. Also, we can look at the logic (the processing of data) and model that processing.

The most common methods (each with their recommended models) that have proven to be of help to the analyst in D.P. are: Process Engineering, Information Engineering, and Object Oriented methods. We can also use State/Transition modeling where business control is an issue.

These methods and their associated models were proffered by their authors to help specify analysis details as well as design and implementation details. Figure 5–1 shows the points of view of each modeling technique and its associated model.

Process Analysis views the Business from the point of view of Data-on-the-Move by modeling:

- ❖ **Data flowing through the organization**
- ❖ **Data stored in the organization**
- ❖ **Processes transforming Data**
- ❖ **Sources and Receivers of Data (the System Boundary)**

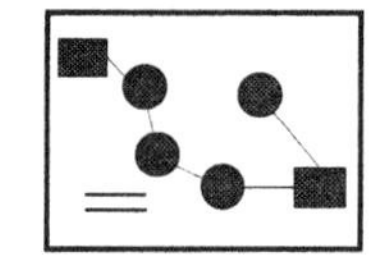

Information Analysis views the Business from the point of view of Data-at-Rest by modeling:

- ❖ **Entities (Data held in cohesive units)**
- ❖ **Relationships between Entities**
- ❖ **Data Elements (facts about Entities)**

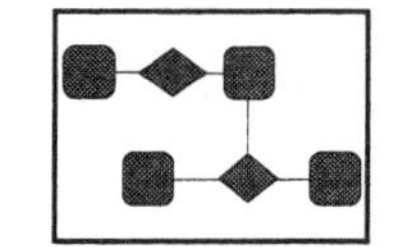

Object Oriented Analysis views the Business from the point of view of interacting Objects by modeling:

- ❖ **Objects (Encapsulations of Data and Process)**
- ❖ **Methods (reusable processes acting on shared variables — Data or States— within an Object)**
- ❖ **Messages (Communication between Objects)**

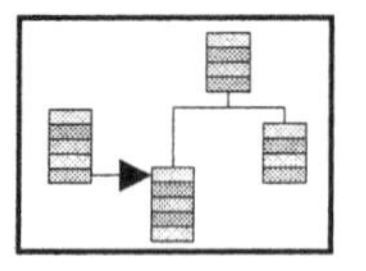

State Transition Analysis views Business Issues from the point of view of control by modeling:

- ❖ **The State of the system being modeled**
- ❖ **Transitions that show what causes a State to change and the action to take based on that change**

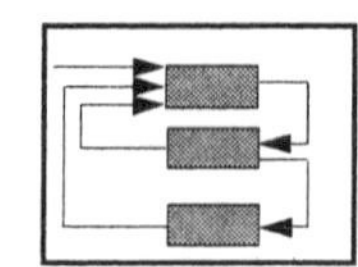

Fig. 5-1: Modeling Business Systems

The common denominator of all the methods employed in Figure 5–1 is that they all use models as a means of communication. The models we draw using symbols simply aid the process of understanding and communicating with the symbols representing abstracted aspects of reality. I do not intend to teach the details of each of these methods and models, but in this chapter let me provide a simple common platform of understanding regarding them.

Figure 5–2 enlarges upon a previous high-level view of a development effort to show where some of these methods and models are used in three phases.

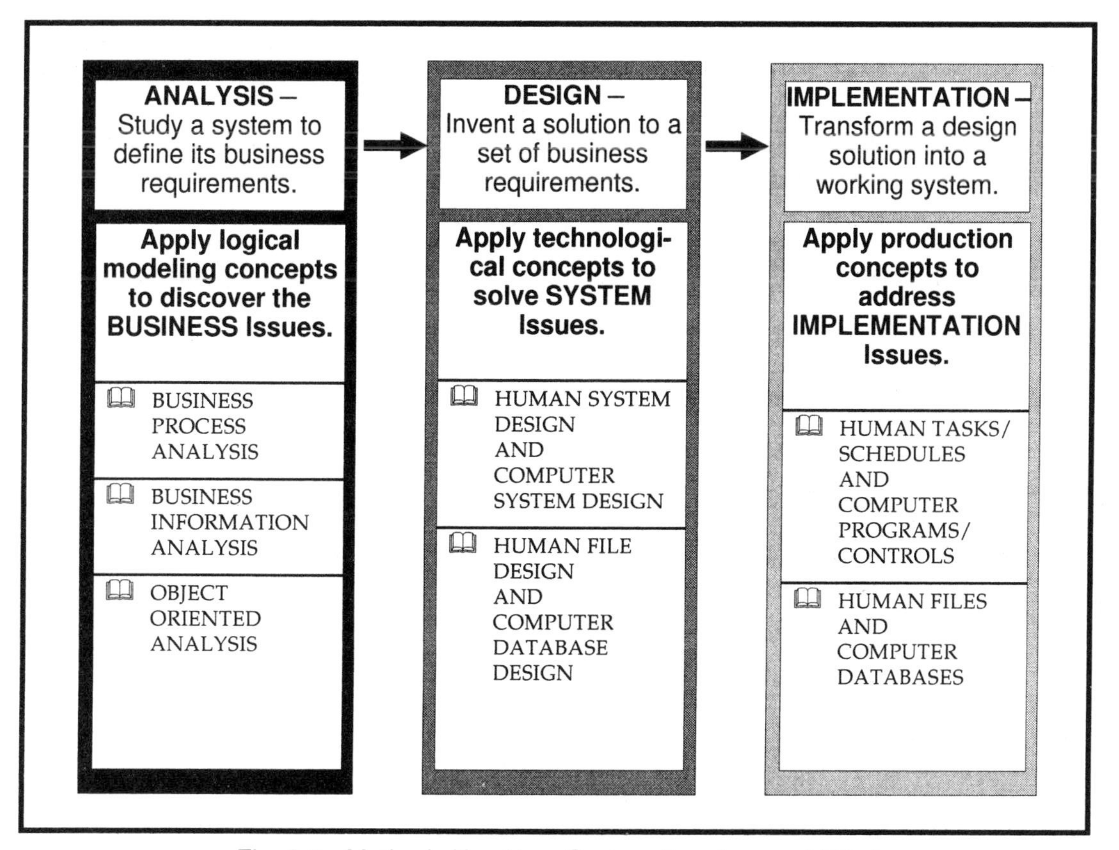

Fig. 5-2: Methods Used in a System Development Effort

You've probably already figured out from the previous chapters that I believe analysis is the most important phase when creating a Customer Focused Organization. However, to conduct analysis we usually have to see through an existing design. Therefore, what follows is a review of a number of models that can be helpful to an analyst. This usually involves understanding the old design when creating an analysis specification of an organization and its systems.

I've categorized these models into four areas: Process Oriented, Information/Data Oriented, Process and Information/Data Oriented, and Control Oriented. Let me enlarge on these types of models in more detail. The following models cover all the analysis models (Figure 5–2) as well as others that may help you understand the old design.

Process Oriented Models

Flow charts like the one in Figure 5–3 are what I'd call a Process Oriented Model.

Flow Charts

It's actually quite rare to find someone who does not know what a Flow Chart is and who does not recognize its symbology. A Flow Chart's intention is to depict processing in a linear (flow-of-control) manner. [2] In this respect Flow Charts and text are very similar in that they both declare a control structure of verbs. Unfortunately, Flow Charts are one model not very well suited to depicting businesses because they are a one-sided view. They show processing only with its flow of control. I call them a Process Oriented Model because they don't show data. This doesn't mean we can't put data in the boxes and say, for example, "IF X = Y, THEN..." (where **X** and **Y** are the data), but the data are not given a symbol in the model. The lines are not the data, they are "GO TOs."

Flow Charts are also detailed diagrams. They are typically used when figuring out detail issues to do with processing/logic. They are not really useful for modeling high-to-low leveling issues or complex control structures. What we need is a model (or set of models) that can show both the data and processing sides of businesses.

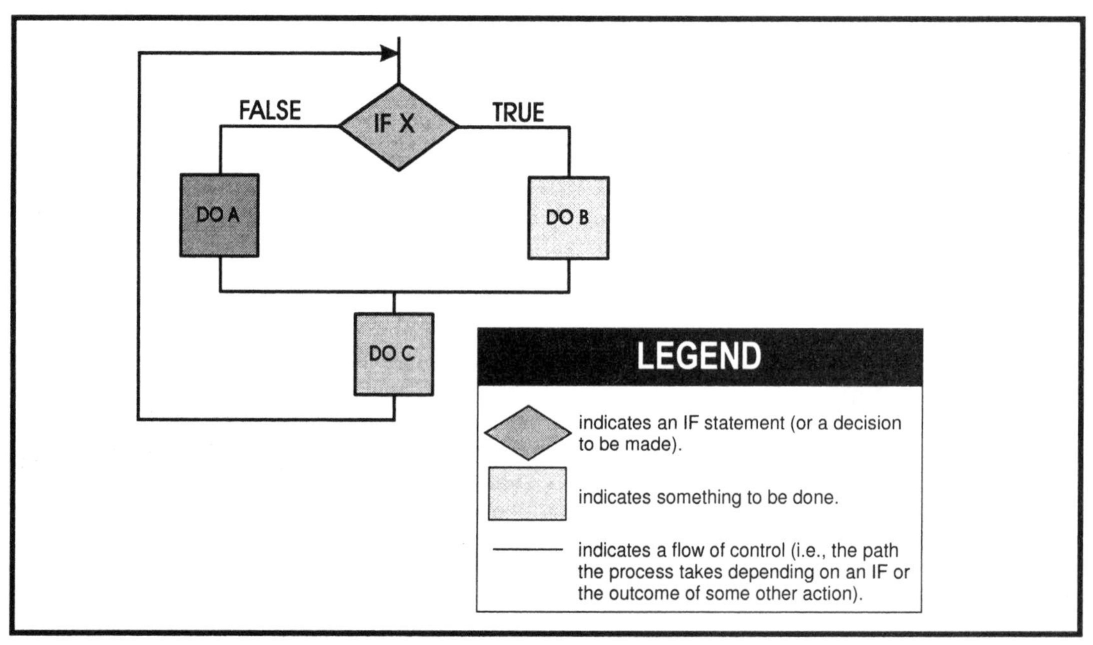

Fig. 5-3: A Flow Chart

2 By the way, that's my excuse for not being able to cook. If you read a linear text recipe from top to bottom, it may say: "Prepare the vegetables by steaming them until you can stick a fork in them." I do just that and get them ready and then it says: "Meanwhile, prepare the meat." Then I ask: "What do you mean by 'meanwhile' prepare the meat? The vegetables are ready?" Of course, if they reversed the steps, the meat would be ready before the vegetables. Text is obviously linear. Even if you give the sets of steps side by side, people would still read them from left to right instead of knowing to perform the tasks concurrently due to the limits of the textual model.

Functional Decomposition Diagrams

Another Process Oriented Model is a Functional Decomposition Diagram (Figure 5–4). These are also a process-only view and are a good way to decompose a process or system and to break down that process or system from a high level to a low level view.

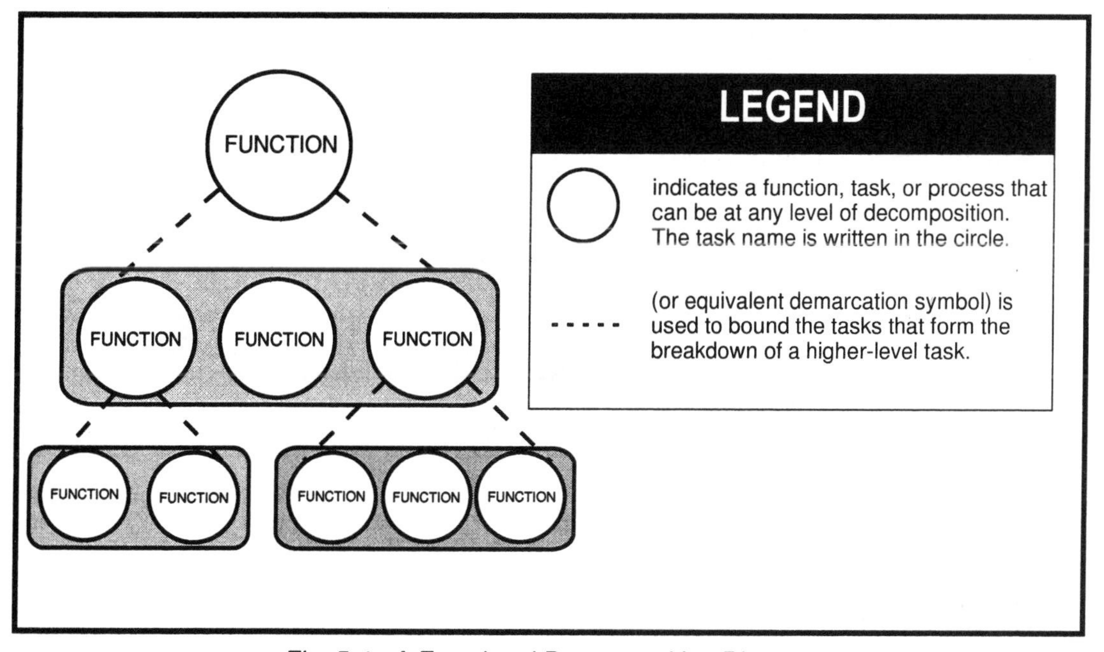

Fig. 5-4: A Functional Decomposition Diagram

Functional decomposition concentrates on processing and its breakdown.

> For example, we might use a Functional Decomposition Diagram to show what's involved in a process called **Calculate Net Pay**. We might break down this process into **Calculate Gross Pay, Calculate Overtime, Calculate Taxes**, etc.
>
> If I didn't understand what was involved in **Calculate Taxes**, I could further decompose (level) that process to **Calculate Federal Taxes, Calculate State Taxes, Calculate Local Taxes**, and so on. Again, if **Calculate Federal Taxes** was a complex process, I could break that down further into **Calculate Taxes for Single People, Calculate Taxes for Married People**, etc.

As you can see from these examples, Functional Decomposition Diagrams are process oriented and do not show data.

Process Hierarchy Diagrams

Many of us are fairly familiar with Process Hierarchy Diagrams (Figure 5–5). Everyone of us is represented in part of one type or another. Everybody has grandparents and parents who are higher in the hierarchy in terms of their ancestry. That is one use of a Process Hierarchy Diagram.

Process Hierarchy Diagrams are good for modeling top-down control to show who is in charge of whom. This is a very common model for organizations where the big bosses are at the top, middle management is below them, and the staff is below middle management in the hierarchy.[3]

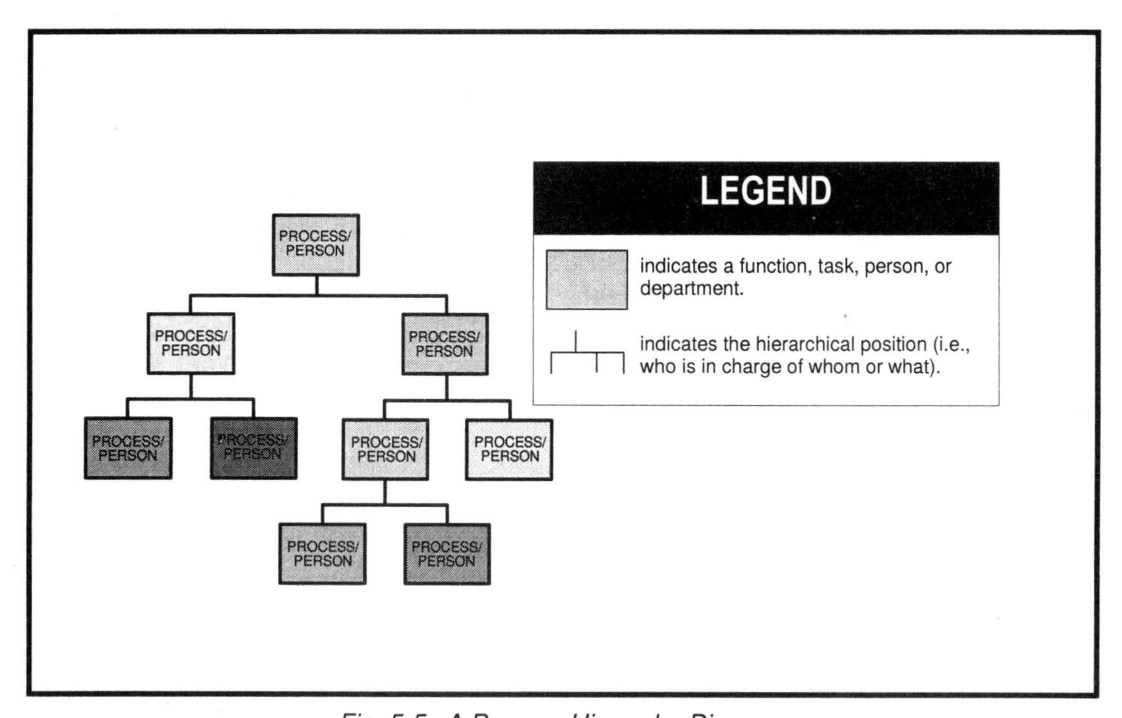

Fig. 5-5: A Process Hierarchy Diagram

We also see Process Hierarchy Diagrams used in the Data Processing Industry when modeling the structure of computer systems to show "boss" modules and "subordinate" modules that are invoked by these "boss" modules.

3 I consider this to be an outdated organizational structure.

Information/Data Oriented Models

There are also models that are completely Information/Data Oriented as opposed to the three Process Oriented Models I just described. Information/Data Oriented Models (Figure 5–6), which are typically referred to as Information/Data Models, show Entities and Relationships (typically two-way Relationships). These diagrams also show the cardinality between these Entities.

For example, one **Employee** may work on many **Projects**, so we have a "one-to-many" Relationship in this case.

Typically, Information/Data Oriented Models are good for seeing cohesive blocks of data and their binary Relationships. They serve as a good design tool for implementing computer databases and human file systems.

Entity Relationship Diagrams

An Entity Relationship Diagram is another model that shows Entities (blocks of cohesive data). However, rather than being limited to showing binary (two-way) Relationships, these diagrams show "n-ary" (multiple) Relationships. That is many Entities can be associated with one Relationship. They also show the cardinality of these Relationships (Figure 5–7).

For example, we may use these diagrams to represent a 3-way Relationship between a **Contract** for a certain set of **Products** with one particular **Client**. In this case the business Relationship can be shown with these three Entities together to avoid a disjointed binary view.

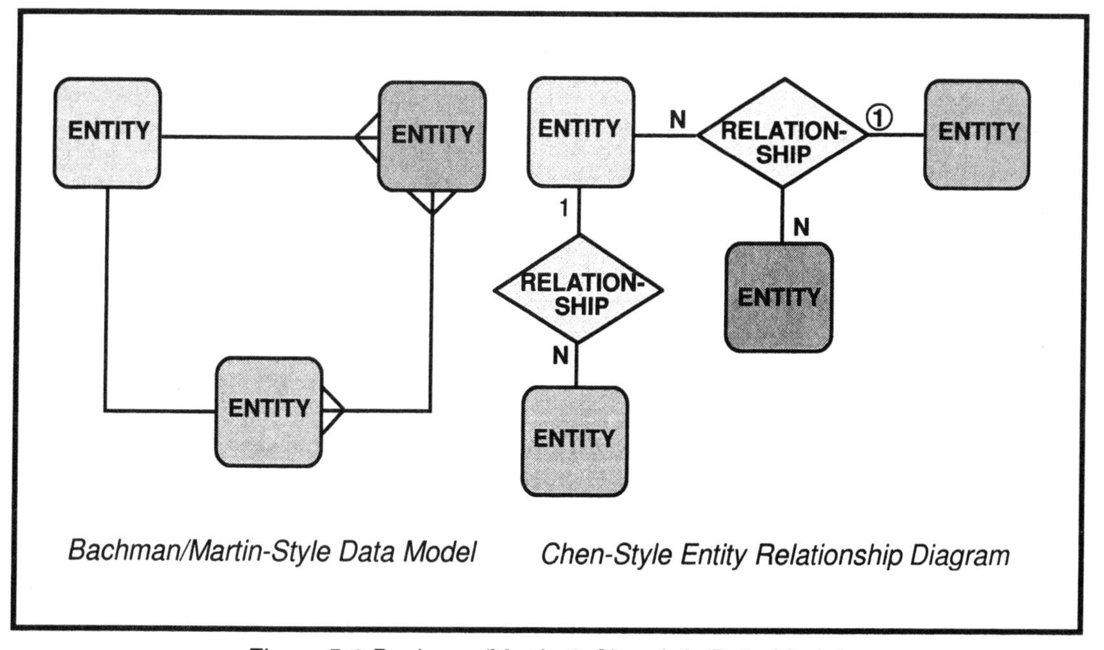

Figure 5-6 Bachman/Martin & Chen Info/Data Models

Figure 5–7 shows the symbols used in a Chen-style[4] Entity Relationship Diagram and in a Bachman/Martin-style Information/Data Model.

Person or Thing		Chen Style Symbol	Bachman/Martin Style Symbol
Product File, Customer, Record (Name, Phone)	Is represented as:	Entity	Entity
Marriage, Index Card – Reference Pointer	Is represented as:	Relationship	Relationship
Relationship Cardinality: 1 teacher & 1 student	Is represented as:	1 — 1 One-to-one	One-to-one
Relationship Cardinality: 1 teacher & many students	Is represented as:	1 — N One-to-many	One-to-many
Relationship Cardinality: Committee to committee	Is represented as:	N — N Many-to-many	Many-to-many
① indicates the point from which to read the Relationship (this will always be singular).			

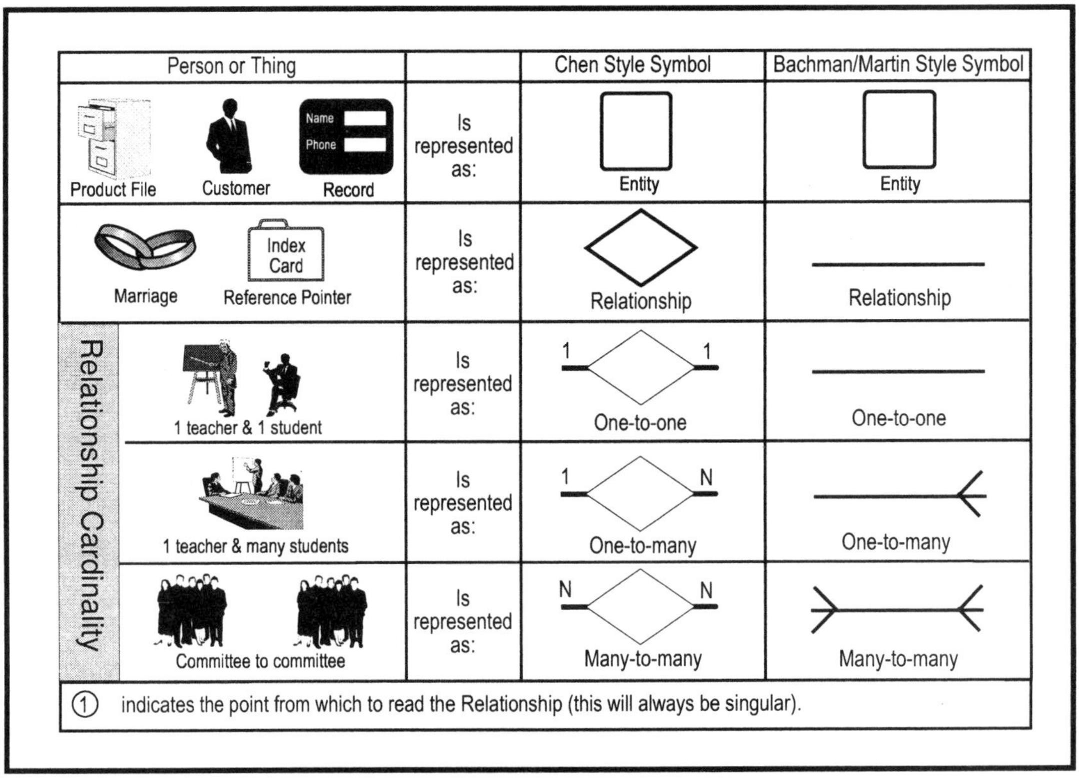

Fig. 5-7: Information/Data Model Symbology

Hierarchical/Network/Relational Data Oriented Models

We have a number of Information/Data Oriented Models to show the design structure of a database or file cabinet system (Figure 5–8). We can use a hierarchical model to refer to a parent-child structure of data.

For example, we can represent a **Customer** (parent) set of data as being "over" all that **Customer's Account** (child) data. This would indicate that we can access all **Account** data given a particular **Customer**.

Another example of an Information/Data Oriented model would be a network structure where we can access one Entity (set of data) from any other. Still another type is a relational data model, typically represented as rows and columns, where we can create dynamic tables that bring together sets of data related to other sets of data.

4 *The Entity-Relationship Model — Towards a Unified View of Data*, Peter Chen — See Bibliography

These last three Information/Data Models are good for modeling a specific type of data structure in an organization.

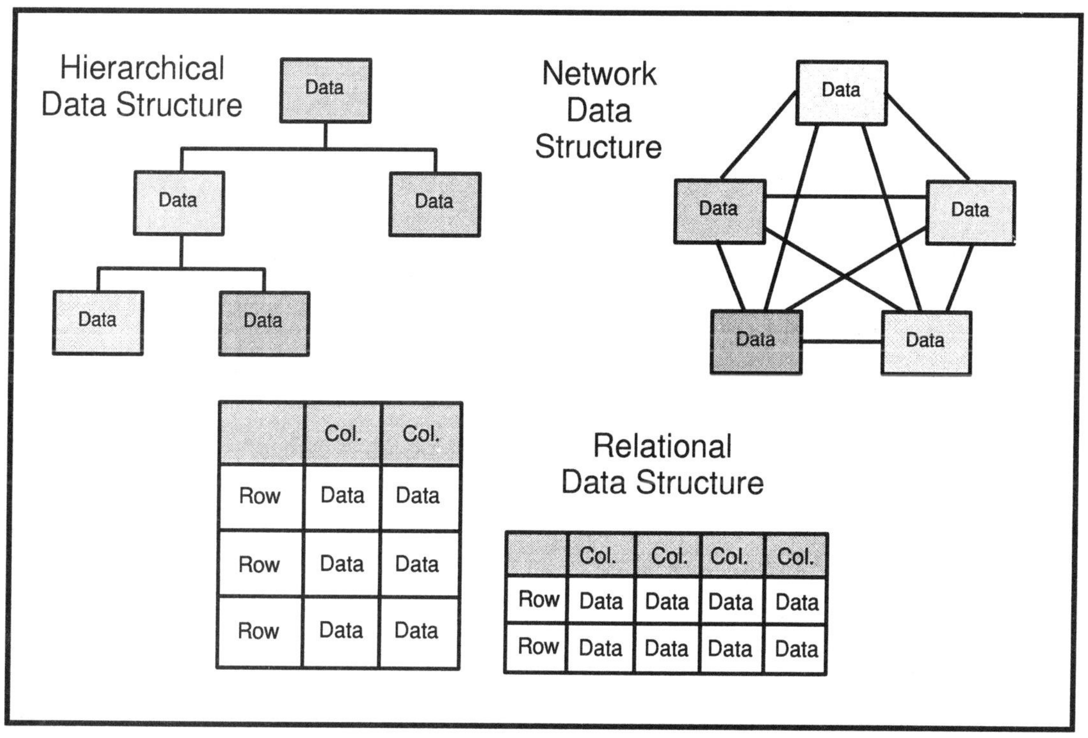

Fig. 5-8: Additional Info/Data Oriented Models

Process and Information/Data Oriented Models

Data Flow Diagrams

There are models that combine Process Oriented Models and Information/Data Oriented Models. A Data Flow Diagram (Figure 5–9) is an example of a model that is both Process and Information/Data Oriented. This particular model allows us to look at the data and the processing acting on that data as it travels through a system. These diagrams can also be represented at various levels of decomposition.

> You can have a high-level Data Flow Diagram that just depicts a single process and its data for what is involved in a payroll system, for example. You could further decompose this single-process view into a low-level diagram that shows a more detailed view with the particular data and processes involved in calculating taxes, pensions, etc.

Figure 5–10 shows the symbology used in a DeMarco-style[5] and a Gane/Sarson-style[6] Data Flow Model.

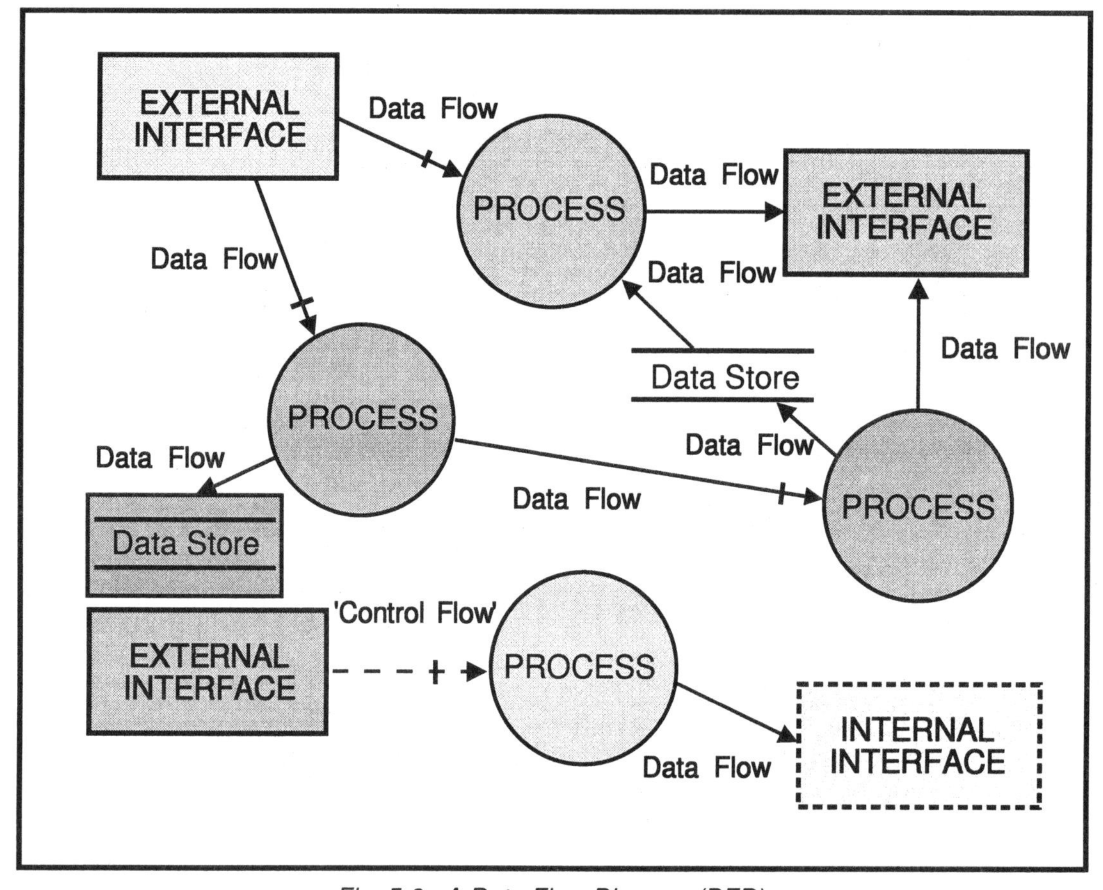

Fig. 5-9: A Data Flow Diagram (DFD)

There is also the need to differentiate between files over which we had total control in terms of the data's creation, retrieval, and deletion (fully conserved data), and external files which we could not conserve. That is why we represent a conserved internal data file with a name between two thick horizontal bars.

When we need to show an unconserved external file, we enclose the two bars and file name within an external interface (a rectangle).

5 *Structured Analysis and System Specification,* Tom DeMarco — See Bibliography

6 *Structured Systems Analysis: Tools and Techniques.,* C. Gane and T. Sarson — See Bibliography

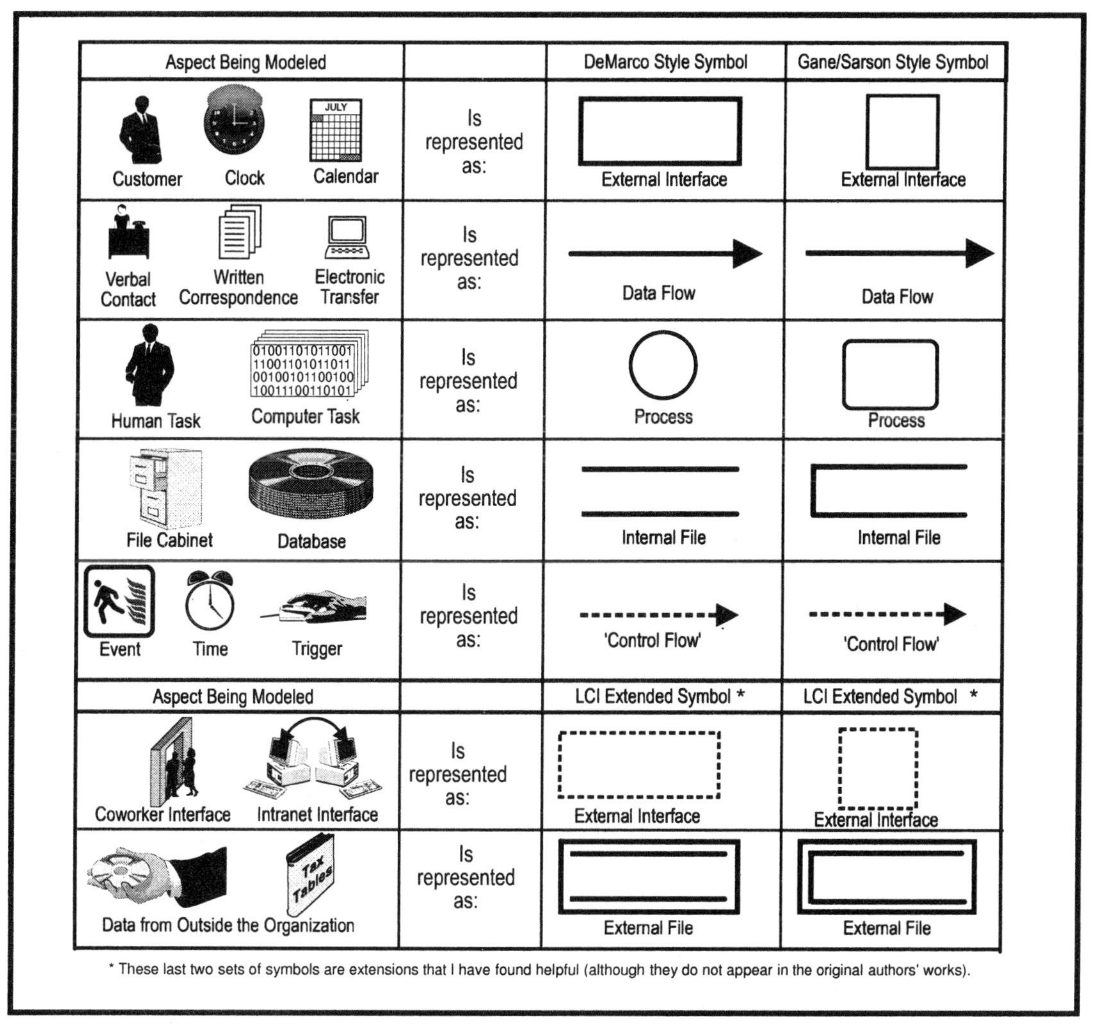

Fig. 5-10: Process Model Symbology

I have found it helpful to show when a data flow or control flow is a stimulus to a process. In addition, on control flows, the flow name is in **'single quotes'** because it typically does not have data content (see Figure 5-11).

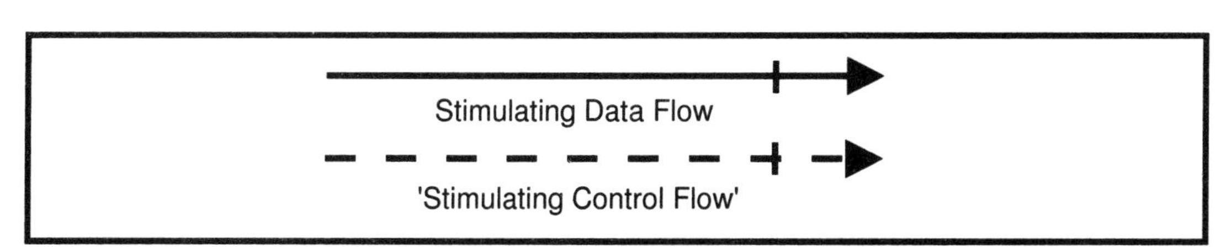

Figure 5-11: Data & Control Flow Symbols

Object Oriented Models

An Object Oriented Model (Figure 5–12) also shows data and processing. The Object Oriented Model is the inverse view of a Data Flow Diagram. They are similar in that they both model data and processing and the stimulation of that data and processing. They vary in that the DFD looks at a specific flow of data (and at just the processes acting on that flow of data) while the Object Oriented Model shows data encapsulated with all of its processes (methods) and the messages that stimulate specific processes within an Object.

An Object Oriented Model focuses on interacting Objects (Encapsulations of Data and Process). Objects can contain Methods (reusable processes acting on shared variables such as Data or States). In addition, Object Oriented modeling tracks Messages (the Communication between Objects). Object Oriented Models are good for modeling systems where the focus is on an Object first and the Object determines the processing.

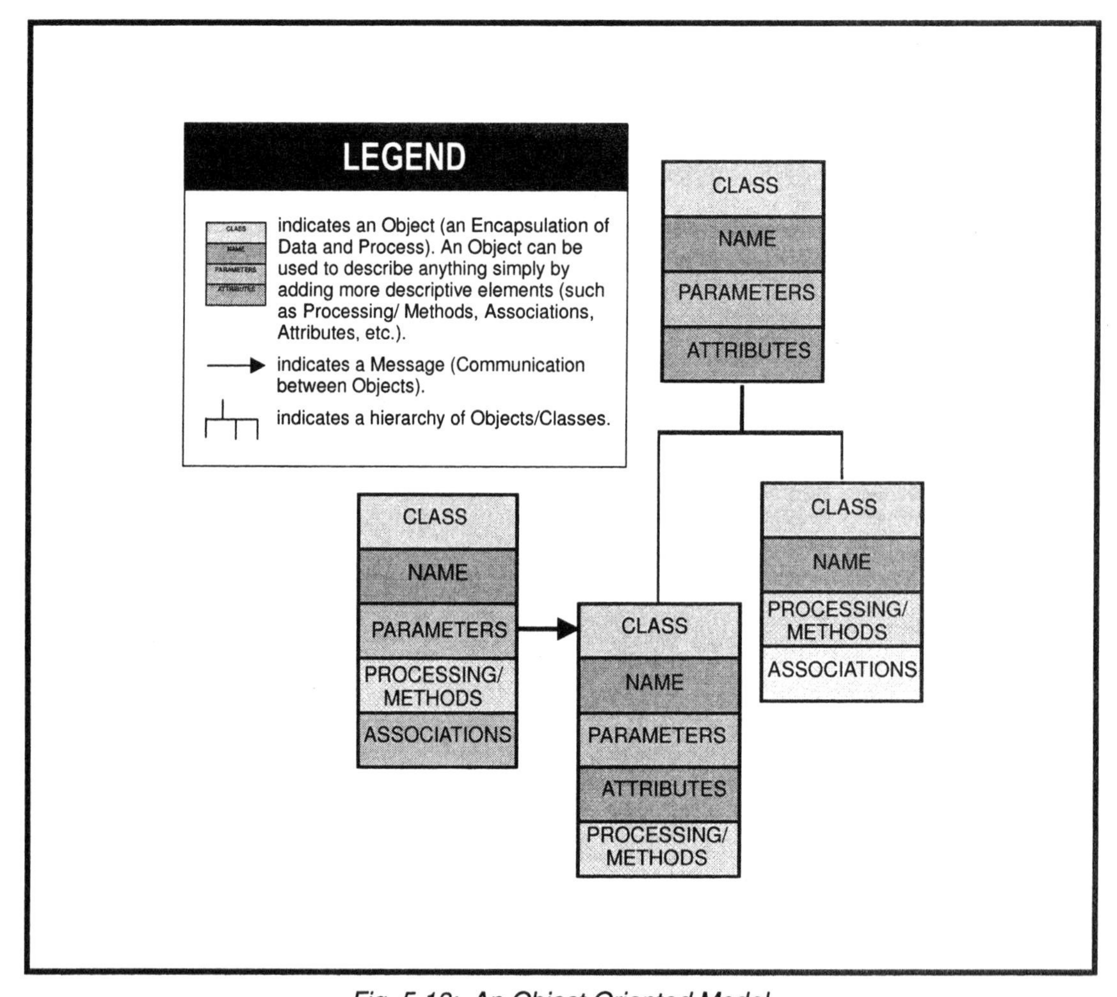

Fig. 5-12: An Object Oriented Model

Control Oriented Models

Control Oriented Models typically depict states and the transitions between those states.

State Transition Diagrams

There are other models we may need when we start to represent all aspects of a system. A State Transition Diagram (STD) is a Control Oriented Model (Figure 5–13). State Transition Diagrams are good for modeling the flow of control between different states. We could look upon a State Transition Diagram as a high-level Flow Chart. This model may show the state that a particular device is in and what causes the device to transfer from one state to another.

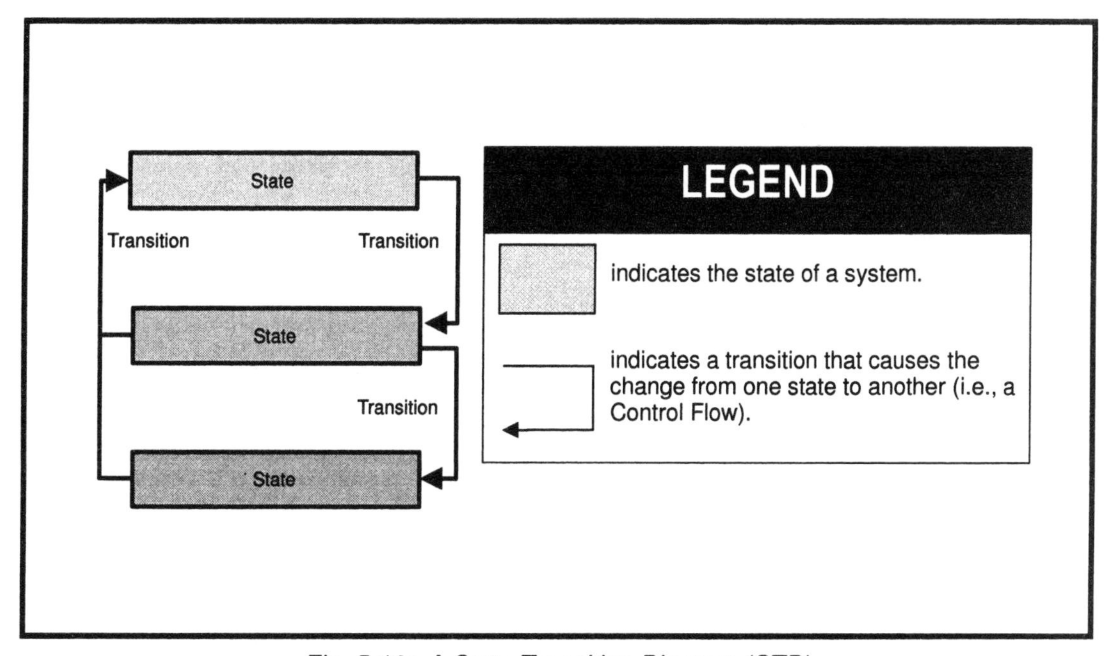

Fig. 5-13: A State Transition Diagram (STD)

For example, when you go to an automatic teller, the device may be in a wait state. When you insert your card into the machine, it transfers from the wait state into a validation state. If you enter the correct PIN, it may enter a menu state in which it asks you what you want to do. If you enter a wrong Personal Identification Number (PIN) three times, it may transfer into a state in which it confiscates your card and goes back into a wait state.

Control Flow Diagrams

A Control Flow Diagram (CFD) is a Control Oriented Model that can also show data movement or is tied to an associated Data Flow Diagram. This model is helpful for modeling the complex control issues of a system (Figure 5–14).

For example, in a system that controls an intelligent set of traffic lights, the decision to change the light and the order of the change can be governed by sensing the flow of traffic, the volume, the time of day, and the number of waiting vehicles at any particular location in the system.

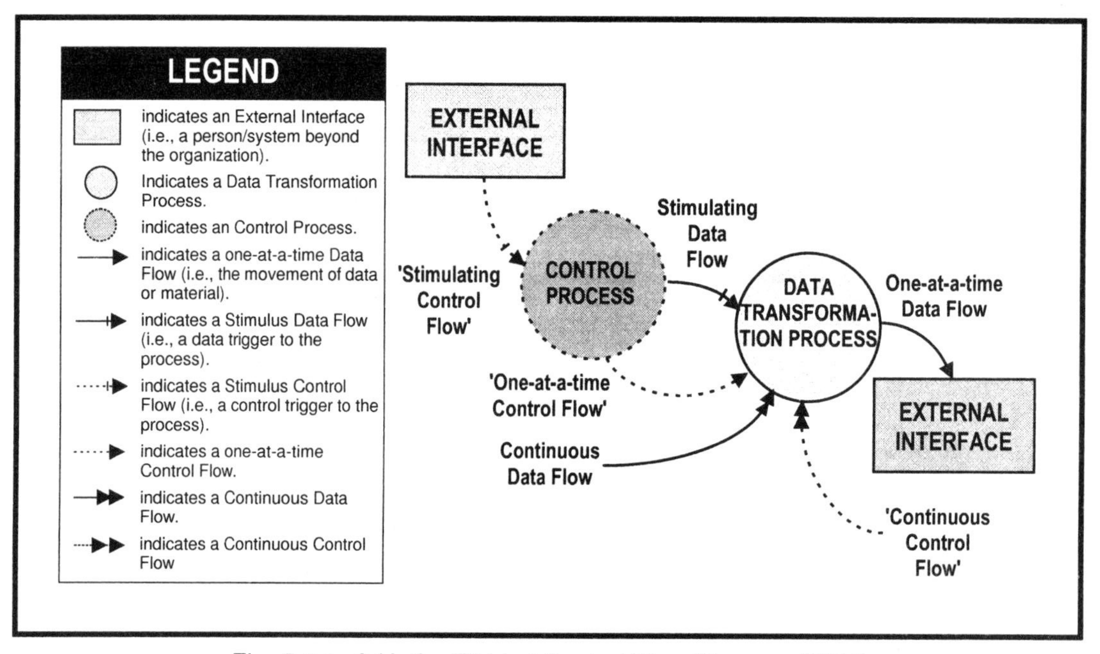

Fig. 5-14: A Hatley/Pirbhai Control Flow Diagram (CFD)

Choosing the Right Model

When we find ourselves analyzing a department and/or a computer system to determine how, for example, a Purchase Order gets through them, we are instantly involved in a "flow-of-data" problem. In this case, an ideal model to use would be a Data Flow Diagram that shows data on the move.

If I wanted to analyze the stored data in that department, or computer system and how one set of data related to another, then I would prefer to use an Entity Relationship Diagram. This model shows data at rest.

This data model might show, for example, that we have **Supplier** information, that certain **Products** come from particular **Suppliers**, that the **Suppliers** deliver to certain **Depots**, and so forth.

If I wanted to study something in the department or computer system (an Object) and see all the processes that can be performed on that object, I would use an Object Oriented Model.

For example, if the department was involved in editing text books, I may want to see all of the activities that can be performed on text (e.g., delete it, change its font, underline it, make it bold).

There are obviously more methods and models available to us for specifying requirements.

A major consideration when conducting analysis is to use the appropriate modeling tool or tools for the particular job. This is especially important if we are conducting analysis with a customer who is non-technical and from whom we require confirmation of the requirements.

If we are trying to gather and verify some information from business people or a customer (and they are not familiar with a particular technical model), try to use the model that best represents the business and train them if needed to understand that model.

Example of Business/Analysis and System/Design Models

Customer Focused Engineering allows us to bring together these different models into one unified methodology. This is where we should produce a progression of specifications (from stating organizational goals in a Mission Statement, objectives in a Project Charter, requirements in an Analysis Specification, solutions in a Design Specification, and on into Implementation Specifications).

Our models must do more than just describe the stimulus-response and process-memory aspects of businesses; they must allow us to represent the issues we talked about in the previous chapter (i.e., business/analysis issues versus system/design issues).

The English Language allows us to use many words to represent the same concept. For example, I've already used the word **what** to indicate business issues and **how** to indicate system issues. So, let me throw in a few more synonyms (Table 5–3).

BUSINESS ISSUES	SYSTEM ISSUES
Analysis	Design
Requirements Oriented	Solution Oriented
Implementation Independent	Implementation Dependent
Logical	Physical

Table 5–3: Business/System Synonyms

Let me show you some examples of Business/Analysis models and System/Design models using Process Oriented and Information/Data Oriented Models.[7]

Let's look at a theoretical warehouse environment in which someone sits at a desk waiting for shipments of **Books** to come in from **Suppliers**. When the **Books** come in, they have notes attached from the **Suppliers**. The person at the desk deals with orders for these **Books** (some of which are **Pending** and some of which are **Accepted**). Their job is to validate the incoming **Book Delivery** with a **Pending Order**, create **Accepted Orders**, and to file the **Accepted Books** on particular shelves. Figure 5–15 shows a sketch of this designed environment.

Fig. 5-15: A View of a Warehouse Environment

If we have the luxury of walking everybody who might ask about this environment through the actual warehouse and have them meet its occupants, we don't need to bother drawing a current design model of it. Or, it would be perfectly valid to create a comprehensive video (and not bother with modeling the actual physical/implementation view) and show this video to anyone interested.

Design models show <u>how</u> a business is, or will be, designed. They show issues such as departmental boundaries or computer system boundaries, and they depict physical details such as media, job titles, computer programs, and form layouts. But, if we then wanted to derive non-implementation (business) views in which we removed the existing design aspects of this human-based system, we could model it using a Data Flow Diagram and an Entity Relationship Diagram.

---

7 These are the ones I've used most frequently over the years and to that extent, I have become somewhat warped by them.

As you can see from Figure 5–16, we might represent this small system (from a business/analysis point of view) in a single-process DFD. The diagram would show **Books** coming in. It may also show us rejecting **Books** if they weren't ordered (by comparing what comes in with **Pending Orders** or by checking our **Supplier** list). If something is wrong, we send back the **Books.** If everything is fine, we change our **Pending Order** to an **Accepted Order**, put the **Order** into a different file, and store the **Books** on the shelves. We would support this model with a Data Dictionary and Process Specification(s).

After we validated this model, we could add any new data or processing needed for the new business view (if any) and cover it with a new design to support this part of the business.

The way we read this diagram is:

- A **Book Delivery** comes in from a **Supplier** which stimulates the **Validate Delivery** process into life.
- The **Validate Delivery** process retrieves the stored information on the **Supplier ID** and the **Pending Order.**
- Any **Unordered Books** and the original **Order** are sent back to the **Supplier.**
- **Accepted Books** are put into the **Books** store and an **Accepted Order** is stored in the **Accepted Orders** store.

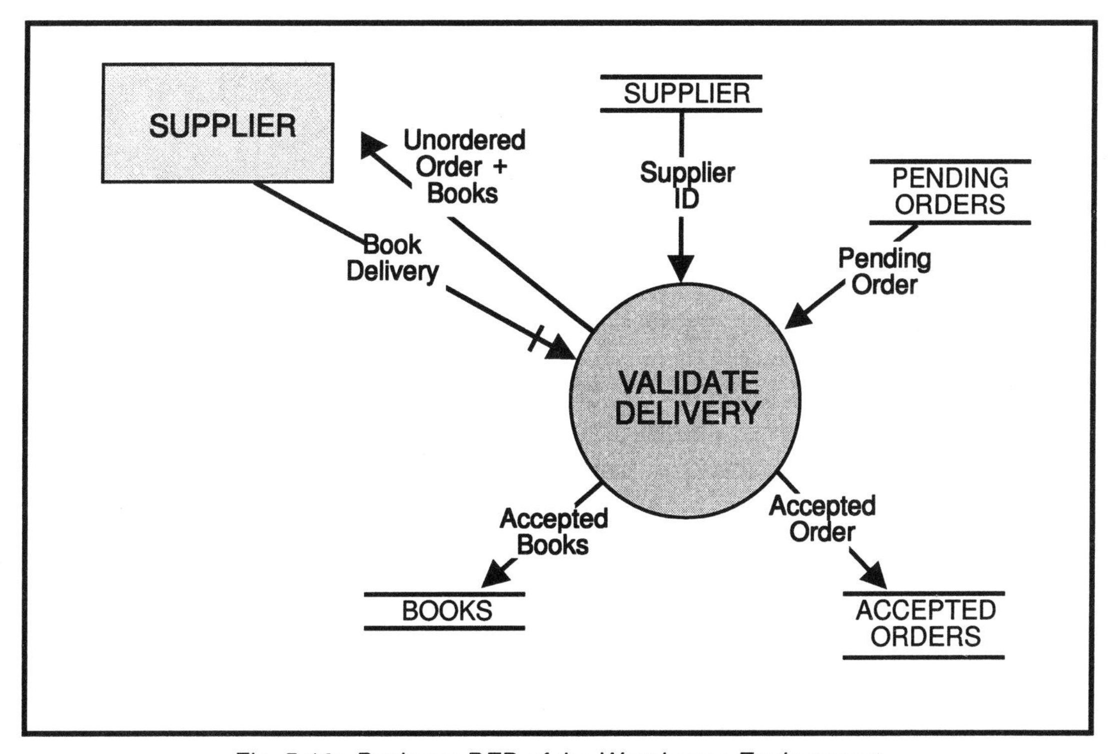

Fig. 5-16: Business DFD of the Warehouse Environment

We could also look at this small system from a "data-at-rest", business/analysis point of view using an Information/Data Model such as an Entity Relationship Diagram. Using this model we would see (as shown in Figure 5–17) that this one shipment of **Books** involved one Relationship where a **Supplier** sends us many **Books**. Part of this **Delivery** Relationship is to retrieve a **Pending Order** and to produce an **Accepted Order**.

We would support this model with Entity Specifications for each Entity on the Information/Data Model (ERD), an optional Relationship Specification for each Relationship between two or more Entities, and Data Element Specifications for each item of information about an Entity.

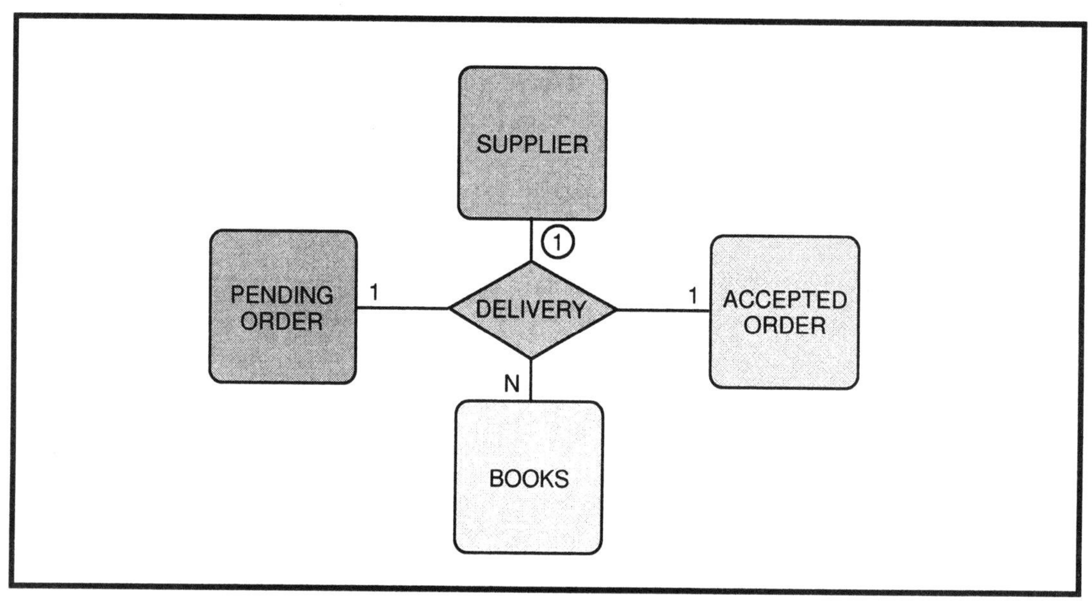

Fig. 5-17: Business ERD of the Warehouse Environment

The way we read this diagram is:

- A **Supplier** (the circled one is an anchor point from which we read the Relationship) is involved in a **Delivery** Relationship with many **Books** and one **Pending Order** and one **Accepted Order**.

After we validate this model, we can again add any new data, or Relationship features, and we can use it to identify a new design. The Process and Information/Data Models should not be separated from each other. In other words, when I have Entities on the Entity Relationship Diagram, they should correspond one-to-one with stores on a Data Flow Diagram. Of course, we can use another collection of models with which we felt we could validate the requirements with the customer/Business Policy Creator. The only constraints are that the models must not be ambiguous and they should satisfy the criteria established earlier in this chapter for an effective model.

The type of model is not what's important. What is important is understanding the business via the model.

I also want to emphasize that if someone comes along with a new model three years after you've read this book, it doesn't make the book (or your use of the old model) obsolete. Simply absorb the new model in your methodology/toolkit if the model is superior to any of the others that have been used.[4]

Summary

The Customer Focused Engineering professional needs to use models just like all other professionals.

System/design models depict the actual details of **how** the system processes and remembers. It declares such things as actual people, locations, files, forms, documents, and materials.

Business/Analysis models depict the *essential* details of **what** the business does and remembers, either currently or in the future. In this model all design and implementation details are removed.

This chapter has proposed some proven models that, with a little training, should be easy for a business person to understand. The important thing is to select the appropriate model for the problem at hand and to partition that model along business lines.

The next chapter will show how to partition these models based on the stimulus-response view presented in the previous chapter. We will also get into the key principles of Customer Focused Engineering. After that, in the *Partitioning by Business Events* chapter, we'll see how we can use this discipline to engineer a portion of, or even an entire organization.

4 I thought that Flow Charts were the greatest thing since sliced bread when I first started in Data Processing. Then I realized that they were inadequate when I was trying to look at concurrent processing and model systems that weren't linear.

Chapter

6

Our Event Horizon — The Boundary of Our Organization

The world thus appears as a complicated tissue of events, in which connections of different kinds alternate or overlap and thereby determine the texture of the whole.

Werner Heisenberg
Physics and Philosophy

We will examine the central concept of Organizational Events in this chapter. We will see that functional partitioning based on some of these Organizational Events provides an objective basis for the structure of an organization.

There are five types of Organizational Events:

- Strategic Events,
- System Events,
- Business Events,
- Regulatory Events, and
- Dependent Events.

However, the only ones that go on the Business Model are Business Events, Regulatory Events, and Dependent Events. From here on in, we'll call these three the Business Model Events. These three types of Events occur at the boundary of our organization. This is our Event Horizon (see Figure 6-1).

We must recognize the Strategic and System Events so they don't cloud our Business Model. These two Event types are somewhat internal; they come from our organization's Strategic Planners and Designers.

We must always keep in mind when determining Event types that Business Model Events are context driven. In other words, aspects that are business issues to one organization may be system issues to another organization. Let me clarify this after we've defined the five types of Events.

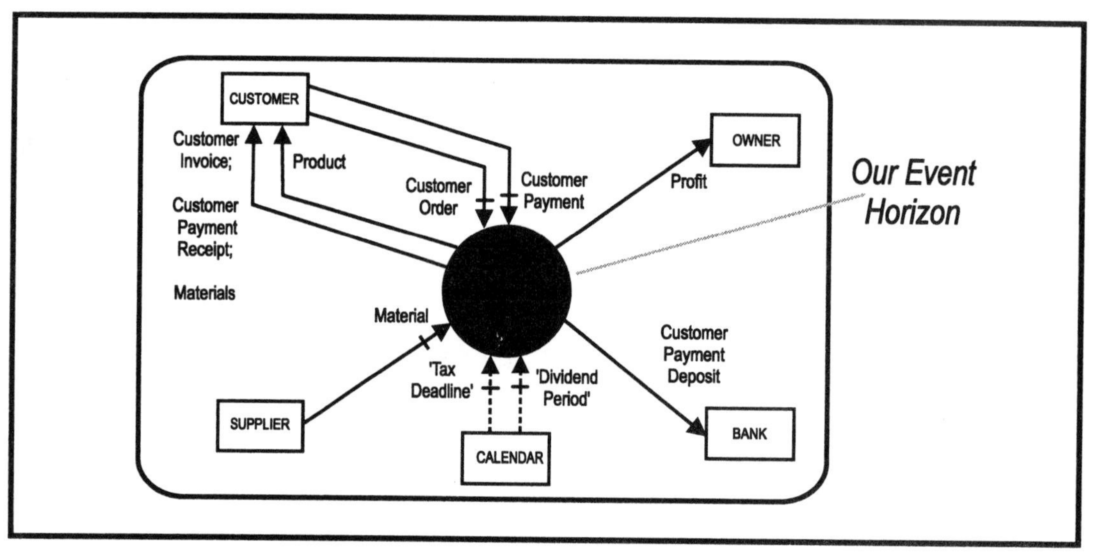

Fig. 6-1: The Boundary of the Organization

I want you to keep in mind the fundamental characteristics of all systems as stated in ***The Nature of Systems*** Chapter. We will use the Stimulus-Response and Process-Memory structure from that chapter as an underlying tenet throughout this chapter.

The Foundation of the Business Event Methodology

There's a multitude of demands people make on the world of business and each requires a different response to satisfy it. On the other hand, any one organization responds to only a small subset of those demands. (As an added benefit, in vertical, non-diversified organizations, we can usually obtain mass reusability in our processing and data across these demands.)

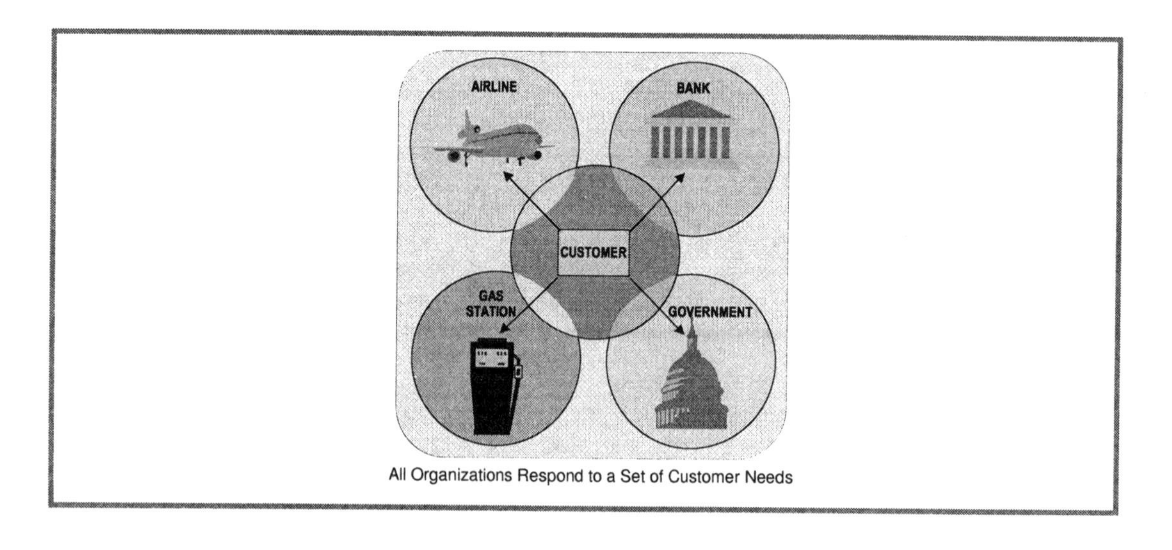

All Organizations Respond to a Set of Customer Needs

From one point of view we could say that any organization has simply decided to respond to a set of customer needs (Events) that occur in the external world. As examples:

- An airline company decides to respond to its customers' travel needs in which they, and/or goods, must be moved.
- A bank decides to respond to its customers' needs for storing and manipulating money.
- A fire department responds to a fire in the environment (the department's customer).
- Logical Conclusions, Inc. has decided to respond to our customers' needs for consulting or training in subjects such as Customer Focused Engineering.

We never know within our organization why the demand was generated. That's why we classify it as *outside* of our organization.

> Why a customer wants to withdraw funds from our bank, for example, is literally none of our business; a funds withdrawal could be made in order to go on vacation, pay a bill, or just have expense money for the day.

A system (a manual operation, a computer system, or even our entire implemented organization) has no control over external customers' needs.[1] The customer has the ultimate control in initiating a stimulus to our organization. At the same time, however, our organization is obliged to respond to stimuli caused by our customer's needs.

Without stimulus from the outside, our organization and its systems remain inactive and are essentially meaningless, and our organization will eventually close down. No organization is self-perpetuating. Even non-profit government organizations respond to external needs as their reason for existing.

All organizations rely on their customer's Events as the reasons for existing.

A Little History of the Concept of Event Partitioning

In the 1970s the Systems Engineering techniques that were being proffered were somewhat disjoint, mainly because the ideas originated with different authors (and their sales staffs). During that same decade, I worked at a young company involved in generating and spreading these new Systems Engineering ideas. In those earlier years of teaching new engineering tools we didn't quite have the techniques down. For example, we knew how to draw a Data Flow Diagram and we understood the usage rules of each symbol, but we

1 This does not mean we should not "think out of the box" and ask our customers if we can assist them beyond our current boundary.

didn't give much advice on how to produce a well-partitioned Business Model. The size of the drafting surface — a page — and the human limitation of keeping track of seven plus or minus two things per page were somewhat the extent of our partitioning rules.

In my seminars I could convince my data processing students that there was a better way to partition systems because almost all existing partitioning of programs and systems was mainly at the discretion of some D.P. professionals. (D.P. was a new industry and almost all programmers were allowed to partition their programs and systems as they saw fit.) This arbitrary partitioning was not so true for the business community.

Of the seminars I taught in the 1970s, the ones in which I taught analysis were the only ones where I would expect a business person to be present. They at least had a longer history and established reasons for how and why they partitioned their departments, people, and tasks. I had to have good reasons for repartitioning these established organizational boundaries.

I and a few of my colleagues who taught analysis seminars at that time used a technique called *Stimulus-Response Modeling* as a means of partitioning a business Data Flow Diagram.[2] This was a means of getting away from any of the physical/design partitioning of the old environment. We used a "string-of-pearls" analogy to illustrate a stimulus data flow triggering a chain of continuous processes connected by intermediate data (indicated by lines and circles on a Data Flow Diagram).

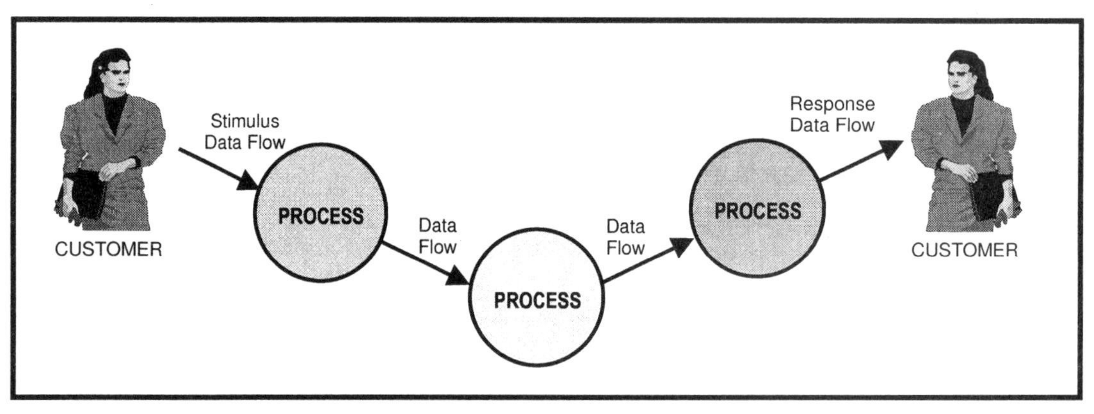

Fig. 6-2: String of Pearls

In Figure 6-2 we see an external stimulus initiating the execution of the set of processes (the string of pearls). The intermediate data and its processing were required to respond to the Event that produced the stimulus. Once initiated, the set of processes runs its course without further stimulus until a response is returned to the outside world. Because all these processes, and *only* these processes, are triggered, we can see that they make up a highly functional partition in that they perform a "single-minded" overall function <u>from the point of view of the external Customer</u>. This basic concept has evolved (over 20 years) into the contents of this book built around the central theme of the Business Event.

2 In 1984 two of my colleagues at Yourdon, Inc., Steve McMenamin and John Palmer, used the term "Event Partitioning" in their book, *Essential Systems Analysis*, for a similar concept.

The Business Event Methodology and Its Players

Before going any further, let me establish a Common Platform of Understanding regarding the term Business Event Methodology. I have heard many discussions/meetings go awry when people use the terms "methods" and "methodology" interchangeably and misuse the word "methodology."

Webster's Dictionary defines "methodology" as:

A set or system of methods, principles, and rules for regulating a given discipline.

So a methodology recommends a cohesive set of methods which in turn usually recommend a set of models (See Figure 6-3).

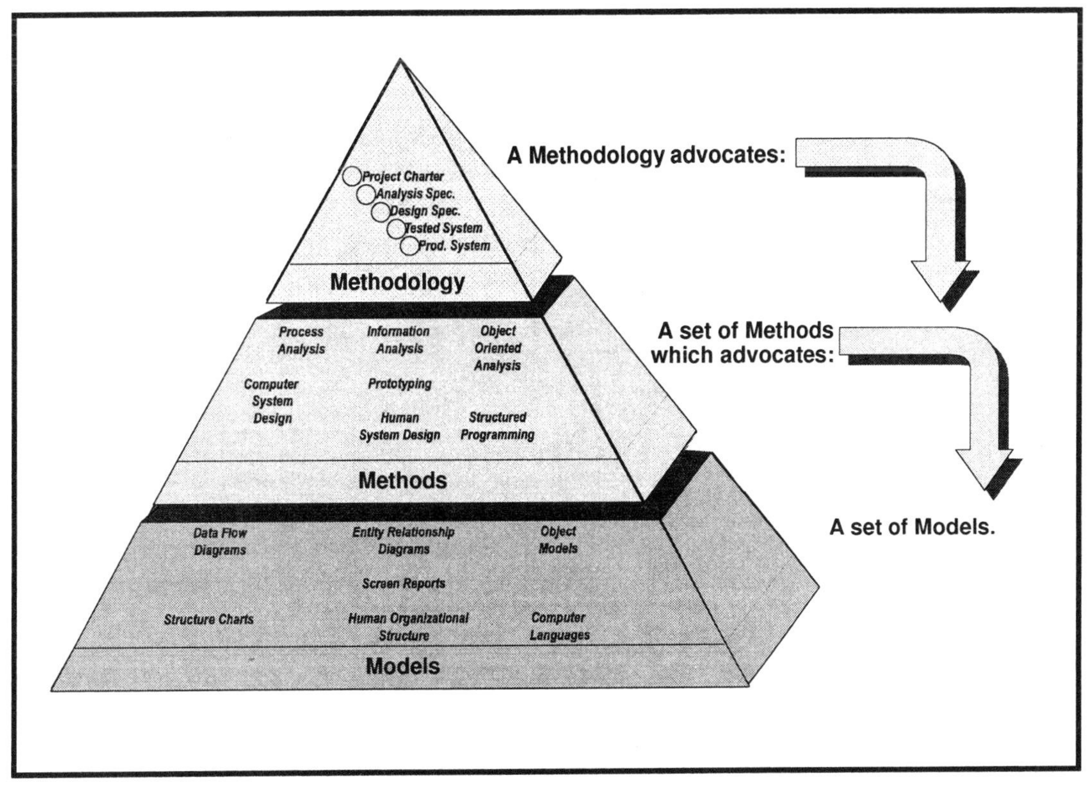

Fig. 6-3: An Example of Methodology, Methods, & Models

At a high level the methods look the same for any project: conduct analysis, design, and implementation, etc. How you perform these methods is different in the Business Event Methodology. The individual methods utilized within the high-level phases are typically selected by each project. Just make sure when creating a Customer Focused Organization that the methods are engineering methods and the intermediate deliverable (the model) from one method flows naturally into the next method.

There are a number of "players" (see Table 6–1) and tasks associated with those players in the Business Event Methodology. Remember, these players could be different people in a development project, or they could be one person wearing different "hats."

- The customer has the need.
- The Strategic Planner identifies if the organization will respond to the need. The Business Policy Creator interprets the need into the specific set of data and processing the organization will use to respond to the need. (In other words, the organization's decision makers determine the business policy.)
- The Systems Analyst models/documents this business policy.
- The Systems Designer invents the design to implement this business policy.
- The Systems Builders (Technical Writers and Programmers) implement this business policy.
- And finally, the Production Systems (People, Technology, etc.) accomplish the customers' need through time.

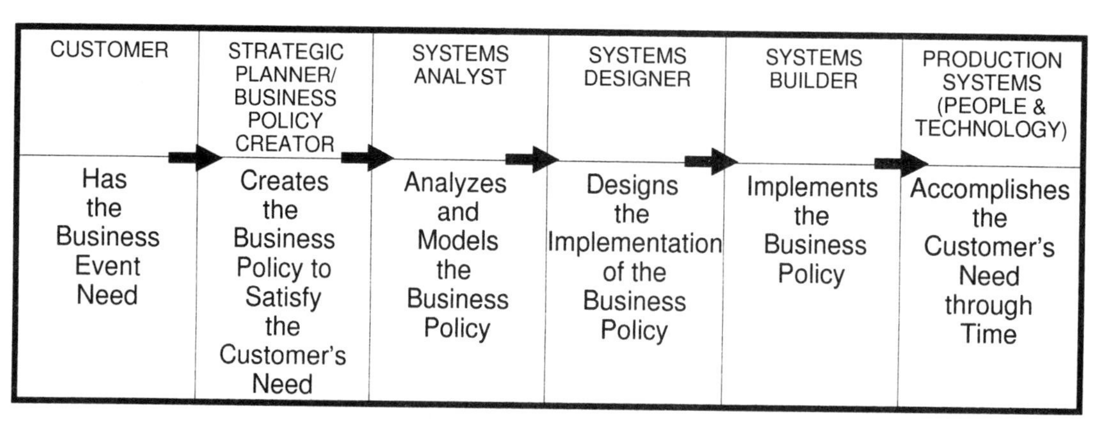

CUSTOMER	STRATEGIC PLANNER/ BUSINESS POLICY CREATOR	SYSTEMS ANALYST	SYSTEMS DESIGNER	SYSTEMS BUILDER	PRODUCTION SYSTEMS (PEOPLE & TECHNOLOGY)
Has the Business Event Need	Creates the Business Policy to Satisfy the Customer's Need	Analyzes and Models the Business Policy	Designs the Implementation of the Business Policy	Implements the Business Policy	Accomplishes the Customer's Need through Time

Table 6–1: The Players in the Business Event Methodology

The Types of Organizational Events

OK. You've reached the most important part of the book. I had to state all of that other stuff in the preceding chapters to set the foundation for this part.

In the Business Event Methodology we identify five types of Organizational Events:

- Strategic Events
- System Events
- Business Events
- Regulatory Events
- Dependent Events

Determining the type of Event we are analyzing is important because we want to concentrate on the ones that make up our true business, i.e., form our Business Model.

Some Organizational Events don't belong on the model at all, while another type actually changes the model.

Before defining these types of Organizational Events, I want to make sure you don't confuse an Event with its stimulus and implementation. A phone call, a piece of mail, a fax, or a customer walking up to a counter are all implementation issues. Phones, the mail, a fax, or a personal visit are just the designed means for communicating the customer's need (the Event).

We need to concentrate on who the external Customer is and on what are their Events to which our organization has decided to respond.

Identifying Organizational Events via where They Occur

Where the Event occurs (i.e., originates) often dictates the type of Event:

- Strategic Events typically originate at our organization's high-level management (although the management may have initiated the Strategic Event based on our competition doing something to which we need to respond). The management may also be responding to an outside agency whose laws affect our organization — more on this later. Strategic Events can't be ignored, but they are not the main focus of the Business Event Methodology.
- System Events originate within the organization. They are invented and imposed on the area of business to meet management and design needs. They will always be associated with an implemented system.
- Business Events originate at our true external customers. Business Events are the most important Event to the Customer Focused Engineer.
- Regulatory Events originate at a government or external regulatory agency. They are imposed on the area of business to meet legal operating requirements.
- Dependent Events typically originate at the vendors used by our organization. Dependent Events are the result of "farming out" part of the business in the past.

Organizational Events also have a hierarchy in terms of their of importance to the Business Event Methodology (see Figure 6-4).

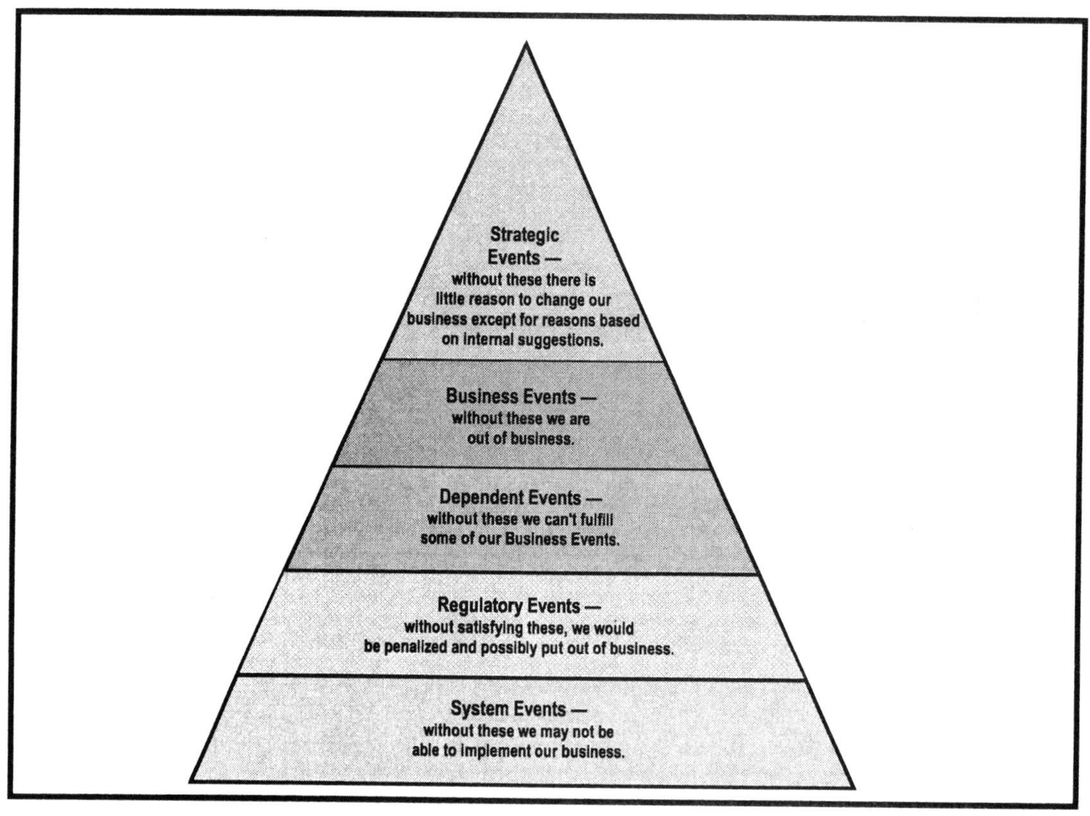

Fig. 6-4: The Hierarchical Importance of Events

Every organization I have studied responds to all these Event types. You may have already guessed that when we create a Customer Focused Organization, we want to focus on the Business Events. They will form the bedrock of our Business Model. However, we have some Event types that get in the way of forming our Business Model and others that we must also include in our Business Model (along with Business Events). Figure 6-5 shows the Customer Focused Engineer's Event filtering and categorizing task.

Notice that this diagram is indicating that we're trying to focus on true Business Events as our organization's mission, but, we have Strategic Events and System Events in the way of our business view. We should filter these out first. We will keep Dependent and Regulatory Events on our Business Model but we should identify and categorize them on our model as such. Even though they appear on the Business Model, they are less important than Business Events because they are not our main focus for "keeping the doors open."

You could create separate models for Dependent Events and Regulatory Events, but I recommend you don't when you're new to this methodology because there will be shared processing and memory between these three types of Organizational Events. Although, it may be a good idea to indicate the different types of Organizational Events that do belong on the Business Model (Business, Dependent, and Regulatory) by using different colors or codes.

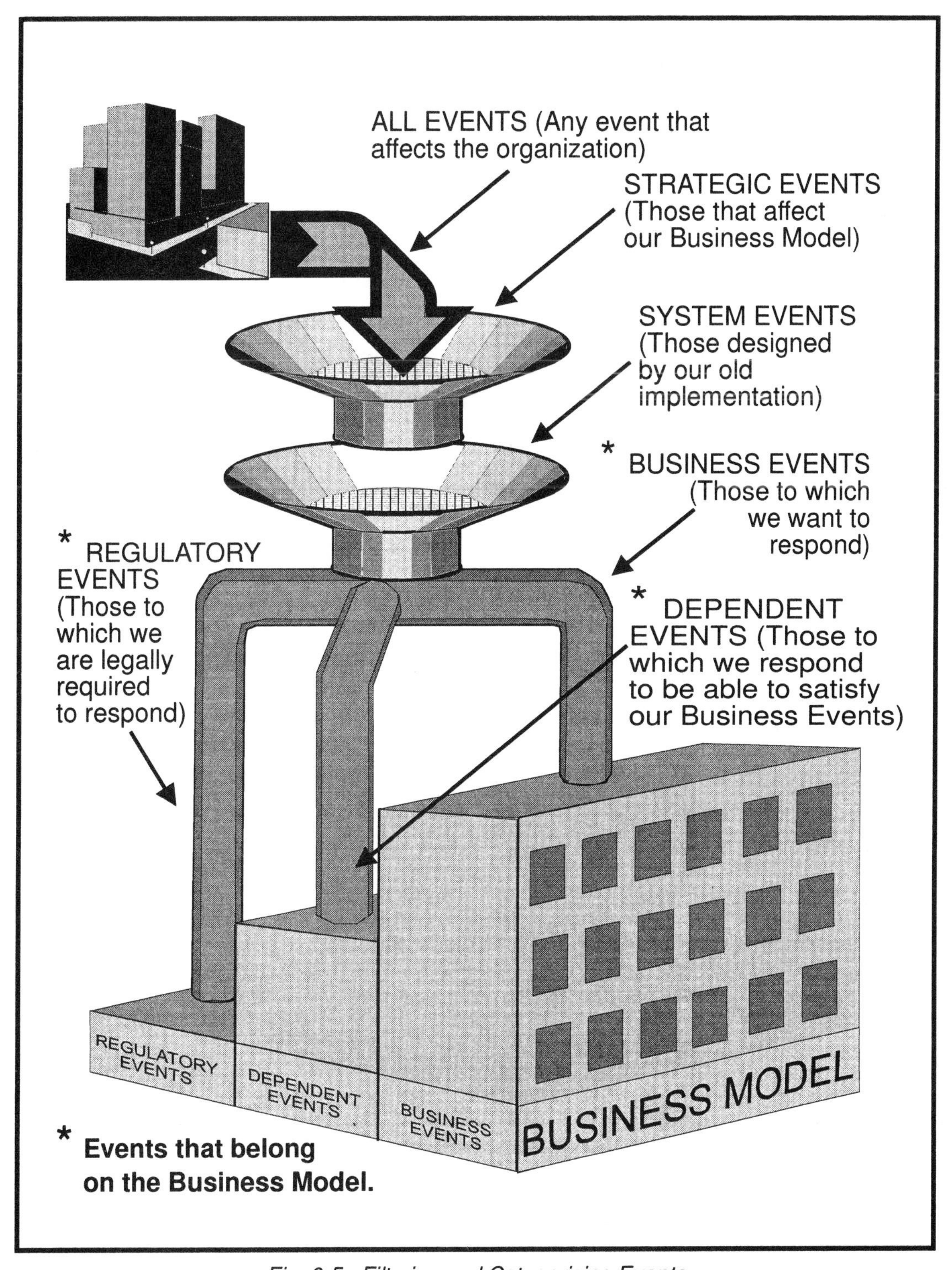

Fig. 6-5: Filtering and Categorizing Events

Summary

The foundation of the Business Event Methodology is its focus on the customer and their Events that stimulate our organization into life. Recognizing the five types of Organizational Events that stimulate all organizations and filtering out the Strategic and System Events is another building block of the methodology. It is only by doing this filtering that we can focus on our true business customers (and their complete needs).

Remember at the beginning of this chapter I said Business Model Events were context driven. I hope you can see after reading this definitions chapter why one organization's Business Events will be another organization's System Events.

> For example, in most organizations issues to do with computer systems and humans are System Event-related and we remove them to form a valid Business Model. However, if we are in the business of creating computer hardware or providing temporary staffing then these are not System Events. When organizations contract with my own business, Logical Conclusions, our services are in fact System Event issues to these organizations. However, within the context of my organization, I'm in business to provide training and consulting and therefore, these are my Business Events.

In the next chapter I will define these types of Organizational Events in detail with the understanding that the Business Events are the most important ones to the organization's success. Please note, however, that everything we say with regards to modeling Business Events can be applied to Dependent Events, Regulatory Events, and even Strategic Events.

Chapter

7

Recognizing the Five Types of Events that Stimulate Our Organization

The customer is the one who supports us. Study his needs; get ahead of him.

W. Edwards Deming

Last published interview in "Industry Week" Magazine

Now that we've identified the different Event types we should define each of the Events in detail. The five different types of Events to which our organization responds are:

- Strategic Events
- System Events
- Business Events
- Regulatory Events
- Dependent Events

Strategic Events

Definition of a Strategic Event

A Strategic Event is an incident, typically originating at our organization's leaders, that makes us change the way we do business and hence change our Business Model in some way.

Recognizing Strategic Events

Strategic Events cause us to add or delete nouns (data) and verbs (processing) on our Business Model. They do not stimulate the Business Model into life as would a Business Event from our customers.

Figure 7-1 shows us changing some logic in the Business Model when a Strategic Event occurs. (Don't get too involved in the Business Model shown in Figure 7-1), we'll cover this in detail in the ***Partitioning by Business Events*** Chapter.

If we were conducting Pre-Engineering (i.e., no existing environment), someone or some group would have to invent the processing logic and data we would need to respond to the needs of our customers. I call these people the Business Policy Creator(s). Our Business Model is created and modified by these people. They are not our customers.

In a private organization the Business Policy Creator would be the upper management (leaders) or the people designated to interpret the Mission Statement of the organization. In a public organization (government) this would be Congress/Parliament or again the people designated to interpret the government programs and laws.

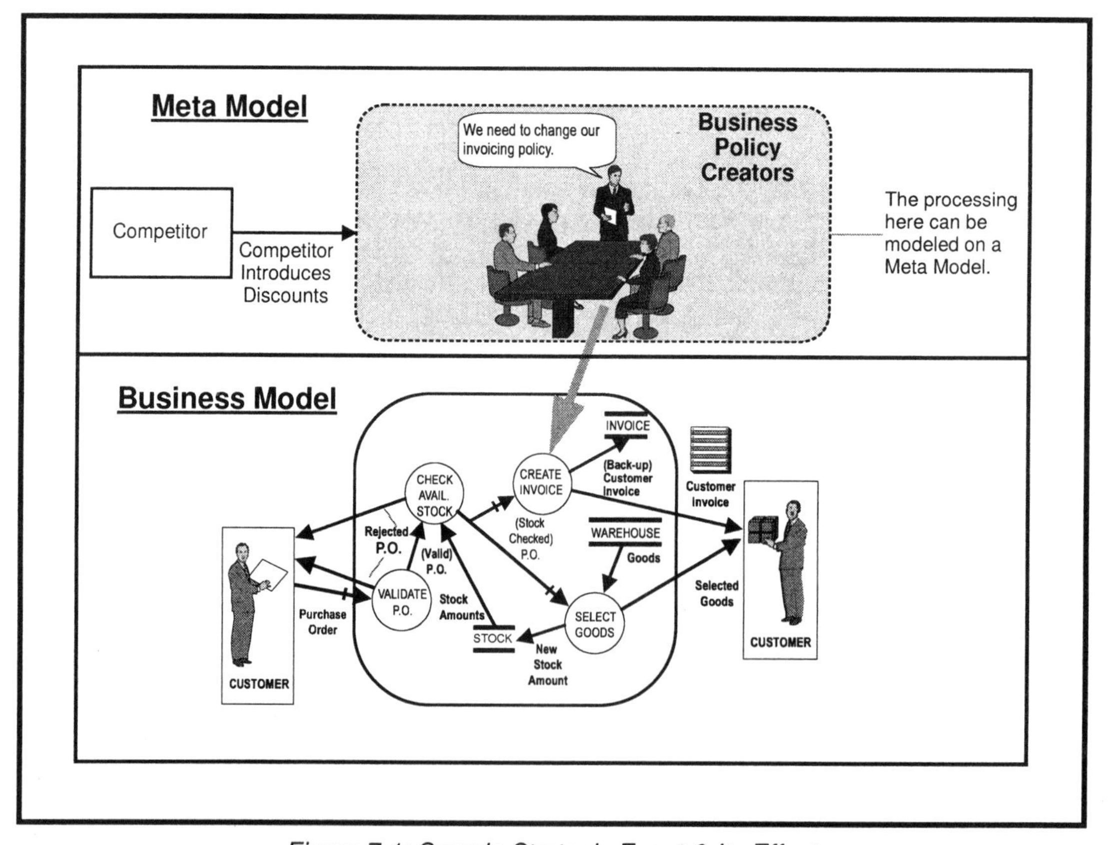

Figure 7-1: Sample Strategic Event & Its Effect

Don't Model the Modeler

Of course, even when creating a Customer Focused Organization with an existing organization in place, there will still be Strategic Events that make us modify our Business Model. The main point to remember about Strategic Events is they do not stimulate our Business Model into life. They make us form the model in the hope that a customer will stimulate it into life with a Business Event.

> For example, the leaders (Business Policy Creators) of an insurance company can create a new type of policy with its logic and Data Element needs inside the organization. However, these will remain inactive until an Event from a customer stimulates this logic and its Data Elements. The Strategic Event forms the business processes and data. The Business Event stimulates, places values in the Data Elements, and executes the processing on these values (see Figure 7-2).

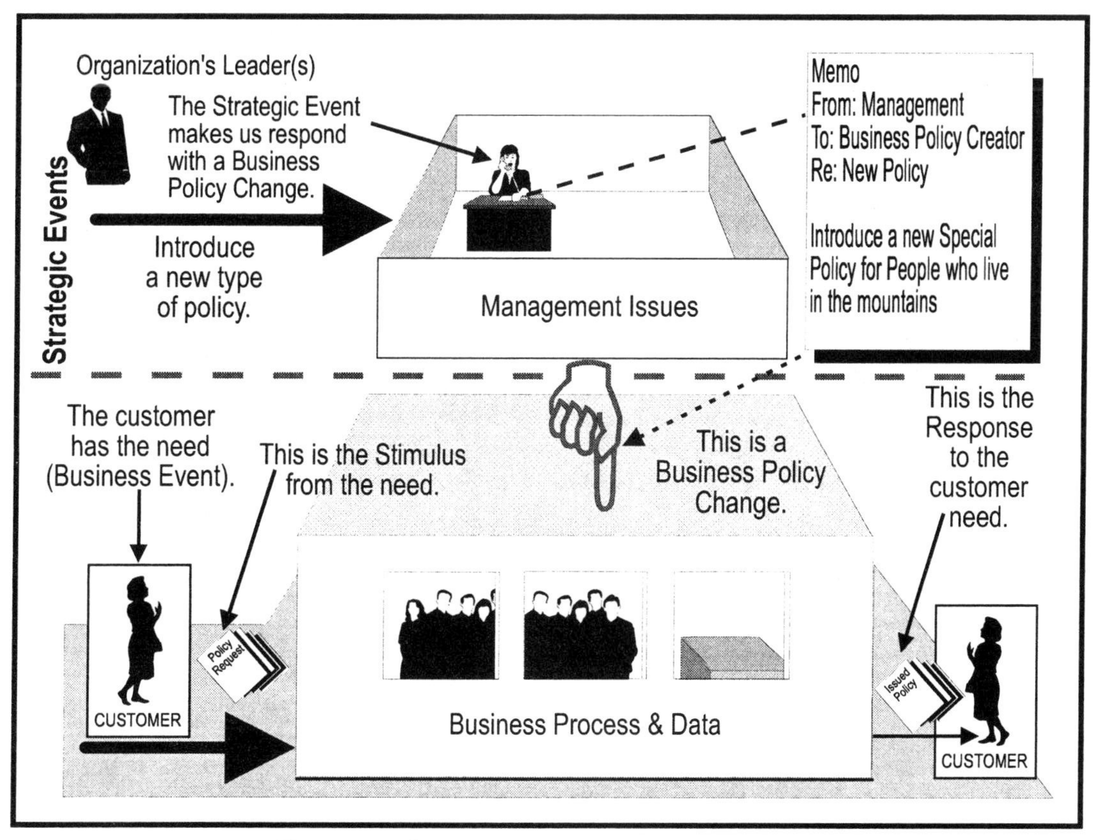

Figure 7-2: Strategic Events vs. Business Events

Another term often used for the data created by a Strategic Event is Meta Data, or the data about data. "Ignatius Pondwater" is Data but "Customer Name" is Meta Data. We can (and probably should) create a separate model (a Meta Model) of just Strategic Events. This would use the same symbology as our Business Model, but would depict the processing and data in response to a Strategic Event.

For example, in the private sector what do we do when we need to introduce a new line of business that requires us to capture additional data elements? Or, in a government agency what do we do when Congress introduces a new program requiring new calculations in our processing? In either case these changes require us to change our Business Model.

Remember you're not in business to respond to Strategic Events, so they do not go on the Business Model.

As I have found Strategic Events difficult to understand, let me use one more analogy for explanation. When we build a system or organization, we are building the equivalent of a radio (circuits, etc.). Our customers aren't really interested in the radio; they want music, news, etc. (see Figure 7-3). We put the radio in place then wait for it's input (radio signals) to create the output (music, news, etc.). In an organization Strategic Events put the business in place and the customer stimulates the business with their input into producing some output.

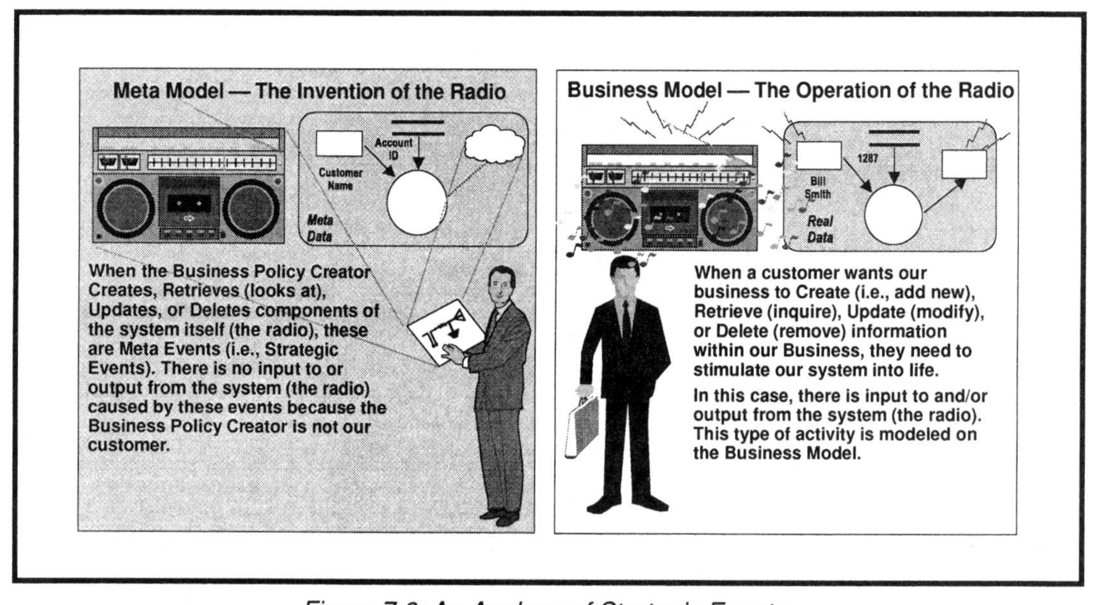

Figure 7-3: An Analogy of Strategic Events

System Customers and System Events

I've found that some organizations get so wrapped up in their historical procedures that they lose sight of the reasons for those procedures in the first place. These organizations seem to be more concerned with satisfying their systems' needs over their customer's needs. The systems are there to support the business, not the business to support the systems.

These systems needs are what I call System Events. This institutionalized importance of System Events over the Business Event needs of the customer leads to what I call the "Clothes Have No Emperor" syndrome. The Emperor (the business need) has gone but the Clothes (the systems) associated with the old need are still being supported.

System Events

Definition of a System Event

A System Event is an internal incident created during the design of our organization's structure and systems in order to satisfy some aspect of technology (human or computer).

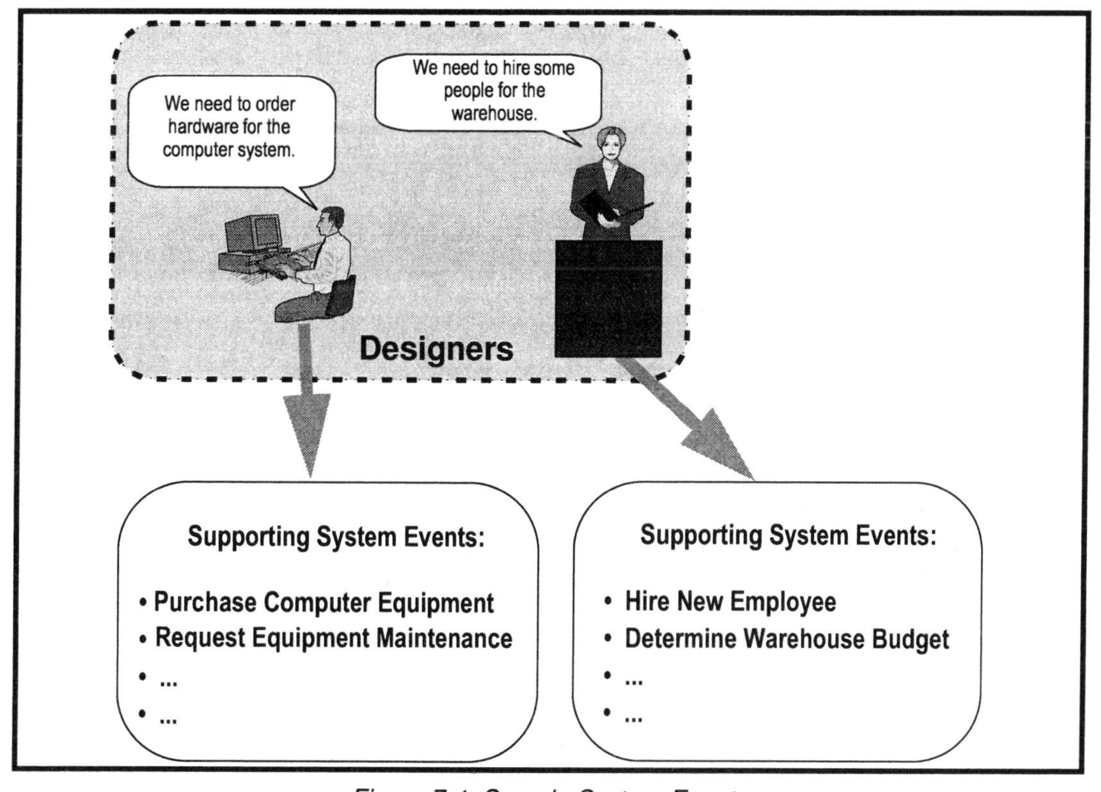

Figure 7-4: Sample System Events

Recognizing System Events

System Events originate with our internal staff, departments, or systems. These entities exist because of the existing structure of our organization. We have control over these Events and they are always designed and implemented within our organization.

Figure 7-4 shows an example of some System Events that support the human and automated designs of a business. System Events are affected by design changes. If the organization (shown in Figure 7-4) decided to automate its warehouse operations, then internal System Events such as employee hiring and firing, that currently support its manual warehouse operations, would no longer be necessary. On the other hand, if this organization decided to provide its customers with a completely manual system, we would not need the System Events to Purchase Computer Equipment or Request Equipment Maintenance.

System Events should not appear on the Business Model. However, to make Business Events run in the real world we extend the Business Model at the time of new design to include System Events (see Figure 7-5).

The historical reasons for dysfunctional partitioning in the ***Systems Archaeology*** Chapter will help us recognize System Events. One significant source of System Events is the existing management inside an organization (for example, management requests are typically System Events).

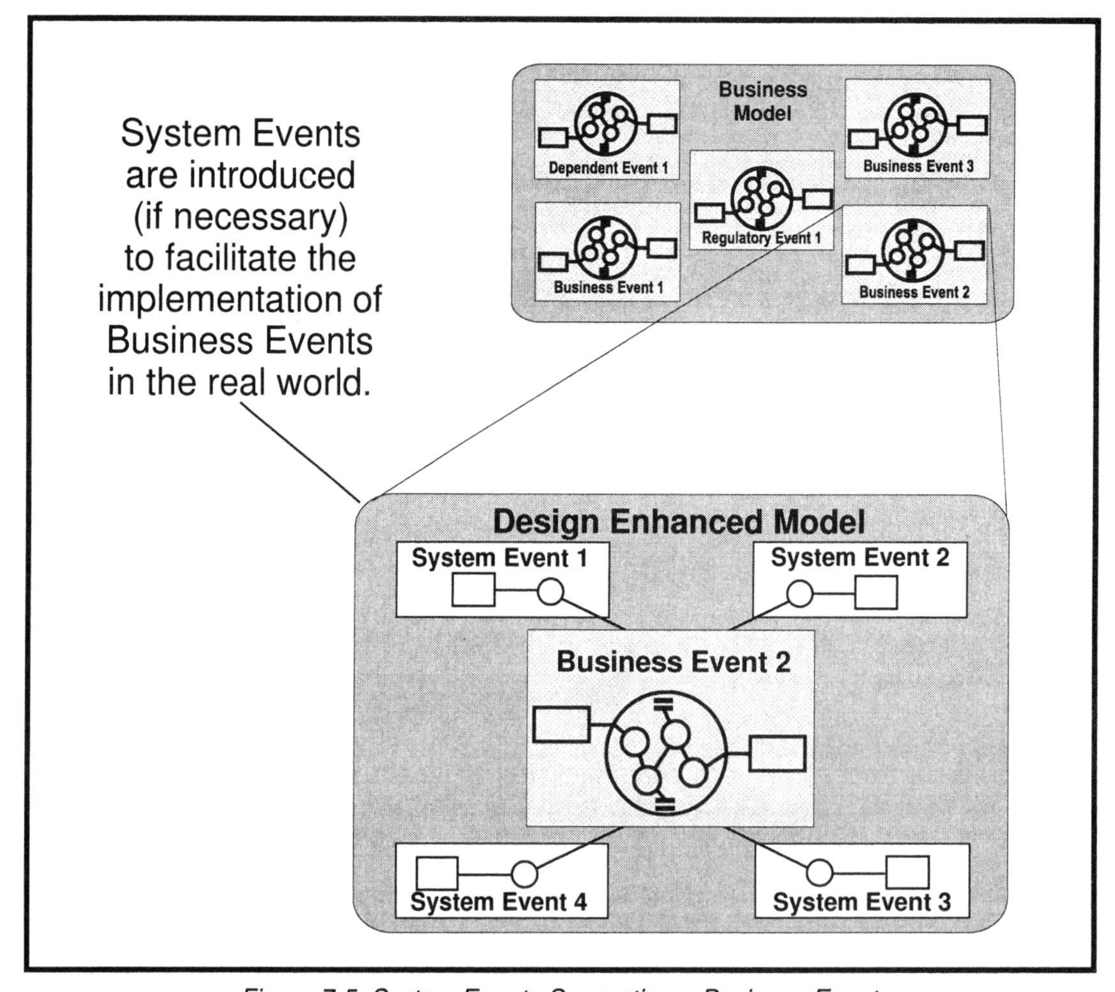

Figure 7-5: System Events Supporting a Business Event

Next let's talk about the three types of Events that make up the Business Model. Understand that the processing and data of these Events (i.e., the nouns and verbs) were originally put in place by Strategic Events. We are dis-covering and modeling these Business, Regulatory, and Dependent Events that were put in place in the past.

Business Events

Definition of a Business Event

A Business Event is an external incident originating at our organization's customer that places a demand on us to which we respond in order to accomplish our strategic mission.

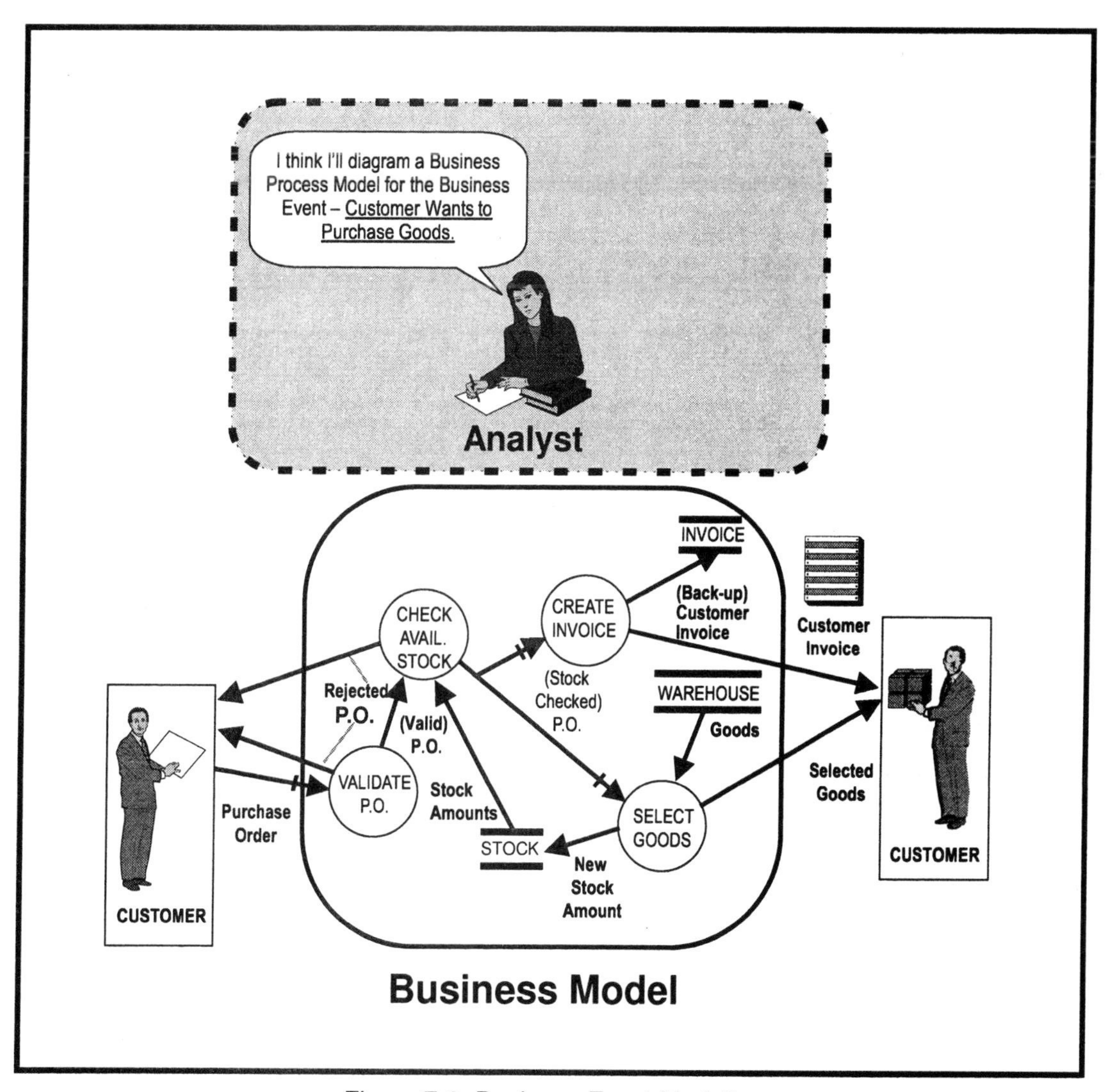

Figure 7-6: Business Event Modeling

Recognizing Business Events

Business Events originate with our true business customers. These are individuals, clients, agencies, and systems that are external to our organization and over which we have no control (that is, the Event does not occur within our organization). Business customers do not regulate our organization in any way.

Business Events are directly associated with revenue generation in a private organization. In a government organization these are the Events that satisfy government programs and laws. In either case, these Events are Business Events as long as they satisfy the needs of whomever or whatever we call an external customer.

When we put on our analyst "hat," we document the organization's internal details regarding a Business Event on the Business Model (see Figure 7-6).

Business Events stimulate our organization (and hence, the Business Model) into life.

In an information organization (banks, insurance and training companies, government agencies, etc.), Business Events Create, Retrieve, Update, and/or Delete the data values.

Business Events can be significant or trivial.

For example, in my organization the Business Event: Customer Wants to Schedule an In-house Seminar is a significant Event. It stimulates many processes and involves the retrieval and updating of a substantial amount of data. A trivial Event on the other hand will involve very little data and processing and sometimes no business processing. For example, the Business Event: Customer Requests Seminar Information just requires a simple retrieval and issuance of data with no real processing.

Even though the latter Event is trivial, we still need to acknowledge these Business Events because they will legitimize why an item of data is gathered (Created) in another Event. As we'll see later in the chapter titled ***Achieving Organizational Process and Data Integrity***, it's important to know where the data are Created, Retrieved, Updated, and Deleted for data conservation purposes.

When I've used this Trivial Business Event example in my seminars, I sometimes get the response: "Couldn't we have an enormous number of Business Events equal to every stored data item in the organization multiplied by four? We potentially have a Create, Retrieve, Update, and Delete Event for each data item, don't we?" In the extreme, this is potentially true; however, you would have to have the customer requesting one specific field as their complete Business Event for this to happen.

One Business Event will usually do any number of Creates, Retrieves, Updates, or Deletes of Data Elements. Also notice in my trivial Business Event example above I labeled the Event: Customer Requests Seminar Information. "Seminar Information" is a significant (cohesive) set of data and the request can retrieve any or all of its contents. This eliminates the need to access one data item with one Business Event. Trivial Business Events usually outnumber the significant ones; however, they also don't take much analysis and documentation effort.

Our response to Business Events (and how we implement them) is what distinguishes us from our competitors.

We are concerned with two aspects when analyzing a *specific* Business Event:

- the nature of the need from the external customer point of view, and
- the policy that our organization has for responding to that need (our organization's internal point of view).

It's vital to our understanding of Customer Focused Engineering and the success of our organization that we have a clear understanding of Business Events.

Business Customers and Business Events

From the customers' external viewpoint we need to know what the customer expects to happen, such as walking away with products or having their service needs met. Taking the external viewpoint is where we get ahead of our customer as related to Deming's quote at the beginning of this chapter. The particular need of the customer (outside person, other system, or external organization), plus the content of the communication, is what affects the particular way our organization responds.

> For example, if you are used to flying on different airlines and they all require the same information from you and process you the same when you fly, there is very little incentive to stay with one airline (assuming the service and routes are the same). If, on the other hand, one particular airline asked for and kept your profile (your meals, seating, and luggage storage needs, reading and sleeping needs, etc.), gave you minimum processing time at the airport, and didn't require you to support their systems with pieces of paper and codes, you probably would want to fly with this airline the most.

After building this methodology and giving seminars based on the Business Event concept, I believe that Business Events and their partitioning is the fundamental view that unifies all the existing methods and models into a Customer Focused Engineering discipline.

Business Event Naming

I recommend a Business Event Name consisting of a whole sentence. I think that it's ridiculous to limit ourselves to only a verb-object name, or worse, to eight characters for our naming standard. (We can invent an additional cryptic identifier during design if necessary.)

For accurate management of our models, the Business Event Name must be unique and associated with only one Business Event.

We should try to name a Business Event from the perspective of what the customer expects of the organization as opposed to what we do in response to the stimulus. If a customer wants to buy our product(s), pay for it now, and take it away, then I would name the Event exactly that:

<u>Customer Wants to Buy Our Product(s), Pay for It, and Take It Away.</u>

The Business Event Stimulus for this may be named: **Regular Order.**

We may also have a Business Event in which a customer wants to buy our product(s), pay for it now, but have us deliver it. Again, I would produce a meaningful name:

Customer Wants to Buy Our Product(s), Pay for It Now, and Have It Delivered.

The Business Event Stimulus for this may be named: **Delivery Order.**

This naming may seem simplistic, but I have found one of the hardest tasks in analysis is to isolate specific Business Events before starting our Business Model. Also, the purpose of producing models is to help us understand a system and then get approval from someone other than ourselves, so meaningful names are essential for stand-alone models.

Our naming convention for a Business Event can be the same as those used for English Language sentence structures, in other words: Subject, Verb, and Object. This is applicable especially as our Business Model is intended for a non-technical audience. This naming standard should be used for all the Organizational Events (Business Events, Regulatory Events, and Dependent Events) on our Business Model.

Regulatory Events

Definition of a Regulatory Event

A Regulatory Event is an external incident originating at a governing body that places a demand on our organization to which we respond in order to comply with legal requirements.

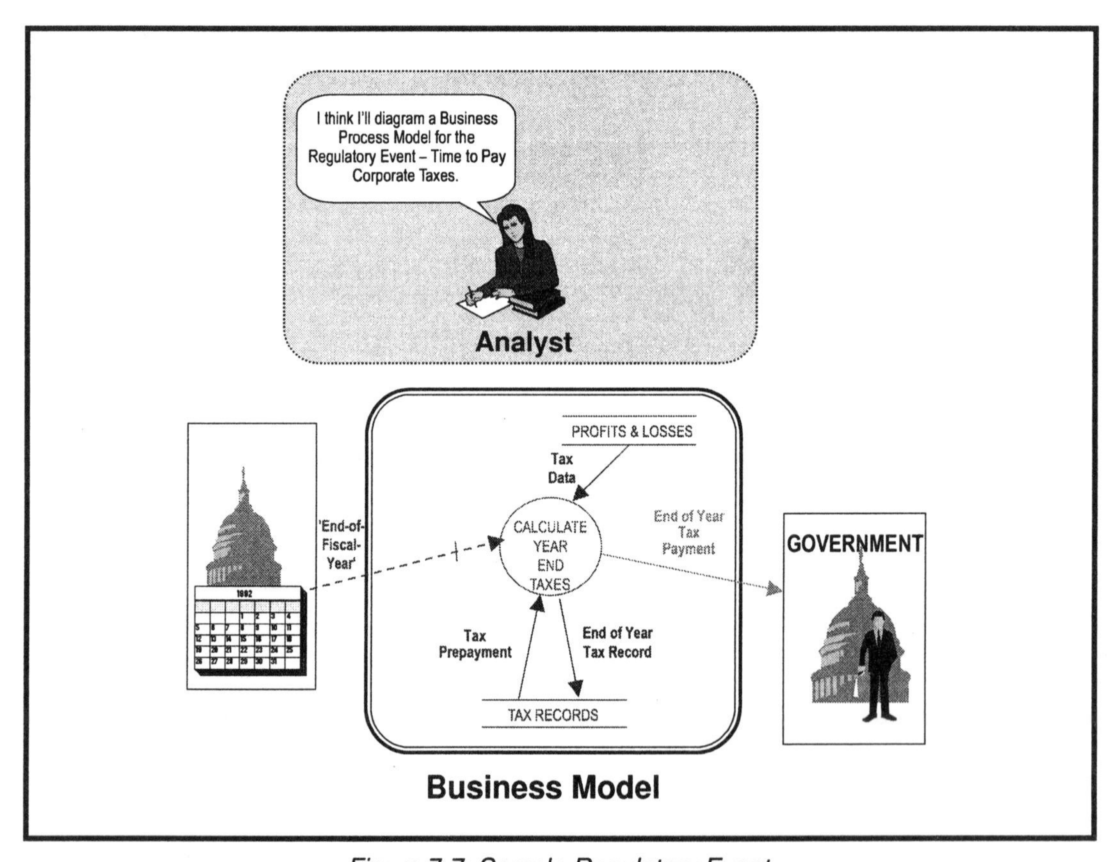

Figure 7-7: Sample Regulatory Event

Recognizing Regulatory Events

Regulatory Events are imposed on organizations to protect the environment, to ensure public safety, or to contribute to the support of the infrastructure within which the organization functions.

Regulatory Events originate at external regulatory agencies and they support laws and regulations that monitor organizations. We have no control over these regulatory organizations and their Events do not occur within our organization.

An example of this would be a corporation complying with "End-of-Fiscal-Year" government tax filing requirements (see Figure 7-7). At the end of the tax year (as set by the governing bodies or selected by the organization), an organization needs to report its tax liability. This Regulatory Event will stimulate the organization to process a certain amount of data and to disclose the results to the government body. This is a "batch" Regulatory Event stimulated by the calendar.

Apart from the obvious Regulatory Events that make us file taxes, we have to respond to inspections by government bodies associated with the type of business we are in and also respond to such things as law suits.

In a private organization, meeting legal requirements forms the majority of Regulatory Events. If our organization is a government agency, the above definitions of Business Events and Regulatory Events seem to be reversed. In a government agency the mission is usually to enforce laws and government programs (i.e., legal requirements). However, these are Business Events to a government agency with their customers being taxpayers (individual and corporate) and such things as the environment being monitored.

Our definition of Regulatory Events still applies to government agencies.

For example, government agencies are still constrained by environmental regulations and by private organizations and individuals.

Note that Regulatory Events are the ones for which our organization can typically purchase "off-the-shelf" software because organizations in the same business have to respond to the same Regulatory Events.

Regulatory Events may be "tagged" onto other Events.

For example, we could **Calculate Sales Tax** (a regulatory issue) as part of processing a purchase order from a customer (a business issue).

I recommend we indicate any part of a Business Event partition that is not actual business data or processing created by our organization's Strategic Planners. We'll talk more about this in the ***Partitioning by Business Events*** chapter.

Dependent Events

Definition of a Dependent Event

A Dependent Event is an external incident originating at our our organization's suppliers/vendors that is in response to an outgoing request made by us in order to satisfy one or more Business Events.

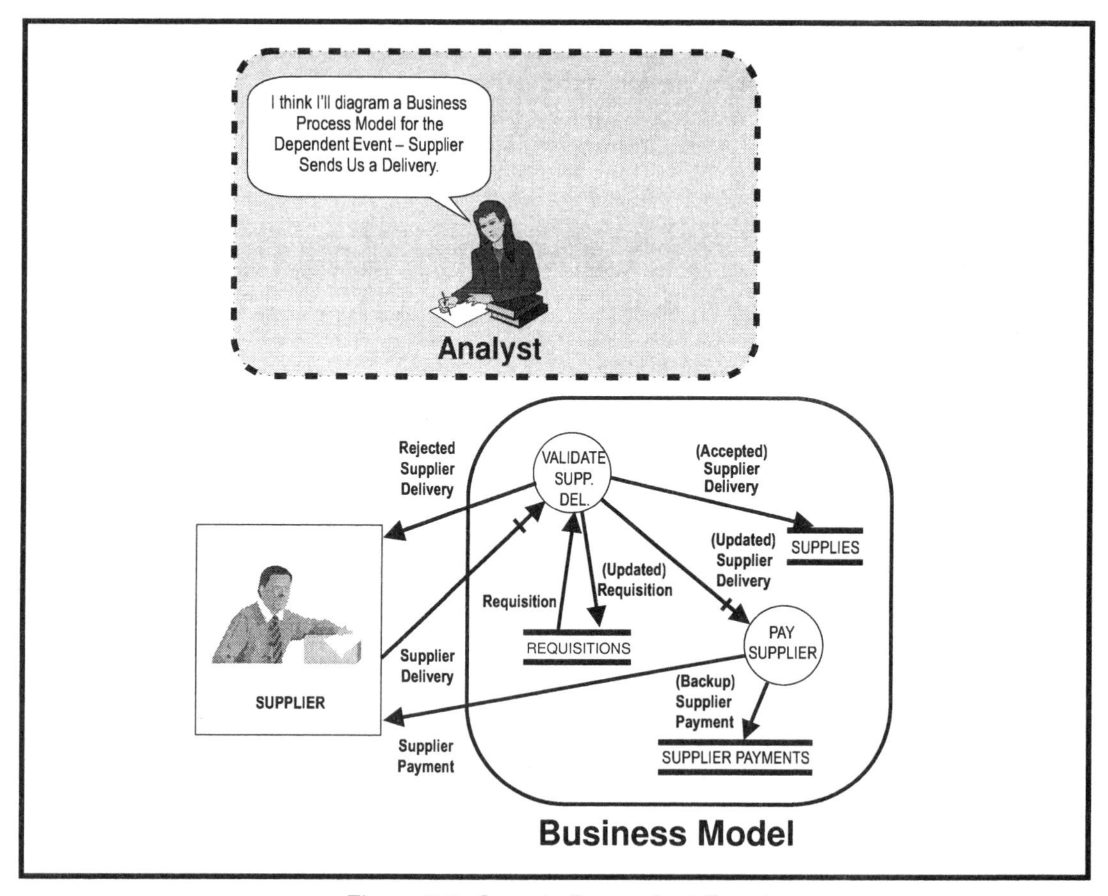

Figure 7-8: Sample Dependent Event

Recognizing Dependent Events

This type of Event is called a Dependent Event because it will be dependent on at least one other Business Event. Dependent Events differ from Business and Regulatory Events in that we have some measure of control over them.

> A common example of a Dependent Event is our organization being stimulated by an Event that originates with suppliers from whom we purchase materials needed to satisfy our Business Events. We could say to a supplier: "Get me these materials by Friday or I'll go somewhere else." This causes them to stimulate us on Friday (hopefully, with a delivery).

This is unlike a Business Event because we're not in control of a customer asking for an order.[1] Some of our outgoing responses from our own Events will form Events for our suppliers. Money doesn't make the world go 'round, Events do (and hopefully our Events make the money).

When we decide to "go outside" our organization's span of control to satisfy one or more Business Events, we typically cause a Dependent Event to occur.

> For example, when we need outside **Materials** to manufacture a **Product** to satisfy our **Customer**, we will issue some kind of **Order** to our **Supplier(s)**. This outgoing **Order** will generate an input to our organization at some point in the future. We are not in control of exactly when our vendor will send us the **Materials**. However, when they arrive, we need to respond to this Event.

Dependent Events are simply the extension of our organization's Business Events into the other organizations that we rely on to satisfy our Business Events.

In Figure 7-8 we need to respond to the Stimulus **Supplier Delivery** so we can stock our **Supplies** store. The **Supplies** can then be sold to our **Customers**. Because we are not in control of our **Suppliers**, we can't guarantee we will always have enough **Supplies** to meet our **Customers'** needs. Also note that the Dependent Event in Figure 7-8 may be associated with a second Dependent Event. The first Event is <u>Supplier Decides to Send Delivery</u> and the second one is <u>Supplier Decides to Invoice Us</u>.

If we outsource part of our business, we create Dependent Events. We also lose control over the Business Event that causes the Dependent Event. Note that any aspect of satisfying a Business Event that occurs beyond our organization leads us to respond to Dependent Events (i.e., we've fragmented a Business Event). This dependence may have been caused at the inception of our organization, where the founders couldn't afford all the facilities needed to satisfy the entire Business Event. This is obviously acceptable when an organization does not originally have the resources to completely satisfy a Business Event. However, you should be very cautious about outsourcing something you already have the expertise in (or the facilities to provide) because of the loss of control in satisfying a complete Business Event.

If we decide to buy out the supplier or produce our own raw materials, then the Dependent Event goes away.

Dependent Events are where we're likely to find cross-organizational, strategic alliances in a Customer Focused Engineering Project. We can look beyond our organization to see how we can best totally support our customers' Business Events. Therefore, a Dependent Event becomes part of its original Business Event when we form an alliance with our supplier.

1 In a government organization this may be reversed when they have legal power to ensure that a Business Event occurs (e.g., filing taxes).

For example, when we're allowed to "pull" goods directly from our supplier's warehouse, as needed, and reconcile payments at the time the goods are removed, we eliminate a Dependent Event and bring back to our organization the control over a cohesive Business Event.

In the above example, the same situation occurs if our supplier delivers the goods directly to our customer.

This is similar to what's happening in the banking industry. You don't have to draw out cash from your account to pay for a product you purchase. You can do a direct debit or electronic funds transfer from your account to your supplier.

Software houses can sell us solutions to many Dependent Events because these fragmented Events are typical to all vendors/suppliers (e.g., supplier invoicing).

Please realize that while evaluating Business, Regulatory, or Dependent Events we should have already filtered out any System Events so any system issues should not even be under consideration at this point.

For example, Events to pay for training or hardware do not support our Business Events (they support System Events) and should already have been filtered out.

The Business Model Event List

The first major deliverable of any Customer Focused Engineering project is a Project Charter. Included in this deliverable is a Business Model Event List (Figure 7-9); a list of all the Business, Regulatory, and Dependent Events for which this project is responsible.

Business Model Event List

Business Events:

B.E. 1 – Customer Wants to Order Our Materials
B.E. 2 – Customer Requests Invoice Copy
B.E. 3 – Customer Changes Their Shipping Address
...

Regulatory Events:

R.E. 1 – Time to Pay Value Added Tax
R.E. 2 – Time to File EPA Reports
...

Dependent Events:

D.E. 1 – Vendor Sends an Invoice
D.E. 2. – Vendor Requests Requisition Copy
...

Fig. 7-9: Sample Business Model Event List

The main point to keep in mind while creating this list is to err on the side of having too many Events. Too many is better than too few because a finer granularity list makes it easier to spot opportunities for the reusability of processing and whole Events. You will also satisfy the customer's needs better with a specific Business Event rather than generic or merged Business Events.

A Business Model Event List allows one to quickly and easily understand a project and an organization rather than relying on system descriptions. Someone new to banking would understand his/her environment more easily with a Business Model Event List containing meaningful names such as: Customer Wants to Know His/Her Savings Account Balance, Customer Wants to Withdraw Funds from His/Her Checking Account, or Customer Wants to Transfer Funds from His/Her Checking to His/Her Savings Account.

For a small, vertical organization with limited products or services, we may have less than a hundred Business Model Events in our list. For a diverse organization, the list may have thousands. It's usual at the beginning of a study to underestimate the Business Model Event count and to discover more Business Model Events during detailed analysis.

I have found in many years of working with projects using this Business Event methodology that the number of Events first identified will almost invariably grow throughout the project. This growth occurs because our initial tendency is to think one order form is the equivalent of one Business Event. In reality, I find an order form contains many different kinds of orders. Also, at the beginning of a project, many project requesters think at a high level and tend to make statements such as: "Our organization just sells materials and that's it." However, by the end of Business Event Analysis, the resulting Business Events will be more specific, more easily understood, and will be truly more customer focused.

Unmasking Events

Even though we understand the five types of Organizational Events, we must recognize they may not be easily discernible in practice.

> For example, we may get a new Regulatory Event from the government requiring our organization to keep track of a new "Excise Tax" and that "Excise Tax" will be set at 4%. This request actually contains two types of Events. The first is a Strategic Event to add a new Data Element (noun). In this case our Strategic Planner(s) would have to recommend changing our Business Model. The second is a Regulatory Event to assign the 4% value to that noun. Remember, Strategic Events deal with Meta Data (the nouns); they don't deal with the data values of those nouns.
>
> Another example would be when the phone company (with whom we have a contract as a supplier to take care of our communications) decides to introduce a new field for satellite communications and they give us new values associated with satellite regions. In this case we have a Strategic Event to add a new Data Element for "Satellite Region" and we have a Dependent Event to Create a set of values for that new field.

Any Event that introduces or deletes a Data Element (a noun) or item of logic (a verb) must be associated with a Strategic Event. Any Event that Creates, Retrieves, Updates, and/or Deletes (C.R.U.D.s) a value in a Data Element must be a Business Model Event (Business, Dependent, or Regulatory). This is why we must unmask these hidden Events.

I have found it helpful to use the questions and answers in Figure 7-10 as a quick reference for determining the type of Event or aspect being analyzed.

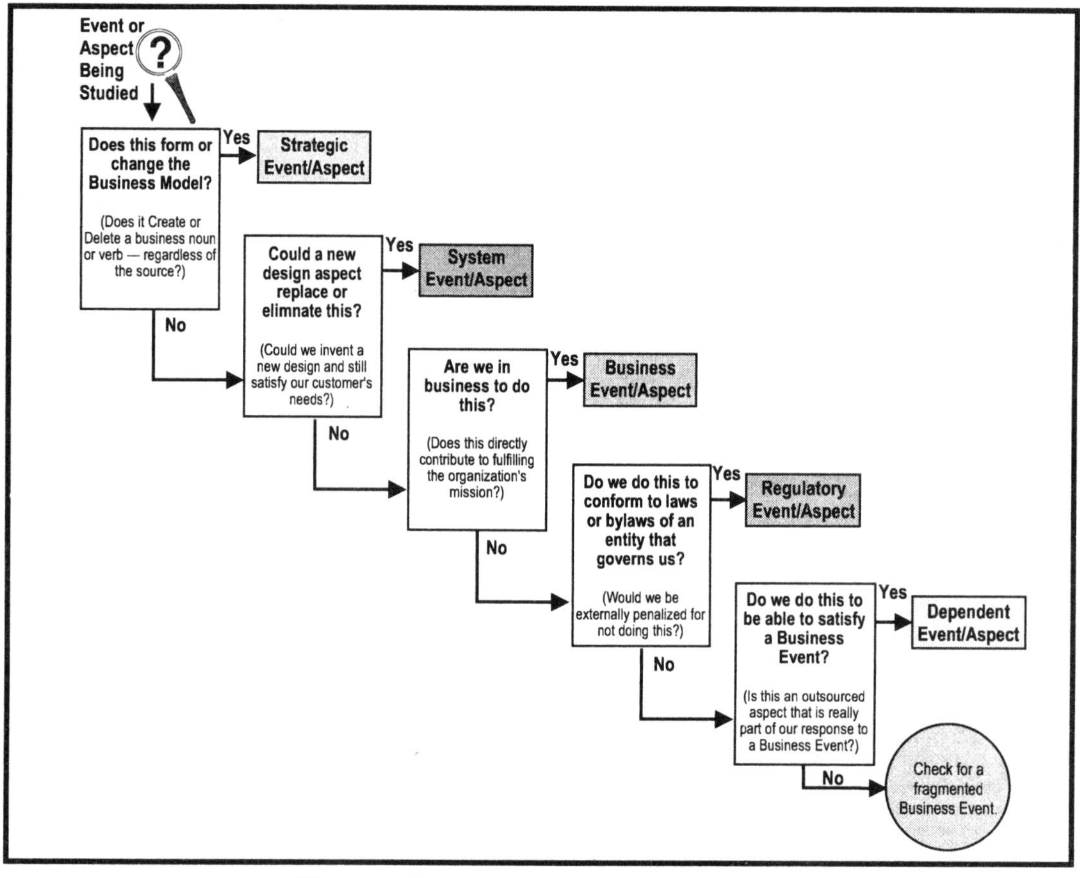

Fig. 7-10: Determining Event/Aspect Types

Summary

We now know that not all Events stimulating our organization's current processing belong on our Business Model. We can apply a sequence in which we identify and categorize Events to form our Business Model.

- **Filter Out Strategic Events** — The first step is to remove Strategic Events. These Events are important for creating and modifying the Business Model, but they do not stimulate it into life. We could create a separate Meta Model for how we address the processing of Strategic Events, but they do not belong on the Business Model.
- **Filter Out System Events** — The next step is to remove those Events that were designed to make the old system run. These Events will be of the least value to us (if they have any value) in our Customer Focused Engineering task. However, we need to be aware that some Events (those formed from fragmented Business Events), masquerade as System Events.

- **Identify Business Events** — These are the Events that keep our doors open. Therefore, they are the Events we should focus upon to obtain the most improvements in our ability to satisfy our real customers.
- **Identify Regulatory Events** — Regulatory Events are external to our organization and we need to model them on the Business Model, but they are probably the ones over which we have the least influence in Customer Focused Engineering.
- **Identify Dependent Events** — These Events are caused by outsourced parts of our Business Events and should appear on the Business Model. Dependent Events will be one of the most important sources for potentially forming strategic alliances as part of a Customer Focused Engineering effort.

A Business Event Partitioned Model captures the true business rules and information of an organization. Thus, Customer Focused Engineering begins by identifying the external Business Model Events that result in stimuli that trigger our organization into action.

The next chapter explains the concept of Business Event Partitions and how to use these as the fundamental basis on which to build manual and/or automated systems.

I believe that the recognition of Business Model Events and Business Event Partitioning (described in the next chapter) used with any of the existing proven System Engineering methods and models allows us to do true Customer Focused Engineering.

Chapter

8

Partitioning by Business Events

In any hierarchy of [metaphysics] classification the most important division is the first one, for this division dominates everything beneath it. If this first division is bad, there is no way you can ever build a really good system of classification around it.

Robert Pirsig
Lila

I have come to realize the most logical/implementation-independent, high level view of any organization is its ultimate organizational goal as stated in a single sentence or paragraph in the Mission Statement. This goal is then broken down into a set of Business Objectives which are still logical.

> If our organization's only goal is simply to make money, any line of business could potentially satisfy that goal and a Business Model for this organization could be as generic as "buy low and sell high." If, on the other hand, we had a specific business goal to be the best training and consultancy organization in the field of Customer Focused Engineering, then we can produce more specific Business Models of our organization and measure back to this goal and its subsequent objectives in systems development and production.

Therefore, our business goal and objectives determine our specific Business Events, versus those of any other organization. In this chapter we'll see that based on partitioning by Business Events we can build systems which address business issues and respond to customers' needs rather than to any historical internal needs.

By definition, partitioning the model of the organization by Business Events keeps Business Events *separate*, making systems which are Customer Focused and much easier to modify when the business policy changes.

The discipline of Customer Focused Engineering is based on partitioning together each stimulus, its responses, and its process and memory that are needed to respond to a specific customer need.

In the ***Systems Archaeology*** Chapter we went on an archaeological dig and saw how the manual and automated systems of an organization evolved based on a variety of factors. We also saw these factors were issues of internal design and not business issues. We found a typical organization is partitioned into departments and computer systems based on historical *(or hysterical)* reasons. Such partitions are seldom related to the way our organization needs to respond to efficiently satisfy our customers' needs in today's environment.

This chapter focuses on what it takes to define a Business Event Partition. This process consists of identifying all of the key components of the Business Event Partition and naming them. Because our task in analysis is building models to communicate with the minds of others, it is important to use business terms and to have a naming standard for a Business Event Partition with all of its constituent elements. So, all definitions in this chapter apply to Regulatory and Dependent Events. I prefer to call the generic forms of these definitions Business Model Events. However, in this chapter I'll focus on Business Events.

In the course of identifying what each Business Event Partition element is and naming it, we also need to define what distinguishes it from all other elements.

Even though we focus on Business Events in this chapter, Dependent and Regulatory Events are made up of the same constituent elements.

Definition of a Business Event Partition

A Business Event Partition is:

The most natural business structure for satisfying one specific need of a customer.

A Business Event Partition consists of a Business Event Stimulus plus all associated processing, stored memory, and outgoing responses that constitute the organization's complete reaction to a Business Event.

A Business Model depicting Business Event Partitions represents the most logical view of any organization. Put another way, if we form any internal views (partitions) of an organization which are not based on outside Business Events, these will be less functional views that restrict the business system in some way.

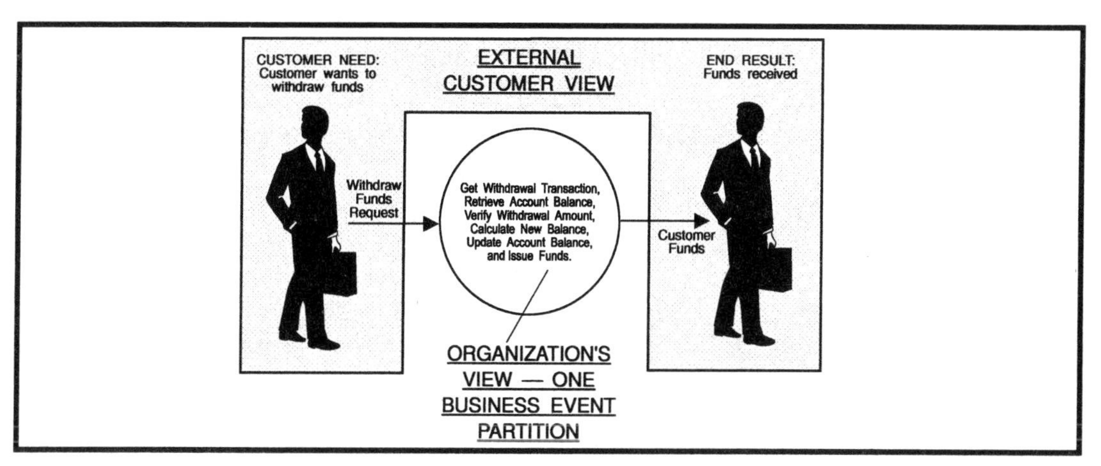

Fig. 8-1: Satisfying One Need of the Customer

Figure 8-1 shows that, ultimately, our task is to bring together the set of business nouns and verbs necessary to satisfy the customer's need. This set of nouns and verbs will make up the Business Event Partition. We can use any model that shows data and processing to capture these nouns and verbs within a Business Event.

Whether we're Pre-Engineering a new organization or conducting Customer Focused Engineering on an existing one, the first major step after identifying the Business Model Event List is to model each Business Event at a high level. I recommend we produce a model declaring one single process for each Business Event and call this the Business Event Context Level Model (see Figure 8-2).

Any further decomposition of a Business Event's processing and data beyond this level will be based on rather arbitrary reasons, such as the reusability of sub-functions or the complexity of individual sub-processes.

The idea is to encapsulate the stimulus, processing, stored memory, and response for one customer's single need together without stopping the flow of the stimulus through our organization. This entire collection of components is needed to accomplish one customer's need within our organization. This leads to what I call true *Business Functionality*.

The most important thing to remember regarding a Business Event Partition is that it consists of everything necessary to satisfy the Business Event from the Customer's perspective; that is, our organization does what the customer expects of us in total within one partition.

The Business Event Context Level Model can be quite large. You may be thinking: "Our organization must have thousands of Events." While that may be true depending on the diversity of your organization and the services you provide, in most organizations it will not be the case. Most organizations are what is termed "vertical" and supply a narrowly defined set of products or services. .

For example, my company supplies training and consulting in Business Engineering subjects. We don't sell airline seats, bake bread, or do any banking for you.

Even if you did have thousands of Business Events, the Business Event Context Level demonstrates the best representation of the complexity of your business; probably many Business and Strategic Planners will be needed to verify such a model.

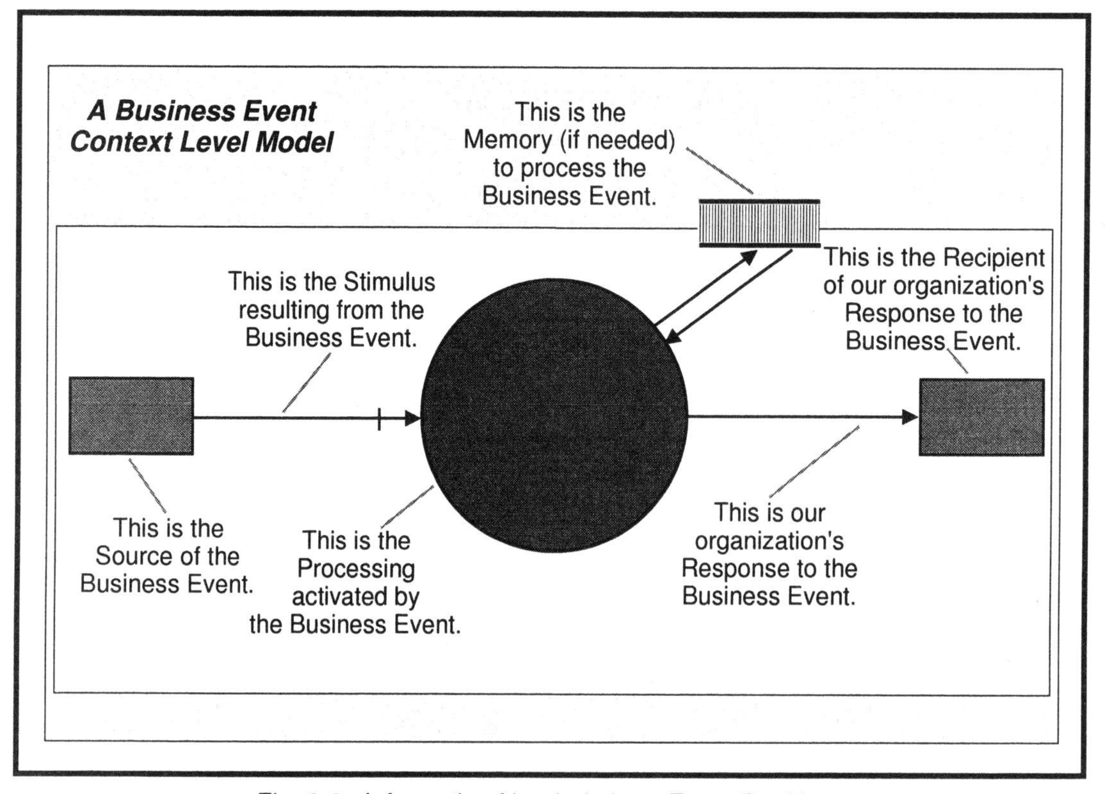

Fig. 8-2: Information Needed about Event Partitions

What Constitutes a Business Event Partition

The elements of the Business Event Partition are:

- The Business Event Source
- The Business Event Stimulus
- The Business Event Processing
- The Business Event Memory
- The Business Event Response
- The Business Event Recipient

Let me elaborate on each of these elements.

The Business Event Source

A Business Event Source is:

A customer — an external individual, agency, organization, system, or other Entity that creates a stimulus to our organization.

The non-shaded area in Figure 8-3 indicates the Business Event Source.

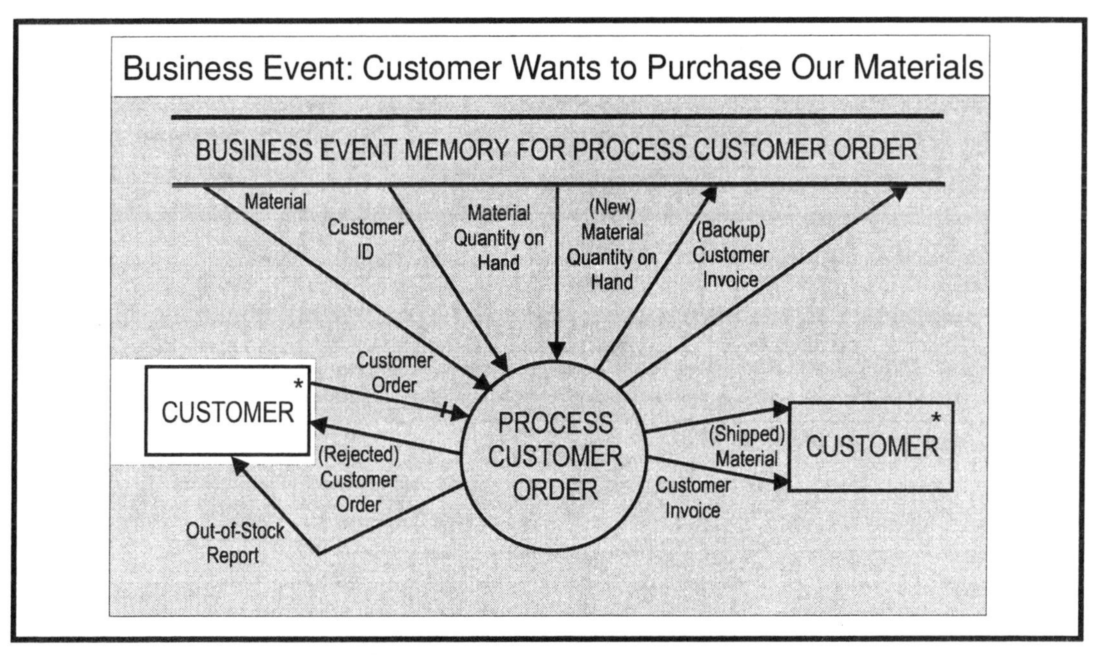

Fig. 8-3: The Business Event Source

The Source is always based on our "Context of Study." If we were limited to only one department or system inside our organization, then the Source of our Events would mostly be other departments and systems. With such a limited context of study, the majority of our Business Events will be fragments of actual Business Events. The best context of study is the actual boundary of our organization.

Naming the Business Event Source

The Source is usually an active Entity where something occurs that stimulates our organization into life. This could be a piece of land where an incident such as a flood occurs or a clock indicating the closing of a foreign stock exchange. The Source could also be any external system including the ecosystem.

> For example, we may respond to Events occurring in the environment such as an earthquake. The Event Source could also be a clock/calendar where we respond to the end-of-the-day or the end-of-the-year. The Source could also be an external device such as a house intruder/fire alarm device where we respond to its signal to our organization.

The naming of this Source should be in singular form and in non-implementation terms such as **Bank** instead of **Joe at the Bank**.

The Business Event Stimulus

A process does not start up on its own — it needs to be stimulated into life by the arrival of data or control. Without such a Stimulus, a process would lie idle. *This is similar to me in the morning; when I'm asleep, I need the alarm clock to wake me into action (or, as a last resort, my internal circadian rhythms kick in).*

A Business Event Stimulus is:

A demand (input data, control trigger, or incoming material) resulting from a Business Event that activates part of our organization.

This could be a customer order/request, a period of 90 days elapsing since we sent an invoice, or unused goods returning from a customer. The Business Event Stimulus is the first thing we see about a Business Event from within our organization. The non-shaded area in Figure 8-4 indicates the Business Event Stimulus.

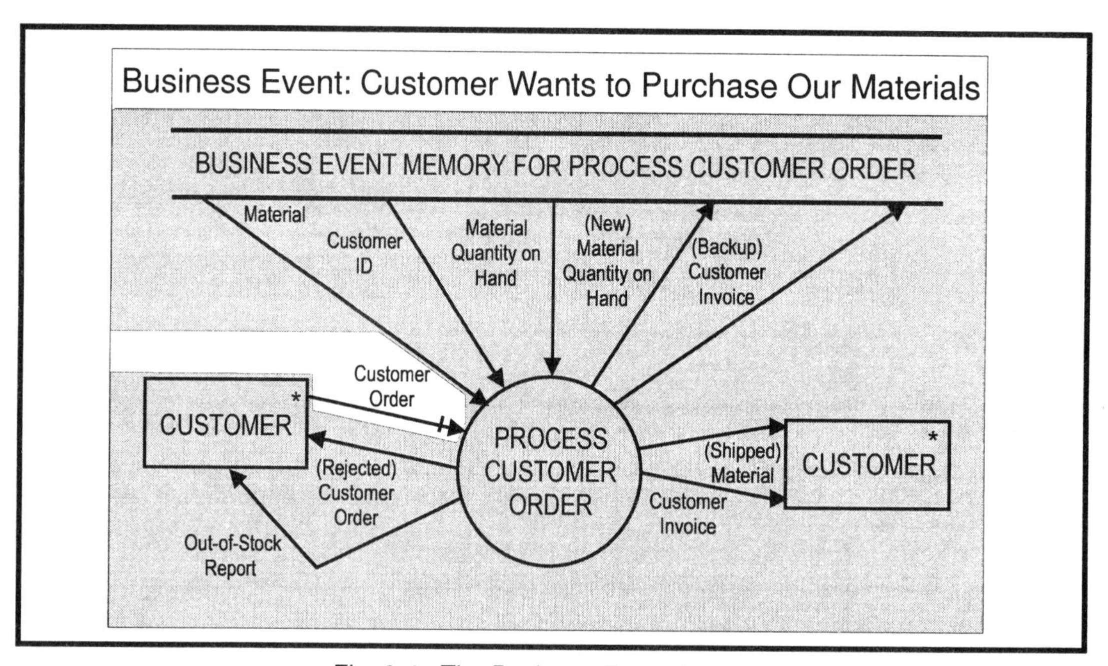

Fig. 8-4: The Business Event Stimulus

Because of the importance of any first division it's vital to identify autonomous stimuli. Therefore, I need to elaborate on this element of a Business Event Partition more than the others.

I recommend using some special symbology to indicate the stimulating data flow or control flow of a process. We can indicate the Stimulus flow with a cross-bar symbol. This is helpful because there are other incoming data and control flows that don't stimulate a process (such as those that are pulled into a process based on the arrival of the Stimulus).

Once we have defined a Business Event, the Stimulus is completely dependent on it. The contents of the Stimulus such as data fields will mostly be dependent on what the customer needs from our organization.

When analyzing a Business Event Stimulus, it's important to identify whether the Business Event Stimuli is Data-, Control-, or Material-oriented.

Data-oriented Stimuli

A Data-oriented Stimuli consists of a cohesive set of data necessary for processing the Event.

> For example, when a **Customer** wants to purchase some of our **Goods**, he or she has an **Order** in his or her head. The **Order** stimulates the system to provide those **Goods**. The **Order** contains data that is essential for generating the Response — which **Goods**, for whom, by when, and so on. The data in the **Order** makes up our cohesive data Stimulus.

Figure 8-5 shows some examples of Business Events with Data-oriented Stimuli.

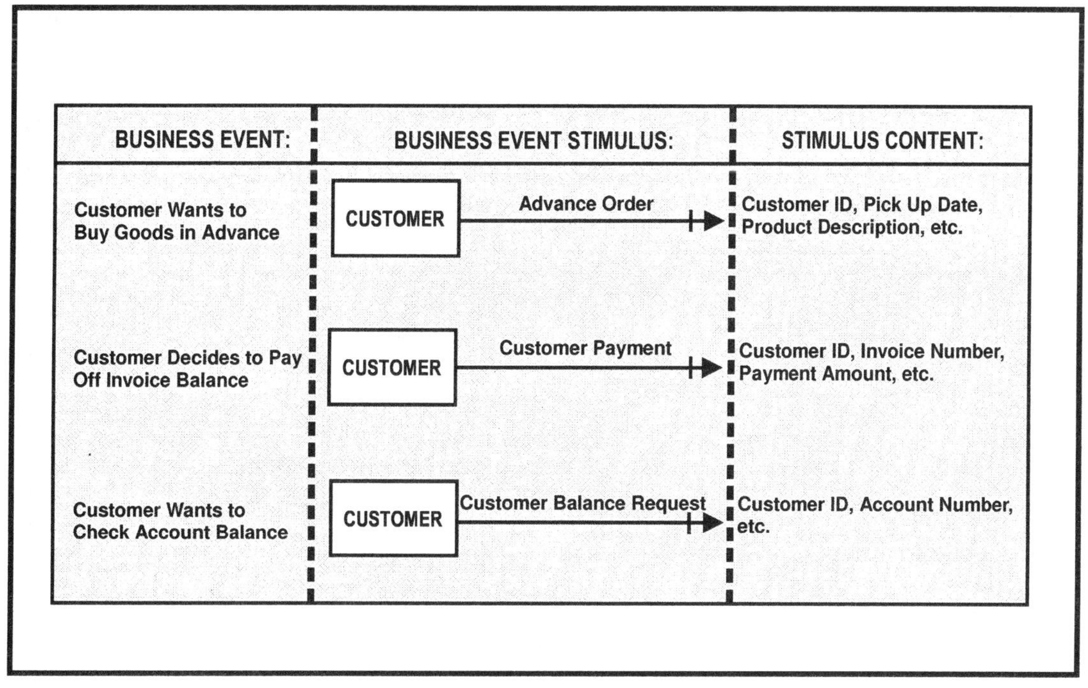

Fig. 8-5: Data-oriented Business Event Stimuli

For a Data-oriented Stimulus, we need enough information in the Stimulus to completely process the need.

> For example, if a customer wants to send a package from point A to B, we'll need a cohesive set of data before we can completely satisfy that Business Event. (Note that all the data does not need to arrive together. It can be input to our Business Event process in increments.) For one organization that delivers door to door to respond to this Business Event correctly, the cohesive data stimulus may contain:
>
> the package size,
> the package weight,
> the package contents,
> the package pick-up address, and
> the package destination address.

For another organization that picks up from one depot and delivers to another of its depots, the cohesive data stimulus may contain:

the customer name and address,
the package weight, and
the destination city.

Using my own organization as an example; when an outside Business Event occurs (such as an organization wants some specific training), I need to know a set of data to be able to respond. To satisfy the training need, I need to know:

the customer ID,
the seminar subject,
the required date,
the seminar location, and
the number of students.

This set of data forms the Data-oriented Stimulus to Logical Conclusions, Inc. For the most part it is the set of data the customer expects to supply to us, but there may be additional data we at LCI need. To satisfy our business policy, we need certain information, such as whether they would allow outside students in their seminar and do they wish to take care of hotel, car reservations, etc. This cohesive Data-oriented Stimulus may also contain data the organization needs in order to satisfy the Business Event.

Another Event outside my organization might be someone wants to train one or more individuals in a particular subject at an open **Public Seminar**. This need results in a different Data-oriented Stimulus to satisfy the policy in my company because this is a different Business Event. In this example, we would need to know:

the number of individuals who need this training,,
the distance their **Students** are willing to travel, and
the **Student Names**.

In this latter case, LCI dictates the date and location of the **Public Seminar**. This information is no longer part of the Stimulus but part of my Response.

If a third caller wants a fully catered seminar, I would tell her that I'd be happy to do the seminar part, but she would need to arrange catering herself. Responding to requests for catering is not part of my present business policy. Although with enough of these types of Business Events I may initiate a Strategic Event and offer catered seminars.

Naming Data-oriented Stimuli

I have previously given some examples of Data-oriented Stimuli names with their corresponding Business Event names. Here I would like to add that the Data-oriented stimuli names will often be aggregate names. A Data-oriented Stimulus would have a name that contains a cohesive set of Data Elements that make up the Stimulus.

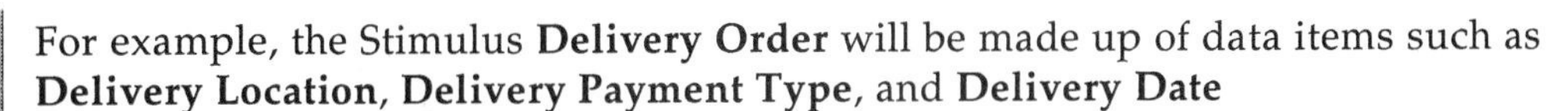

For example, the Stimulus **Delivery Order** will be made up of data items such as **Delivery Location**, **Delivery Payment Type**, and **Delivery Date**

Material-oriented Stimuli

These are actual products or physical items that arrive at our organization. However, for a governmental disaster relief agency, the Material-oriented Stimulus could be water overflowing the banks of a river.[1]

Figure 8-6 shows some examples of Business Events with Material-oriented Stimuli.

Notice in the first example that I may not have sent any request to my propane supplier. I may have a contract with them to fit their schedule (set up via another Business Event), so there is no data with the Stimulus. However, the gas arrival may trigger me into performing some action.

---

1 Note, I find that Data Processing people have a hard time associating with Material-oriented Stimuli on a Business Model because they tend to want to turn everything into data. To deter this I use an example of a warehouse where incoming goods must make it to the actual inventory not just a quantity field stored in a database. These goods must be shown on the Business Model because they are not an aspect of design, they are an aspect of the business.

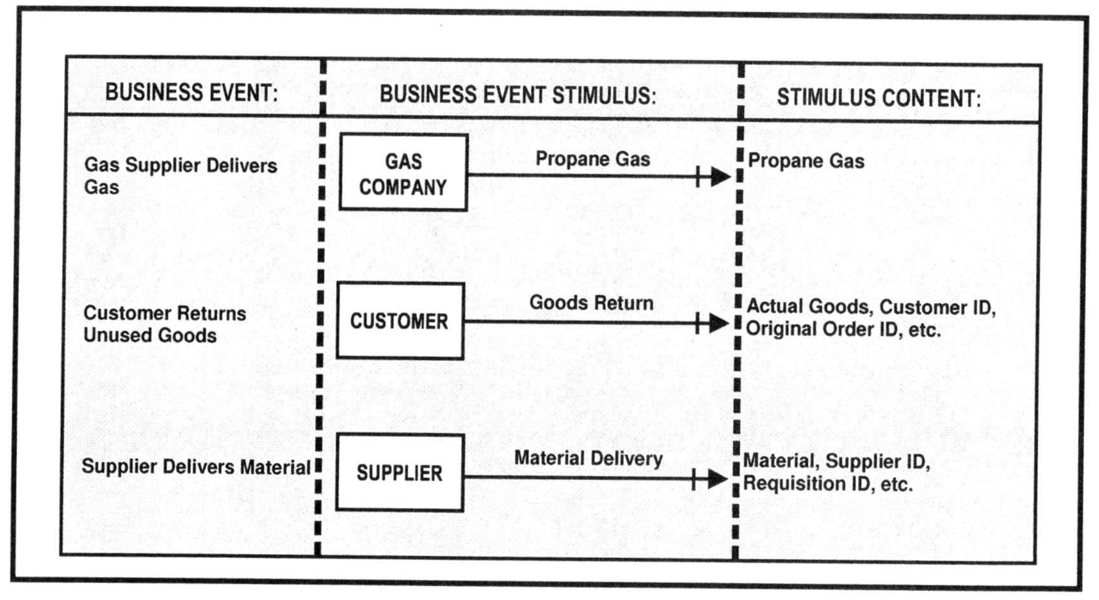

Fig. 8-6: Material-oriented Business Event Stimuli

Naming Material-oriented Stimuli

Material-oriented Stimulus names may or may not contain data, but they will always contain a description of what actual material is in the stimulus flow such as money, products, or legal documents.

Control-oriented Stimuli

Control-oriented Stimuli prompt the system to do something and have no data. Figure 8-7 shows some examples of Business Events with Control-oriented Stimuli. In these examples the Regulatory Events that produce these Stimuli are, in fact, the calendar (or clock) reaching a predetermined point in time. Not all Business Event Stimuli are so straight-forward.

> For example, one Stimulus might be 30, 60, or 90 days of elapsed time after an item is purchased from us on credit. These durations may stimulate us to produce various overdue notices.

In this example, our first analysis task, of course, would be to question the 30, 60, or 90 day periods as design issues. Always question control issues in business systems, as control is primarily a design issue; however, control can also be a business requirement issue (especially in real-time systems). If these duration's are found to be essential business policy, then we have Control-oriented Stimuli (e.g., **'Review Invoice Time'**) that makes us review every invoice looking for a 30, 60, or 90 day difference between the invoice's date and today's date. It sounds strange, but in these kinds of Business Events the calendar is the customer because it is the passing of time indicated by the ever changing calendar that produces the Stimulus.

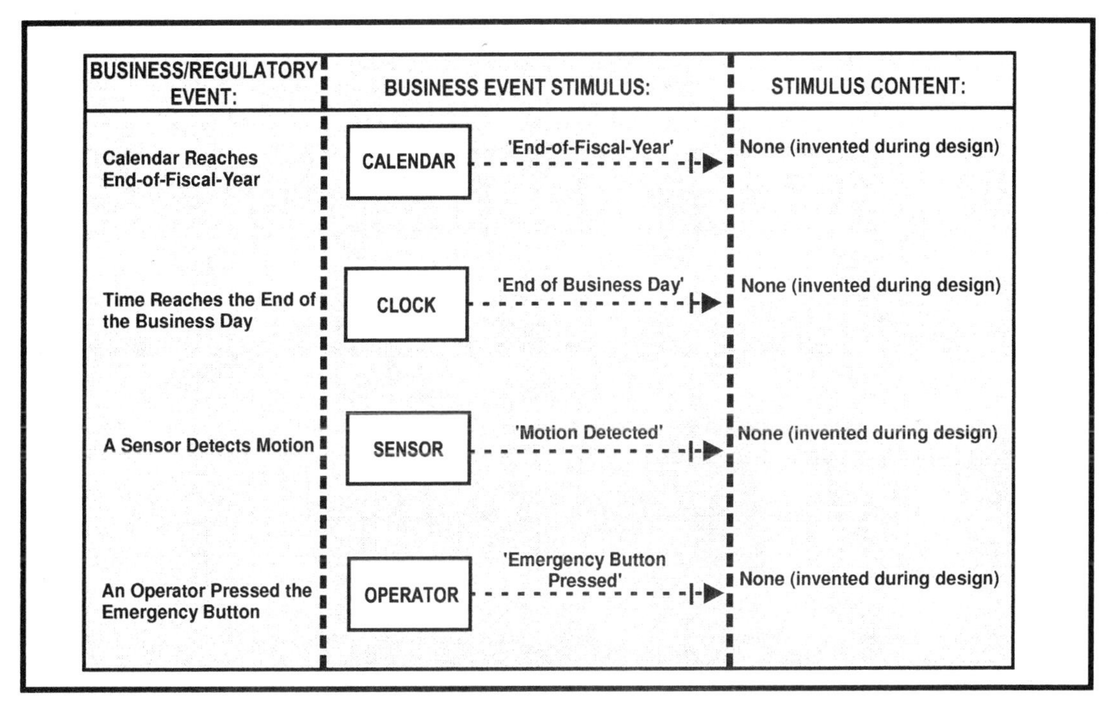

Fig. 8-7: Control-oriented Event Stimuli

Be aware that we are always looking for Events beyond our organization.

For example, the marketing staff of an organization may dream up ways of trying to get people outside the organization to initiate Business Events, as when a phone company advertises: "Have you called your mother lately?"

But we should not look upon the marketing drive in the example above as a Business Event because it's internal. The marketing drive will be a response to some other Event, and the Stimulus that initiates a marketing drive may be poor monthly sales figures or a scheduled time of the year such as Christmas. The Stimulus in these cases came from the calendar. The customer who happens to respond to the ad has the Business Event.

Naming Control-oriented Stimuli

Control-oriented Stimulus names do not contain any data, so their names will be associated with what is happening, what has happened, or a threshold that has been reached. So we name the Stimulus with a descriptive statement in single quotes such as **'Temperature has exceeded acceptable limits'**, **'Time to restock'**, or **'End-of-Fiscal-Year.'**

The Business Event Processing

Now let's talk about the processing component of a Business Event Partition. Business Event Processing is:

> ***All the business logic and its associated transient data required to produce the total response to a Business Event.***

The non-shaded area in Figure 8-8 indicates the Business Event Processing.

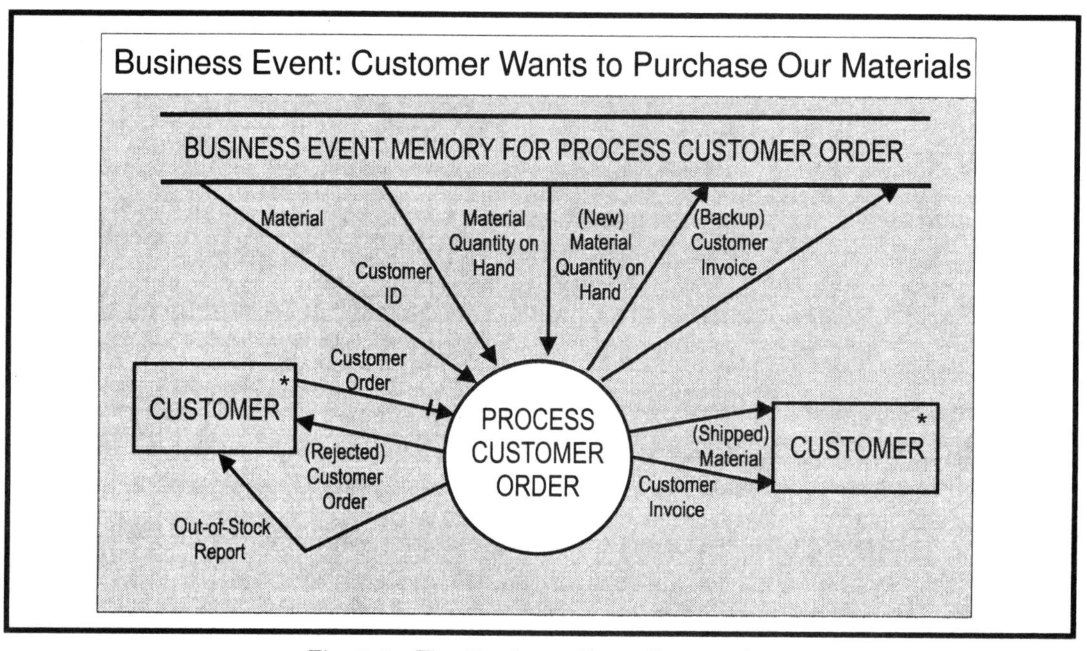

Fig. 8-8: The Business Event Processing

This logic is put in place by our organization's Business Policy Creator(s) or Strategic Planner(s). When a Business Event Stimulus arrives at the boundary of our organization, a number of instructions/statements are stimulated to perform whatever logic is needed with whatever data are needed to satisfy that Stimulus and to execute the business policy associated with that Stimulus.

> Using a Dependent Event as an example, if a **Delivery** comes in from a **Supplier** we should do everything necessary to deal with the **Delivery**, such as verifying the **Delivery** against the original **Purchase Order**, restocking **Material**, and storing **Accepted Delivery** data (for a future inquiry, or when the **Supplier's Invoice** arrives). We may also immediately pay the **Supplier** based on what we accepted.

All the processing we do and all the data we access is in response to the external Event of the **Supplier**. It is this collection of processing (and its transient data) associated with an Event that we should logically package together into Business Event Processing.

The business processing may be trivial such as responding to a customer's Business Event to check his/her account balance. This will probably require validation of the **Customer ID** and **Account ID** and the retrieval and issuance of his/her **Balance Amount**. The business processing could also be significant, such as in my organization when a customer wants a training seminar in-house. This Business Event triggers a few hundred instructions requiring the retrieval and updating of many data items and the output of the data in a dialog.

When the processing is significant and involves hundreds or thousands of instructions then we will obviously partition the processing further (not the Business Event). We will discuss this later in ***Chapter 10***.

I include the Business Event Partition's transient data in my definition of processing. This is data you expect to "burn up" after the processing is complete. It is equivalent in the design world to a person's scratch pad notes or a computer program's working storage/Random Access Memory.

Naming the Business Event Processing

Unlike the Business Event name, we want the Business Event Processing name to be relatively cryptic because it will always be supported by more detail in either the form of a lower-level diagram/model, or a detailed process specification/procedure. So, we use a verb-object description to sum up the process (for example: **Order Goods, Process Customer Order, Control Fire**). The names may be more descriptive, such as **Calculate Special Customer Product Discount**. The important thing to note here is that we're not looking for generic names such as **Process Stuff, Handle Data, Do Everything Else**, etc. Rather, we're looking for quite specific names that truly sum up what's going on in the process.

The Business Event Memory

A major factor that affects our Business Event Partitioning is how we deal with stored data or data at rest; the Business Event Memory. Due to its importance, this topic is covered in some detail here.

This book is aimed at the business person, so I do not pre-suppose that the reader already knows Information Engineering techniques or such methods as Normalizing data structures. Therefore, I have found it useful to teach the concept of grouping stored data via a logical collection of Data Elements that I call Business Event Memory.

The Business Event Memory is:

> ***The collection of all the stored (i.e., non-transient) Data Elements necessary to accomplish the processing for one Business Event.***

Business Event Memory is not strictly contained within any one Business Event Partition's boundary, but rather is shared by two or more Events. The non-shaded area in Figure 8-9 indicates the Business Event Memory.

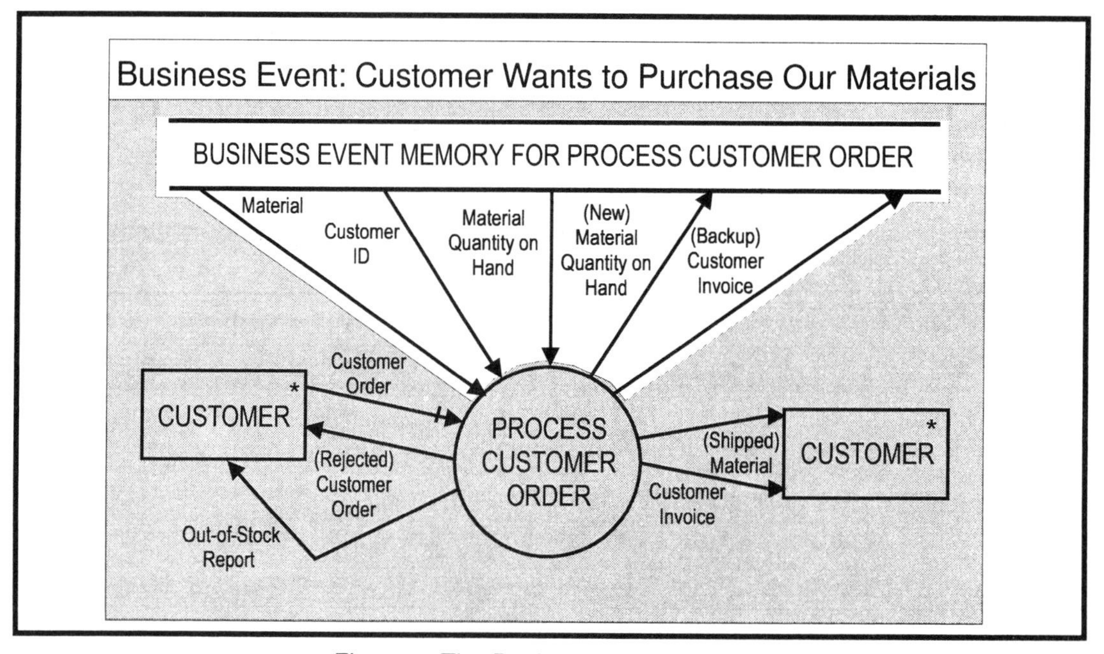

Fig. 8-9: The Business Event Memory

Beware of Design/Convenience Stores

Before explaining Business Event Memory, we need to identify the current design source for producing these sets of data. We also need to be aware of certain things from the old environment that may corrupt our collection of this data.

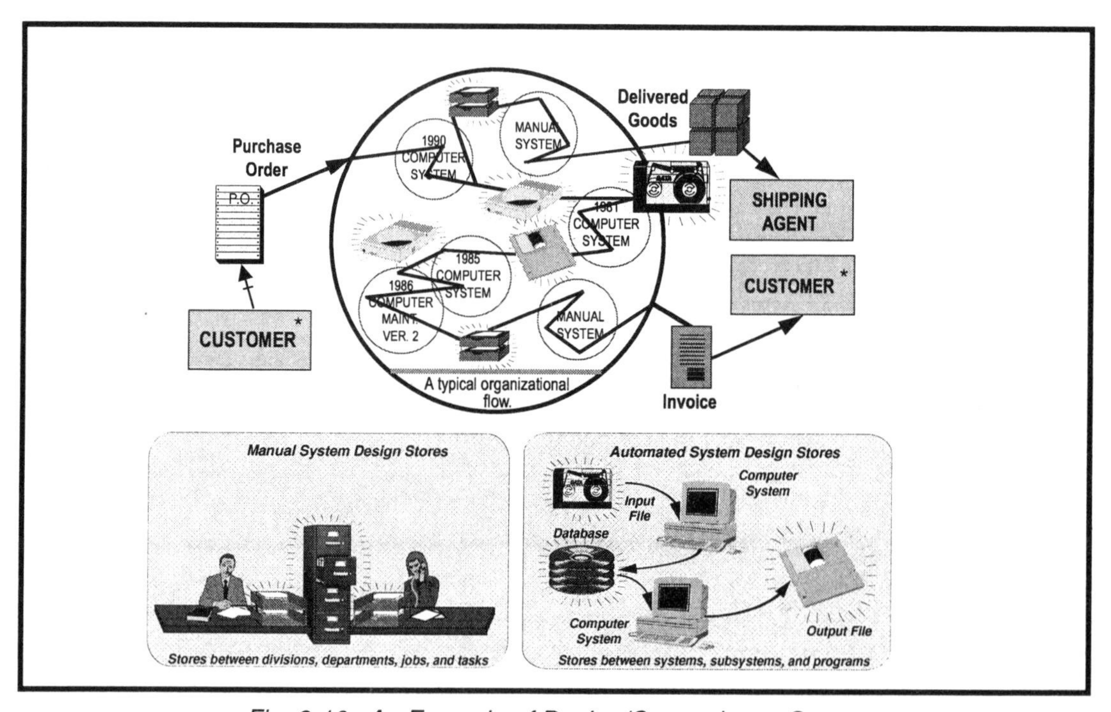

Fig. 8-10: An Example of Design/Convenience Stores

The big problem we have when looking at stored data is that a significant number of the stores/files in the existing design of an organization should not be there at all. They are what I call "Design or Convenience Stores." They are there to support the old system's design. I call these Convenience Stores because they are created as a matter of human or computer access convenience between old system boundaries. These stores will turn out to be transient data flows on our Business Model.

Unfortunately, in many existing environments, we will find Designer Stores between divisions, departments, etc. in manual systems and between systems, subsystems, and programs in automated systems (see Figure 8-10).

When we look at any false computer batch systems, such as the one shown in Figure 8-11, we see Convenience Stores: the **Transaction File** input to the batch run, the **Edited Transaction File**, and the **Updated Data File**. These Convenience Stores were needed only because of the batch design partitioning into three programs. We will need to analyze for their stored data the three Business Event Stimuli (**Close Account**, **Regular Order**, and **Special Order**) to see if any of their processing requires data to be held across these Business Event Partitions.

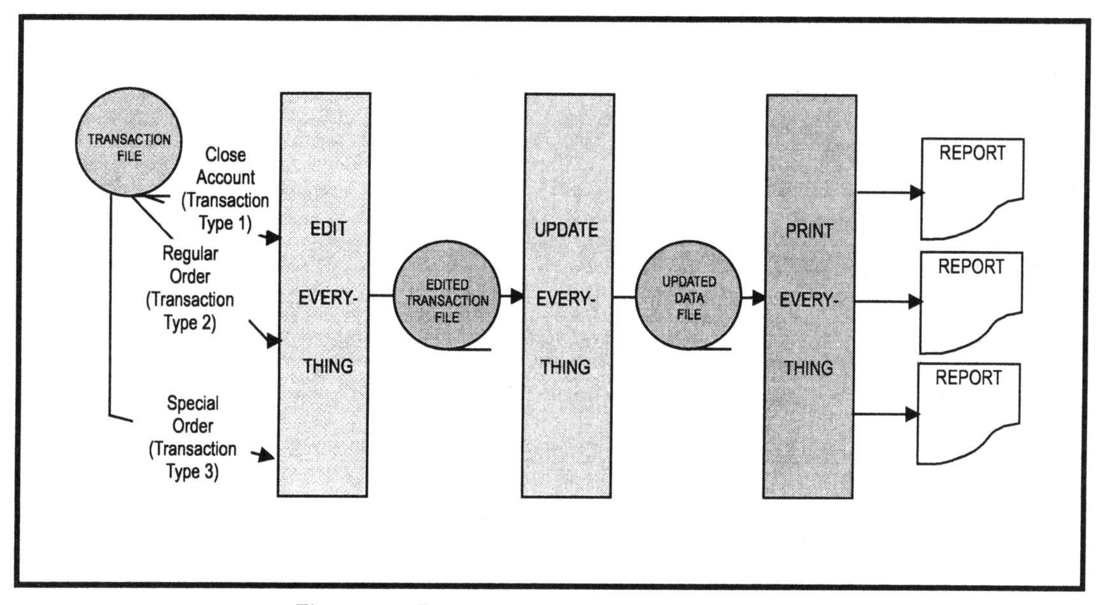

Fig. 8-11: Traditional False Batch Partitioning

As stated in the ***Systems Archaeology*** Chapter, we have a problem when we try to form a cohesive data store (an Entity) because its Data Elements will be typically scattered or bundled across old Design Stores.

For example, information on a **Customer** could be scattered across **Sales**, **Accounts Receivable and Payable**, and **Invoicing** files within one organization. Also, customer data will be typically bundled with other data in these files such as **Product Data, Order Data**, and **Invoice Data.**

Also note that any Convenience Store will make us form an Internal Interface and its false stimulus to restart the flow of data for the fragmented Business Event Partition.

Figure 8-12 is a generic model showing a Convenience Store — **Orders**. The designer had to create this so that the **Order Department** could accumulate orders, waiting until the scheduled time for the **Accounting Department** to pick them up. Note that whenever the data in one Business Event Partition crossed an old design boundary such as **Order Department** to **Accounting Department**, a Convenience Store would have been created. Between each and every old design boundaries there will be some kind of false stores. These stores are usually unnecessary in the business world, and the analyst must remove these during analysis.

Convenience Stores always will be associated with the old design and if the analyst does not recognize this characteristic, the old store is very likely to become part of the requirements for a new manual file/computer database. This will perpetuate unnecessary stored data and the housekeeping processes associated with it, such as backups, restores, and audits.

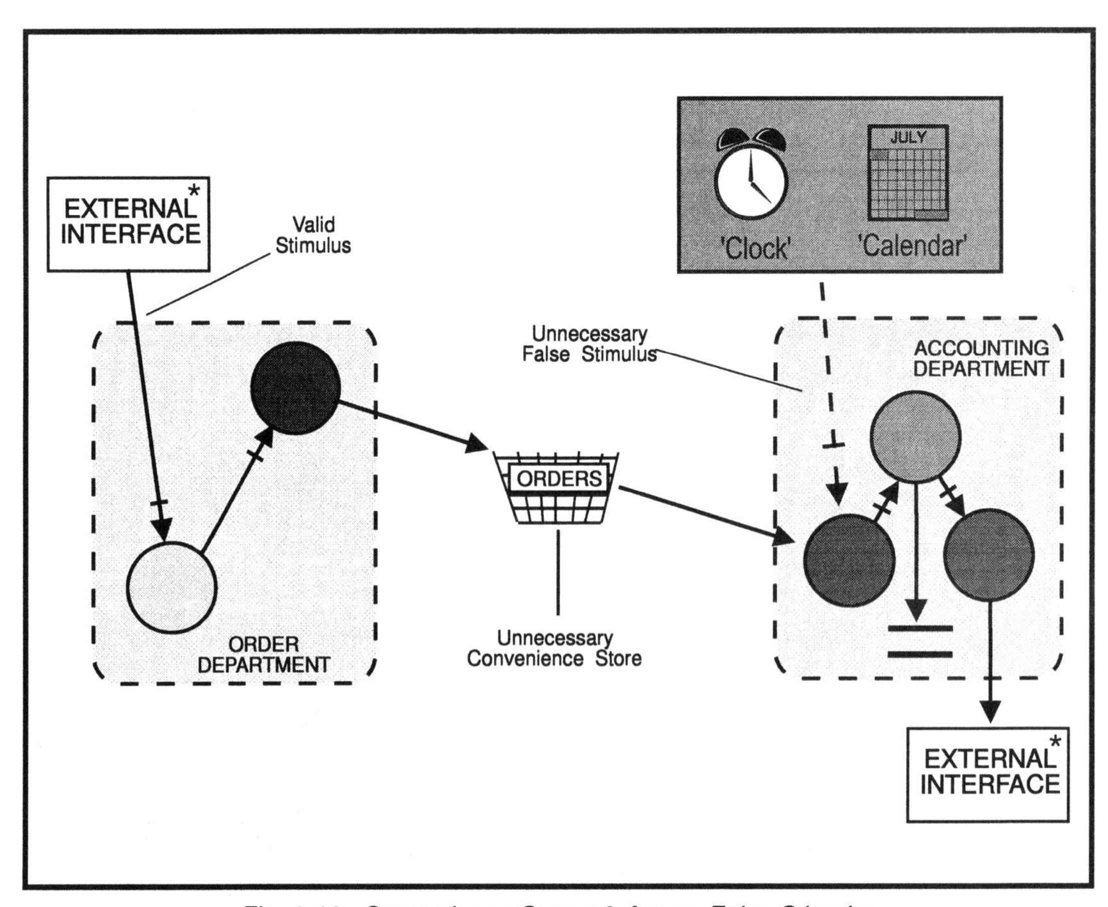

Fig. 8-12: Convenience Stores & Assoc. False Stimulus

These bundled and fragmented files are the source of our data for repartitioned memory in our Business Model. We need to identify "essential" stored data from all of these old Design Stores and then gather together associated Data Elements into cohesive units for processing in our Business Events.

Why We Need Stored Data

Now we can identify the need for business data stores.

With Business Event Partitioning we can see the true business need for data stores and state a fundamental rule of Business Event Partitioning:

We need to store data from one Business Event Partition to await retrieval by another Business Event Partition. In fact, this is the only reason why stored data exists, to respond to external (Business) Events occurring independently.

Data stores are only necessary when data needs to be shared across two or more Business Event Partitions or across two occurrences of the same Business Event Partition. All other data are transient and "burn up" after completion of the Business Event Processing.

This means that stored data will only exist at the intersection of Business Event Partitions (see the Essential Data Stores in Figure 8-13).

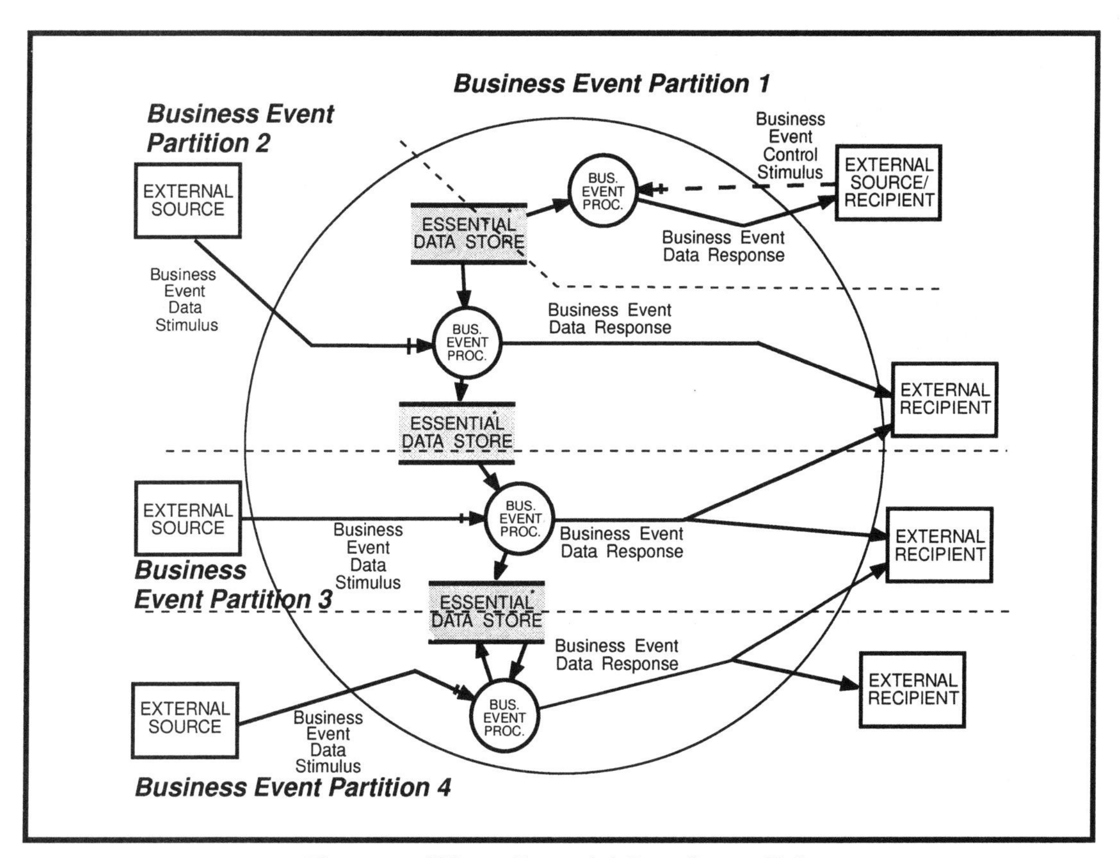

Fig. 8-13: Where Essential Data Stores Exist

These "Essential Data Stores" are the only ones that belong on our Business Model. We can support these using a an Information/Data Model such as an Entity Relationship Diagram with support specifications. Later we will talk about the need to make each Data Element within these Essential Data Stores, Data Conserved. The graphic in Figure 8-13 shows Essential Stores conceptually; in reality, many Business Events can Create/Update data that can be Retrieved/Deleted in many other Business Events.

Logical Stores — the Concept of Business Event Memory

When we think about stored data logically, we could think about the way the human brain assembles stored data. The human brain seems to work by dynamically linking global Data Elements.

> If I asked you to think of all your friends who are male and who have red hair, you would bring together from somewhere in your memory that set of male, red-haired friends. The brain does this so quickly and effortlessly that we don't usually stop to think how we did it. It's as though we have a "Data Element pool," a large set of facts (each Data Element being a separate fact). We don't have a list in our brain of only male, red-haired friends waiting to be accessed when this request comes along. We just logically bring together this set of data. The Business Event Memory for each Business Event Partition is the logical set of data needed to satisfy the Business Event with one mini Data Element pool for each Business Event Partition.

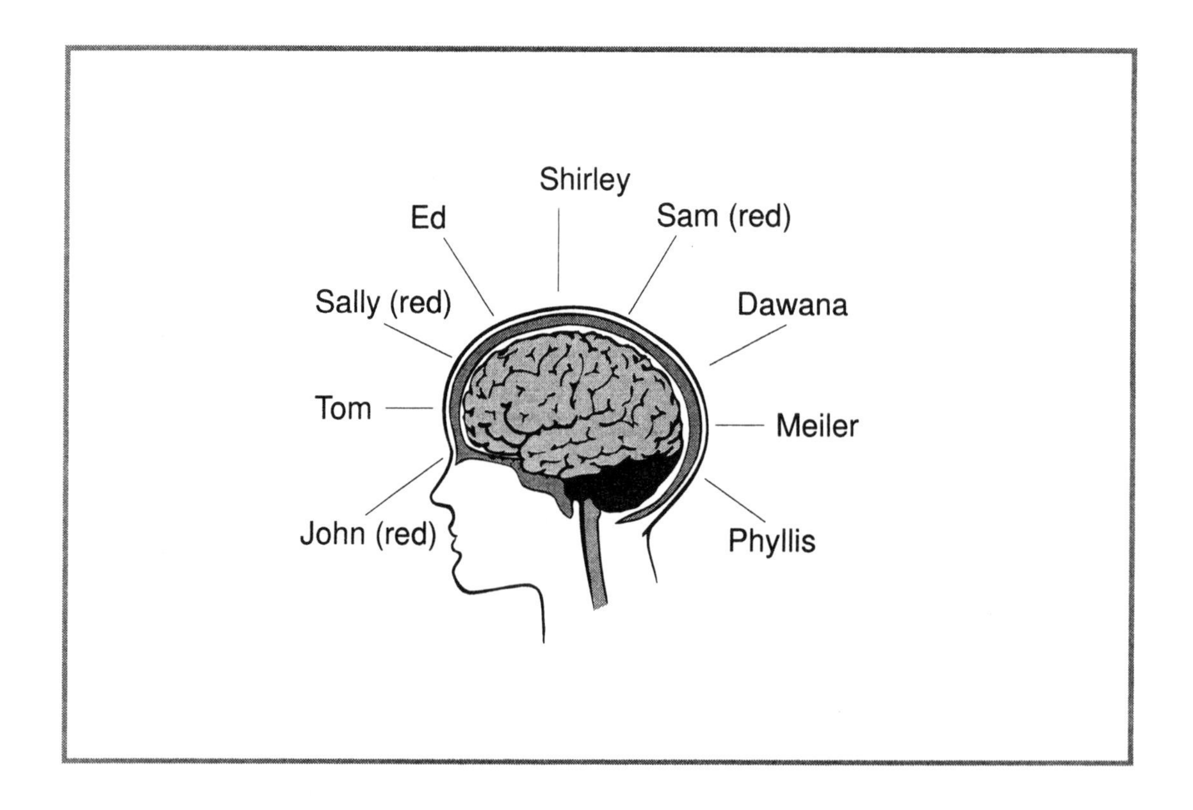

Obviously data would be duplicated across individual sets of Business Event Memory because stores are only necessary when data is shared across Business Events. But, as with the example of a data table on the wall between two people, data really doesn't appear twice, and each Business Event Partition can Retrieve the actual data it needs.

Of course, it's very likely that, for any Business Event other than those that just Create, Retrieve, Update, or Delete a specific set of data, the Business Event Memory will contain a diverse set of Data Elements. In Figure 8-14 we see an Event Context Process Model of a Business Event, Customer Wants to Purchase Our Materials, resulting in the stimulus Customer Order. This Business Event Partition requires a Business Event Memory store which contains information about a **Customer**, a **Customer Invoice**, and **Material**.

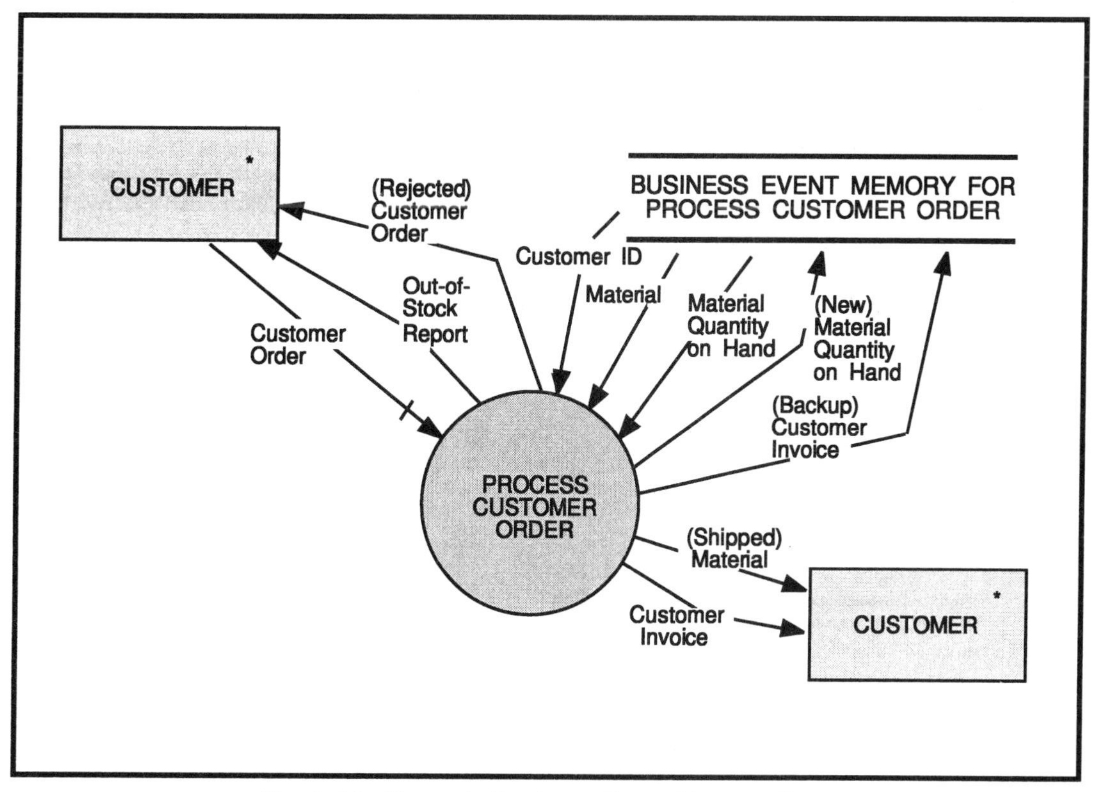

Fig. 8-14: Example Business Event Memory Store

If we add another Business Event to this Process Model for Customer Pays Our Invoice resulting in the stimulus **Customer Payment**, the result is as shown in Figure 8-15.

Note that we now see the redundancy between the stored data of the two Business Events: Customer Pays Our Invoice and Customer Wants to Purchase Our Materials. The two stores both contain some identical data (**Customer ID** and **Customer Invoice information** but this is only "modeling" redundancy.

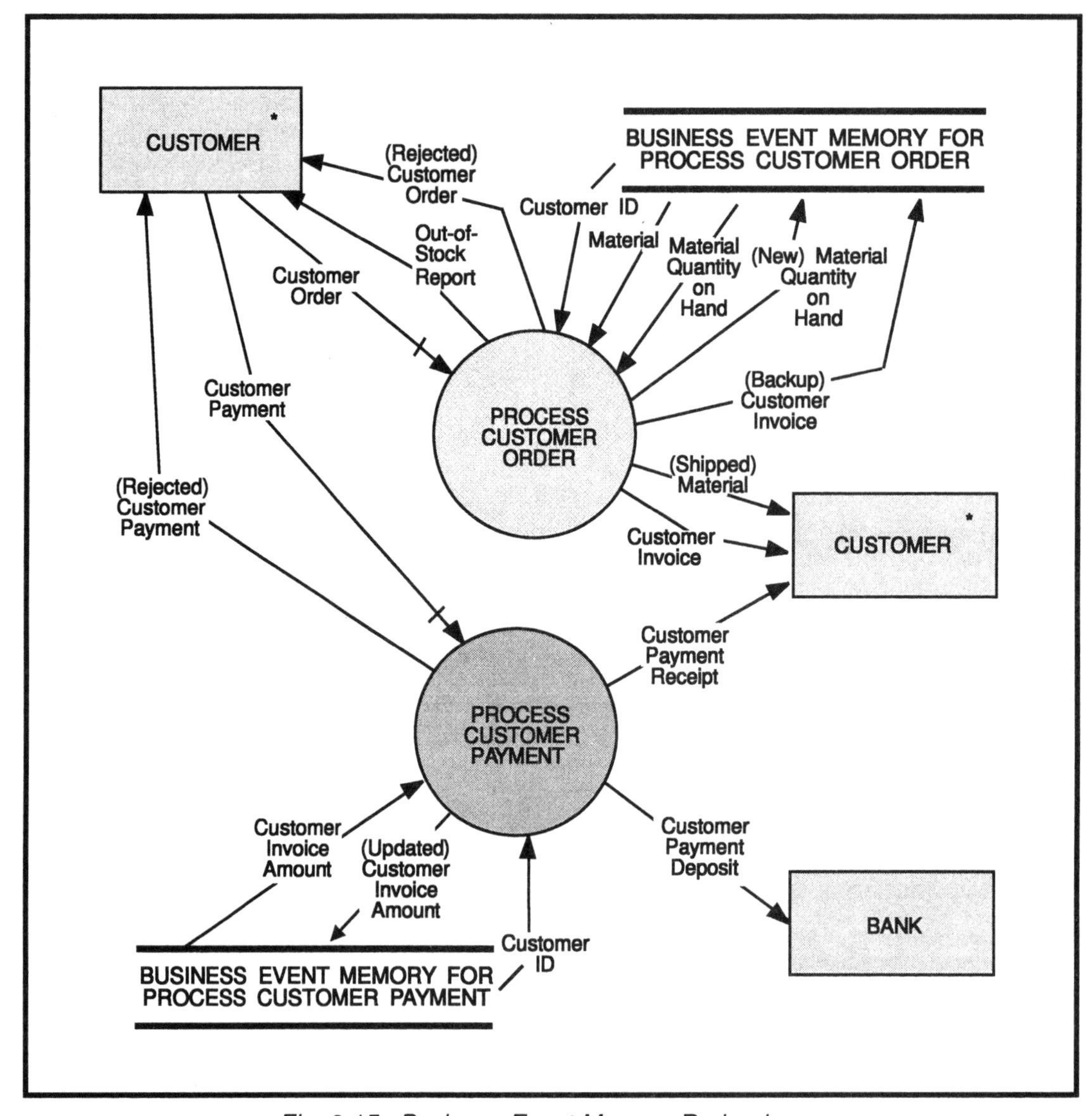

Fig. 8-15: Business Event Memory Redundancy

Naming the Business Event Memory

The naming of the Business Event Memory can simply be based on the summary process name for the Business Event (see Figure 8-15). This summary name can be used because this memory is only going to be used by this one Business Event. It is important, however, to specifically name the data flowing to and from the Business Event Memory. The data flow name must list exactly what is being used in the memory. After we apply Information Engineering techniques to the total resultant set of Business Event Memory stores, we can replace them with stores that reflect Business Entity Names.

The Business Event Response

A Business Event Response is:

All output, such as data, control, products, and/or services resulting from one Business Event.

The non-shaded area in Figure 8-16 indicates the Business Event Response.

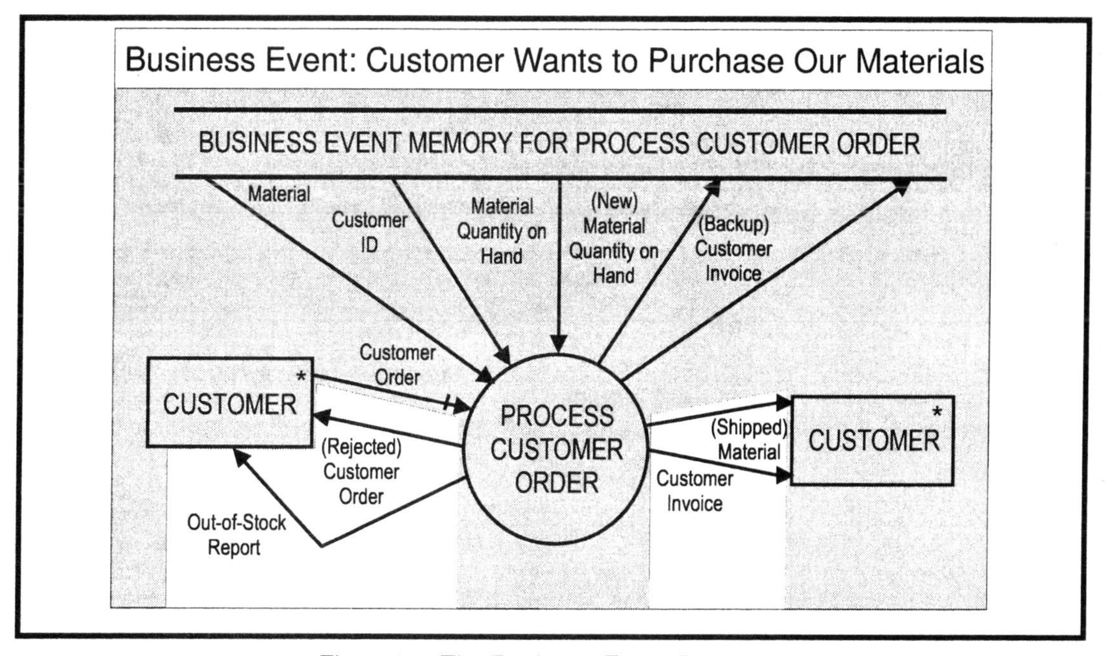

Fig. 8-16: The Business Event Response

A Business Event Response might be delivered products, or the switching on of some equipment.

Naming the Business Event Response

These names use the same naming convention used for Business Event Stimulus names.

The Business Event Recipient

A Business Event Recipient is:

An external individual, agency, organization, system, or other entity that receives the Response of a Business Event.

The non-shaded area in Figure 8-17 indicates the Business Event Recipient.

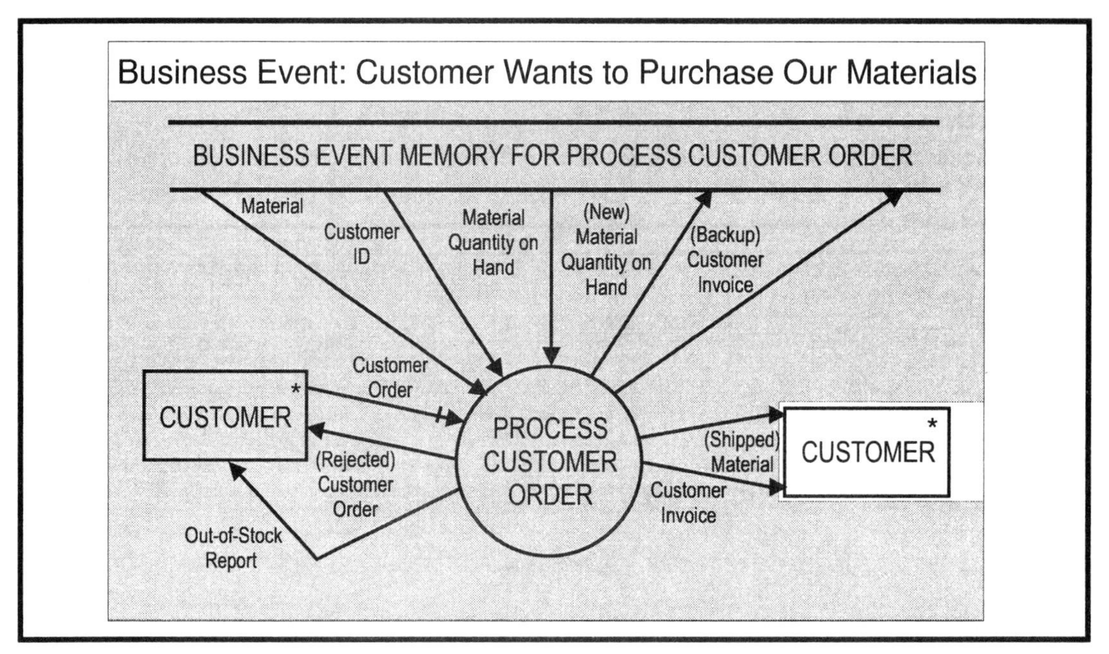

Fig. 8-17: The Business Event Recipient

This may be exactly the same as the original Business Event Source (e.g., the Customer), but it doesn't have to be.

> For example, when I initiate a charitable contribution I don't expect any Response (except good karma), but I do expect my selected Recipient to get the benefit of my contribution.

There may also be many Recipients of one Business Event Partition.

> For example, when an organization requests a seminar from my company, many Recipients receive Responses from this Business Event such as the original customer gets a confirmation, the instructor gets an updated schedule, a shipping company gets a delivery request, a hotel may get a reservation, etc.

Naming the Business Event Recipient

These names use the same naming convention as used for the Business Event Source.

The Beginnings of the Business Library

The result of producing the context-level Business Event Specifications, as recommended in this chapter, will be the beginnings of our Business Library. This library is the model of our business (the organization's repository). That is, it is the description of the what the organization currently does, or will do from a functionally partitioned, customer-oriented (Business Event) point of view. Figure 8-18 shows a sample entry at this stage in our Business Library.

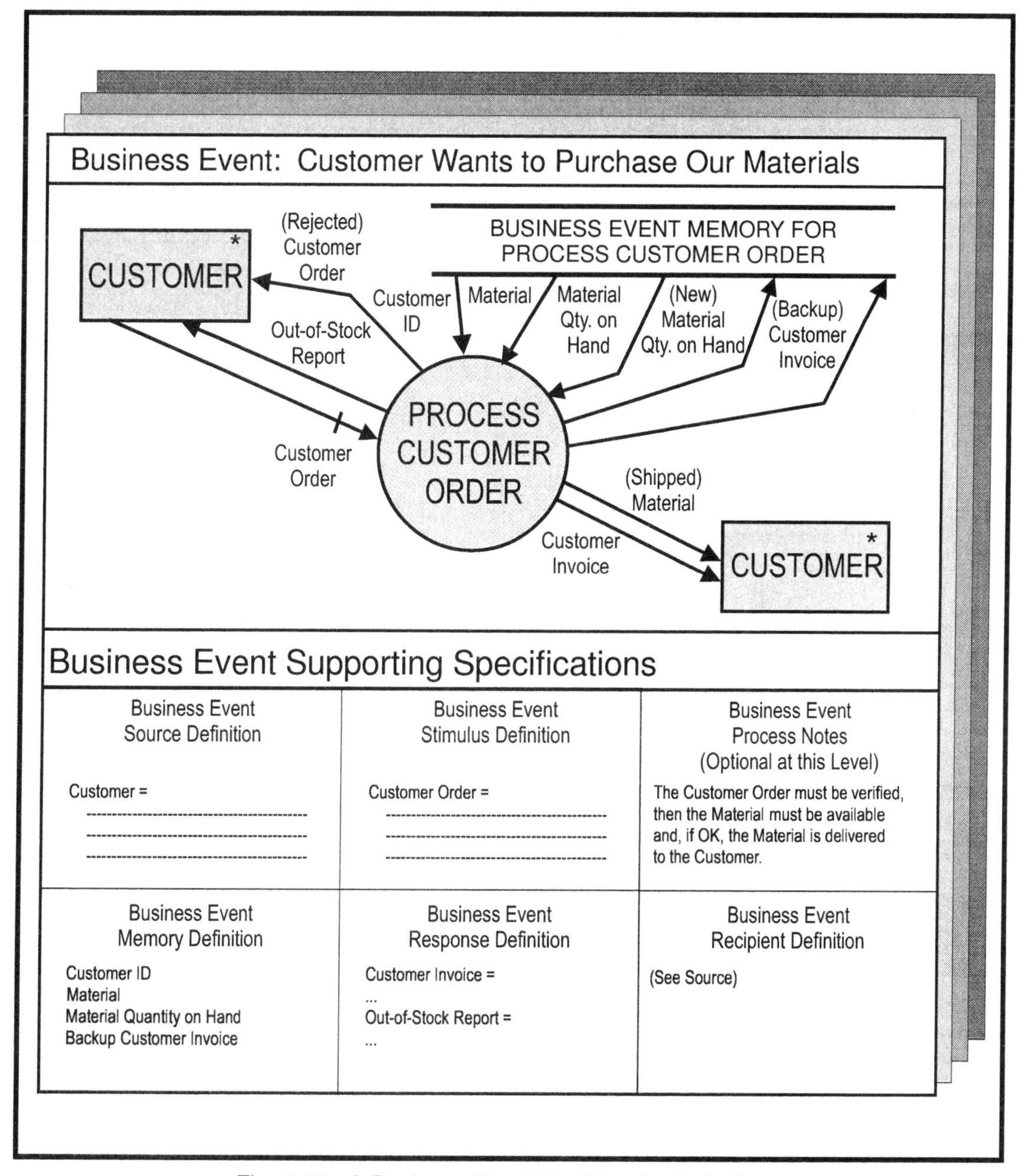

Fig. 8-18: A Business Event Partition Analysis Spec.

Summary

Understanding Business Events is the most important part of creating a Customer Focused Organization.

The analyst's task is to identify individual Business Event Partitions and define the six aspects of each Business Event Partition (Source, Stimulus, Processing, Memory, Response, and Recipient). Any other basis for partitioning will reflect some aspect of design and therefore will not be a functional Business Partitioning as I've defined it.

The result of identifying and documenting our Business Model Events is the foundation of the beginnings of our Business Library, which is the only way we're going to see the real business without being warped by its arbitrary implementation. The remainder of the analysis portion of Customer Focused Engineering is to flesh out the Business Library in more detail.

In the next chapter we'll discuss some traps to avoid when analyzing our organization to identify the constituent elements of a Business Event Partition. In the chapter after next, we will bring these Business Event Memory stores together to form a single Information/Data Model (for example, an Entity Relationship Diagram) which would contain no redundancy of data.

Chapter

9

Unfragmenting Events from Old Design Traps

One day through the primeval wood, a calf walked home as good calves should.
But made a trail all bent askew, a crooked trail as all calves do.
Since then three hundred years have fled, and I infer the calf is dead.
But still he left behind his trail, and thereby hangs my moral tale.

The trail was taken up up next day, by a lone dog that passed that way;
And then a wise bell-wether sheep pursued the trail o're vale and steep,
And drew the flock behind him, too, as good bell-wethers always do.
And from that day, o're hill and glade, through those olds woods a path was made.

And many men wound in and out, and dodged and turned and bent about,
And uttered words of righteous wrath, because 'twas such a crooked path;
But still they followed — do not laugh — the first migrations of that calf.

And through this winding wood-way stalked, because he wobbled when he walked.
This forest path became a lane, that bent and turned and turned again;
This crooked lane became a road, where many a poor horse with his load,
Toiled on beneath the burning sun, and traveled some three miles in one.

And thus a century and a half, they trod the footsteps of that calf.
The years pass on in swiftness fleet, the road becomes a village street;
And this, before men were aware, a city's crowded thoroughfare.
And soon the central street was this, of a renowned metropolis;
And men two centuries and a half, trod in the footsteps of that calf.

Each day a hundred thousand men, follow this zigzag calf again,
And o're his crooked journey went, the traffic of a continent.
A hundred thousand men were led, by one calf near three centuries dead.
They followed still his crooked way, and lost lost one hundred years a day;
For thus such reverence is lent, to a well-established precedent.

A moral lesson this might reach, were I ordained and called to preach;
For men are prone to go it blind, along the calf-path of the mind.
And work away from sun to sun, to do what other men have done.
They follow in the beaten track, and in and out, and forth and back,
And still their devious course pursue, to keep the path that others do.

They keep the path a sacred groove, along which all their lives they move,
But how the old wood-gods laugh, who first saw the primeval calf.
And many things this tale might teach — but I am not ordained to preach.

Sam Walter Foss
The Calf Path[1]

---

1 Forgive me for this long opening poem, but it was so perfect for this chapter that I couldn't pass it up because I guess I am "ordained to preach."

Of the five types of Events identified in the previous chapter, Business Events are the most important in our effort to create a Customer Focused Organization. It's relatively easy to spot our organization's Business Events since they always originate at our true external customers. However, it's usually not so easy to follow the Stimulus from an Event through its Processing and Memory in a typical organization. The larger and older an organization, the more the trail will resemble a long, winding calf path.

There are two stages to unfragmenting Events from old design traps:

- Dis-covering the real business processing and data by removing the cover placed on these by the old design, and
- Unfragmenting our response to a Business Event into a cohesive partition to satisfy the customer's need.

The fragmentation of Business Events will typically occur across many automated and manual boundaries. If we look upon department and computer system partitioning as vertical slices in an organization (see Figure 9-1), then Business Event Partitioning will probably take the opposite view, that of horizontal partitioning.

Notice in Figure 9-1 Business Event #5 is completely subsumed within an existing designed system boundary. Hopefully, we will find some of these, but they will usually be coincidental.

For example, the Regulatory Event for End-of-Year Tax Reporting is completely within the Accounts Department because they happen to have all of the financial data and this Event only requires financial data and processing.

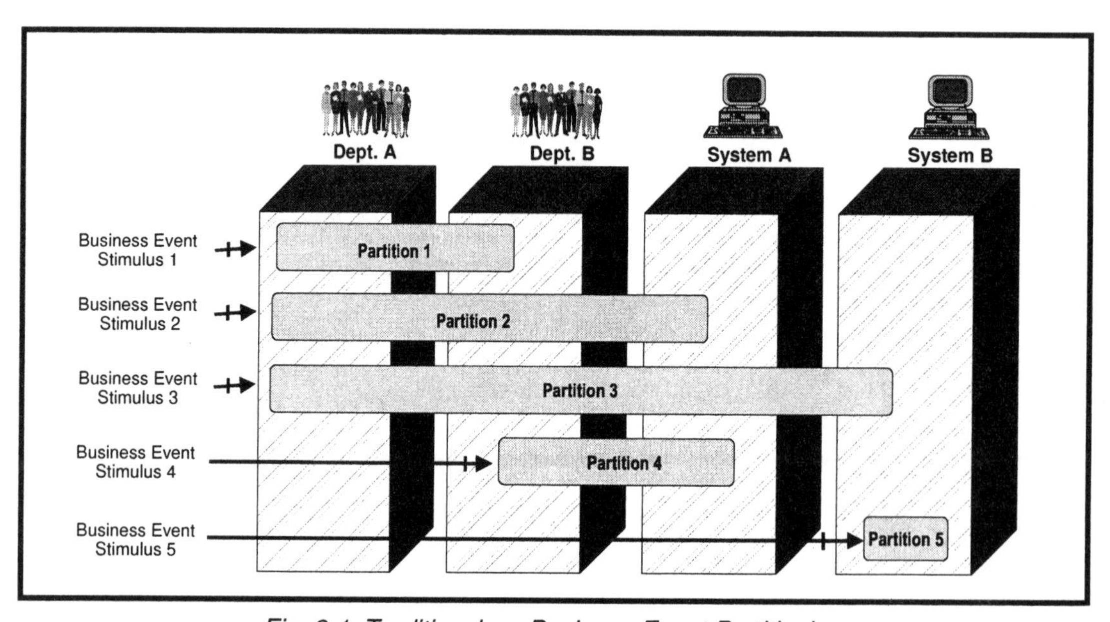

Fig. 9-1: Traditional vs. Business Event Partitioning

Because a Business Event will probably have been broken up by the traditional design of an organization, we will find a number of "traps" that will make it difficult for us in "dis-covering" our true (and whole) Business Event Partitions. Let me identify some of these traps.

Beware of Fragmented Events (Design Trap #1)

As a Customer Focused Engineer we must be able to recognize "fragmented" Business Events that occur at internal design boundaries. These are often places where a Business Event stopped in relation to a past area of study, and hence, a point where a Business Event got fragmented. This Internal Interface may be an internal person (such as a clerk in Accounting), or it could be a calendar that stimulates a batch of data to be input into a portion of the business.

In the past, many designers believed they had to break a Business Event into many individual transactions based on the restrictions of batch processing, central mainframe capacity, or, from a manual point of view, on human skills and memory.

> An example of this latter case would be to send the customers to another department to take care of part of their request rather than have one employee or a team empowered to service their need.

When performing analysis and identifying an interface where a Business Event occurs, we must first ask whether the person or system stimulating us is external to our organization and not just to the portion under study. In other words, are they truly a customer — someone to whom we need to respond in order to satisfy our strategic mission? (Remember, this customer could be a place on Earth where lightening starts a fire, or even the calendar reaching a significant point in time to which our business must respond.) If the origin of the Event is internal and it is there only because of project scope or for design reasons, we must be careful not to dismiss this fragmented Business Event as a System Event because it looks as though we are in control of it. We must also make sure we bring back together fragmented Business Events when we create our Business Model.

It's when we study a subset of our organization that we start to confuse System Events with fragmented Business Events. This also can happen when we break up our project into different project teams to study subsets of the organization.

If you start from the organization's true outside boundary with a Business Model Event List, then any Business Event should be modeled in its entirety without any fragmentation. Therefore, we should not confuse System Events and fragmented Business Events in an organization-wide Customer Focused Engineering project context. Figure 9-2 shows a complete Business Event that was fragmented. This may make us look upon the processing triggered by the System Event Stimulus **'End-of-Day'** as unnecessary System Event processing; however, it is really part of the fragmented Business Event triggered by the customer.

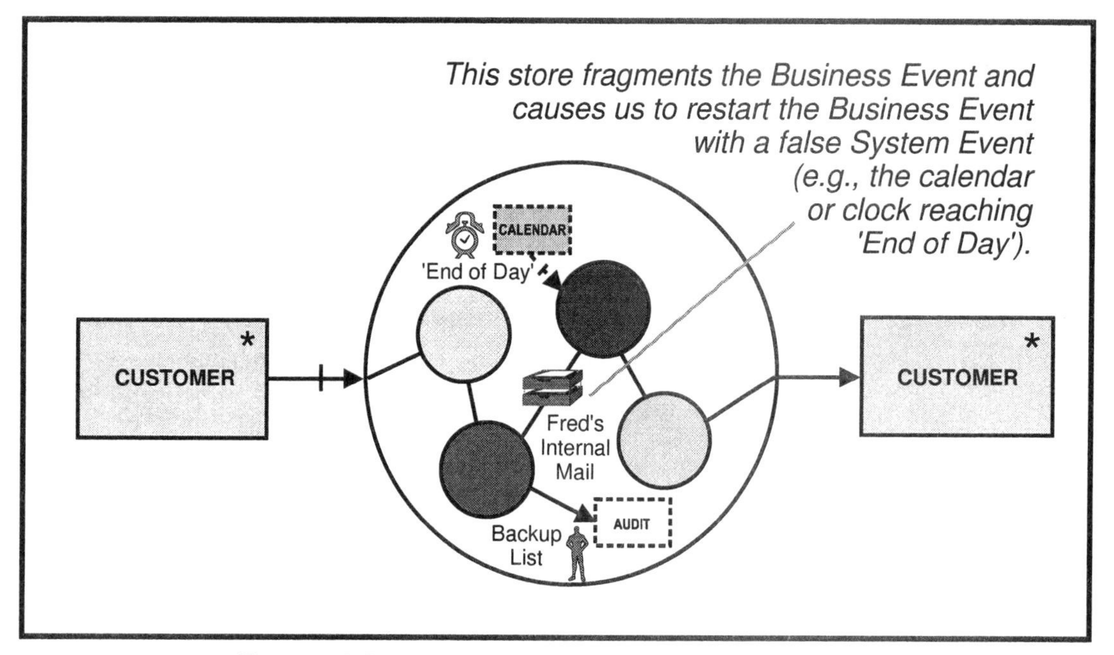

Fig. 9-2: A Fragmented Bus. Event Causes a Sys. Event

The classic "real-world" example of this is a stock control system that is triggered by the **Calendar** (appearing to be a System Event); however, ordering **Materials** (in an organization that sells **Materials**) is not an aspect of design. This is where reordering got fragmented from a real Business Event such as: Customer Wants to Purchase Our Materials resulting in **Materials** being removed from the **Warehouse**. This is the point at which **Materials** could fall below the necessary reorder levels. Therefore, it is the most natural time and place to reorder **Materials**.

Transaction Types and Fragmented Events

Transaction type fields and codes alert us to look for fragmented Business Events.

For example, in one of my seminars I had a Data Processing student ask, "Isn't a Business Event the same as what we call a transaction in batch systems?" For the Customer Focused Engineer it's important to keep the notion of Business Events clear of transaction issues that pertain to design and implementation. A Business Event is not a transaction. Transactions belong to the world of the designer/programmer.

Typically, "transaction thinking" is a single Business Event broken down and lumped together with other kinds of Business Events to be processed in transaction batches. A dead give-away that a Business Event has been broken apart and/or batched is when we see "transaction type" or "request number", etc. fields. A transaction-type field is used to keep track of each Business Event within a batch and its associated processing (e.g., a person or program has to interrogate the transaction type field before they or it knows what type of processing needs to be executed).

We call these broken apart events Fragmented Business Events. The important task for the Customer Focused Engineer is to bring together the fragmented parts of a Business Event. (See the ***Systems Archaeology*** Chapter, for why these bundled/fragmented events occur.) So, some "system" stimuli will be in place to support fragmented Business Events. These stimuli typically will be associated with a transaction type field or a colored form.

For example, we may have a batch process triggered by the start of the business day and this process may start at that time because that's when people show up to work in the Accounts Department. This process really should have been part of our Response to a Customer Order where the flow triggers the processing in both the Order Department and the Accounts Department and ignores the department boundaries entirely.

This transaction view is one of the hardest habits to break. People who have this batch-oriented view of manual or automated systems want to put a distribution process at the beginning of their Business Process Model (a system processing issue rather than a business issue). This process manifests itself in the manual system world as the Customer Service/Help Desk and in a computer system as a Transaction Center module at the beginning of the program. In non-Customer Focused systems this distribution process is needed to separate out the lumped together Business Event Stimuli (see Figure 9-3).

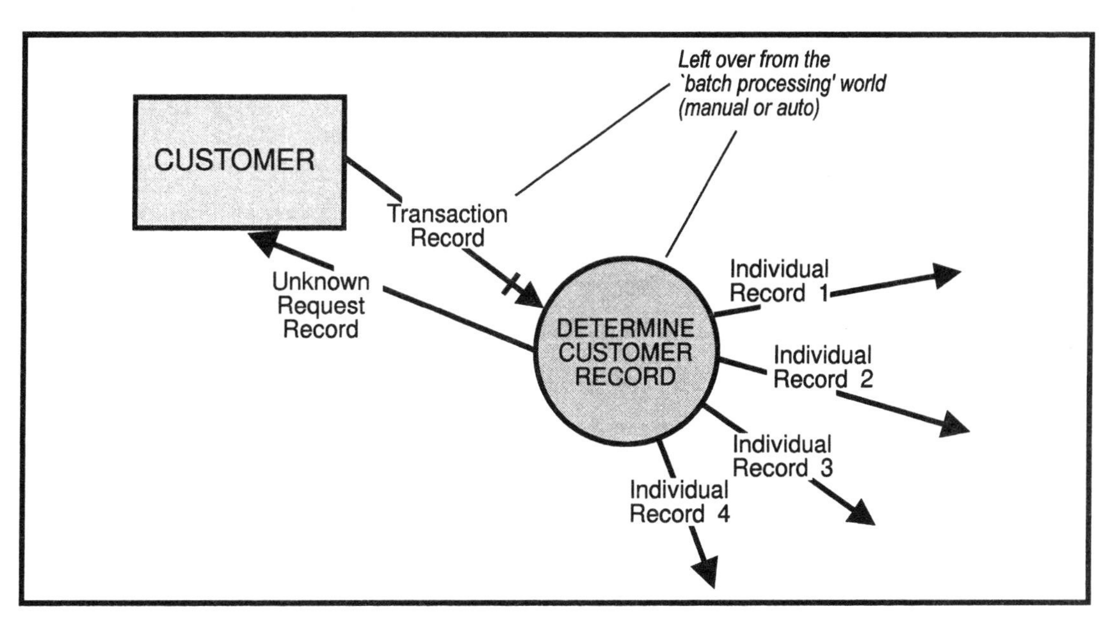

Fig. 9-3: Erroneous Process Model "Transaction Warp"

I always ask these "transaction warped" students to try to write the business logic inside that distribution process, invariably a fruitless task because the logic needs a transaction type indicator that was invented by the designer, rather than requested by the business or the customer, and all of the logic is system/design oriented.

There is a need to differentiate between Internal and External Interfaces on a model. If we're using a Data Flow Diagram to model our business, then an External Interface is a solid-lined rectangle; if the interface is Internal, I recommend a dashed-lined rectangle. I introduced this into the Data Flow Diagram symbology in ***Chapter 5***. If you use another model, then I believe it would be helpful to extend that model's symbology to address this need.

For example, on an Object Oriented Model you may have Messages that are internally stimulated represented differently than externally stimulated Messages.

My reason for putting forth the idea of an Internal Interface and for recommending the introduction of new symbols is to point out where future projects need to link to other parts of a fragmented Business Event caused by the data flow terminating at an Internal Interface. At some future point we would eliminate these Internal Interfaces and implement complete Business Events across the whole organization; this should be part of the Strategic Plan of our organization. The timing of this future point depends on the implementation of a Strategic Plan based on Business Events. After the Strategic Plan is completely implemented, there will be no Internal Interfaces on our organization's Business Model. (More on this in the ***Strategic Planning via Business Events*** Chapter.) Also, under this unifying discipline we would not continue our old, historical error of implementing fragmented Business Events. Note that if we do Pre-Engineering (as discussed in the first chapter), we wouldn't introduce these fragmented Business Event driven Internal Interfaces in the first place.

Beware of Bundled Events (Design Trap #2)

Just as "Edit-Update-Print" is a poor basis for partitioning because it fragments business policy, so is collecting all similar processes together, such as Create, Retrieve, Update, and Delete to form monster partitions.

For example, we might collect all Create Events (such as Create New Customer or Create New Vendor) together to form a generalized Create Program. This view, shown in Figure 9-4, could come from focusing only on data instead of both data and processing together. This is not a specific Business Event view, and if implemented as four design partitions (programs), we would have similar maintenance problems as with the Edit, Update, and Print partitioning.

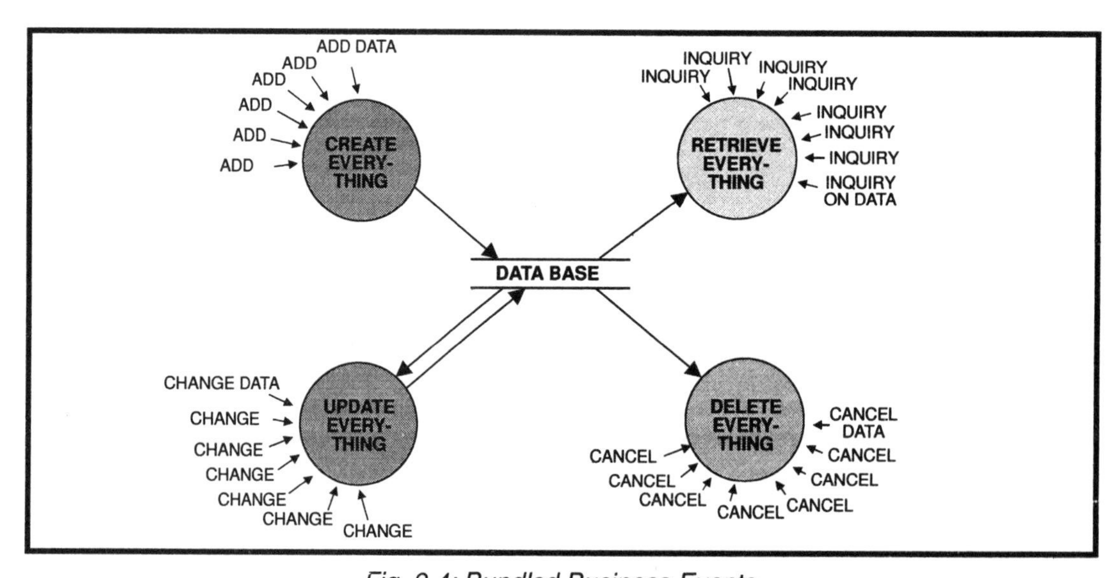

Fig. 9-4: Bundled Business Events

Carried to its extreme, we could incorrectly multiply the number of Data Elements/Fields and Entities in our business by four to incorrectly derive the total number of Events in our business. (I hope that you can see that this does not represent a *business view* but rather a *data-only oriented* point of view.)

If we talk about quality as "conformance to customer requirements," to obtain the ultimate customer satisfaction, we need to satisfy the customer's specific requirements. If we are lumping together different types of orders from customers into a general Customer Order, we will have bundled together a set of separate Business Events. This leads, in design, to what we find in most businesses today — the standard order form/screen and the standard processing of that order form/screen. The processing of a generic order may be implemented using human beings or an automated system. In either case, we end up with a form or screen containing data fields that should never be filled out for a Special Order, but which definitely need to be filled out for a Delivery Order. Also, different parts may never have to be completed for an Advance Order while other parts may have to be filled out for a Pre-paid Order. What happens is, we end up with a global form or screen and the global processing needed to satisfy that form or screen. Often the person who fills out the form or screen has to conduct a long-winded procedure to overcome the design of the form or screen.

A problem also occurs when shared logic is initiated for many different Events that were usually bundled together. Unfortunately this makes it difficult to modify procedures and/or computer code when changes are needed because there is so much procedure and code (design logic) associated with the fragmentation. This design logic gets in the way of the business logic.

The logic (in computer code and procedure manuals) is usually strewn with what I call designer "IFs." The Business logic is cluttered with "IF" statements where we try to figure out which one of the bundled Events use this logic (see Figure 9-5).

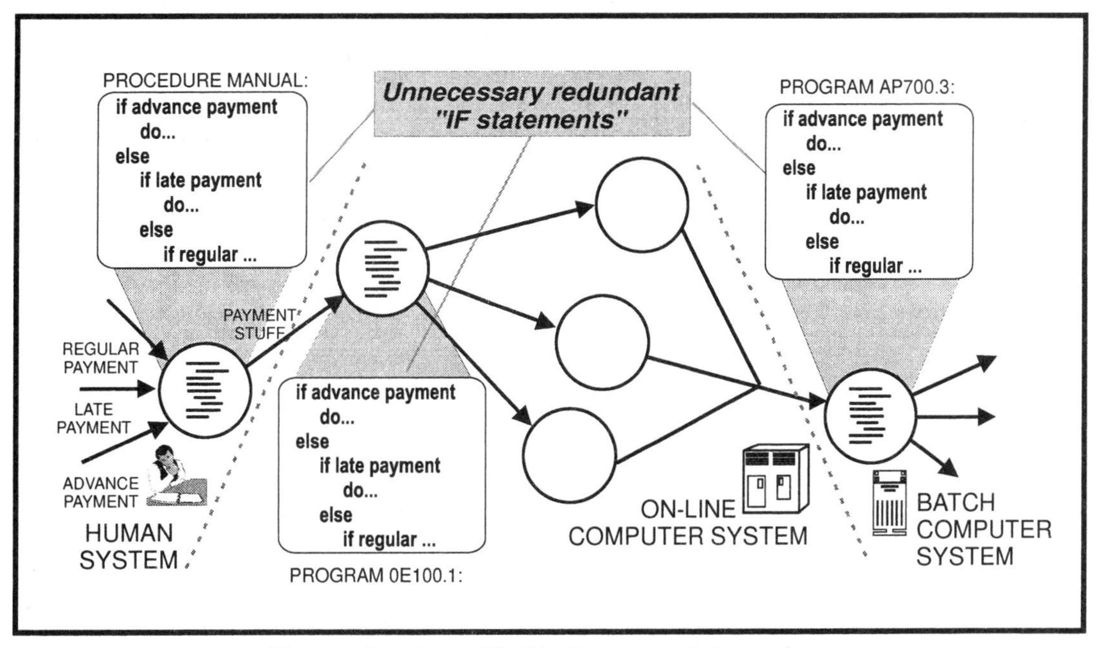

Fig. 9-5: Designer "IFs" in Programs & Procedures

It gets worse, of course, when we have Fragmented and Bundled Events combined. This is what we find in the classic, historically/hysterically partitioned Edit All — Update All — Print All computer programs. We also find the equivalent historical partitioning in a manual environment in areas such as Order Entry, Accounting, and Invoicing (see Figure 9-6).

Figure 9-6 shows the evolution of Business Event modeling that would occur in the process of conducting Customer Focused Engineering on an organization that was originally historically/hysterically partitioned, causing fragmentation and the bundling of Events. This is represented in the model shown at the top of Figure 9-6. The middle model shows a partially Systems Engineered organization, and finally the bottom model shows a Customer Focused Organization completely restructured with complete Event Partitions.

Note that the Internal Interfaces (shown as dashed rectangles on the interim model) are the result of not being able to conduct Customer Focused Engineering on the entire organization at one time. Also, the first two models in Figure 9-6 show where the designer complicated and slowed down the response to the Customer by bundling and/or fragmenting the Business Events.

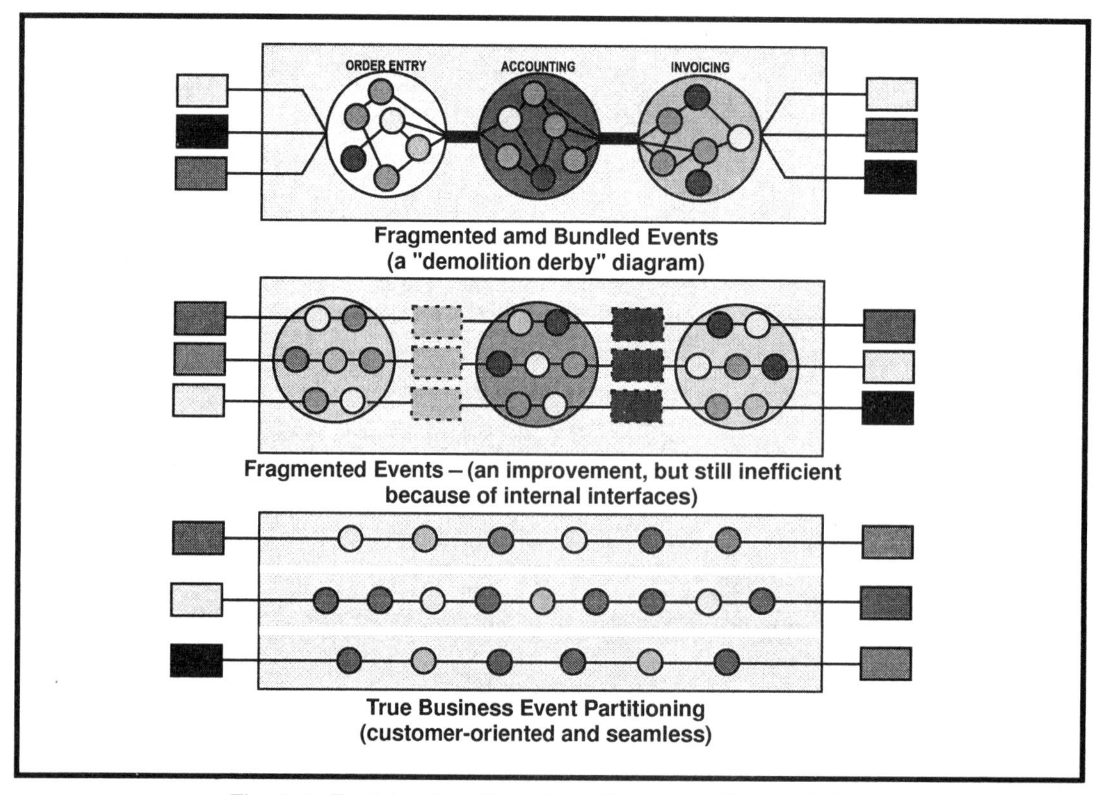

Fig. 9-6: Evolving to a Seamless Customer Focused Org.

Beware of Historical Events (Design Trap #3)

We should even guard against believing that what was a Business Event in the old system is really a Business Event that was initiated by the original customer. The last analyst (or Strategic Planner) may have dictated what a customer was allowed to do with the organization's old systems. If the inventor of the business logic in place today had not done some type of customer survey when putting the old logic in place, then our customers may just be putting up with our limited Events until something better comes along. We don't want to create Customer Focused Systems in which our customers aren't really interested. We need to "think out of the box" on this one.

For example, no one in the phone company asked me if I wanted to keep a thick directory of Business Information (the Yellow Pages) in my house; nor was I asked the order in which I would like the directory to be arranged. The phone company also makes me access their systems (manual or automated) with a false design key — a phone number.

Of course, there is a problem with asking a customer what they require of our organization. Many times our customer is trapped in a design world also and we may get a request to simply improve our existing designs — not new requirements.

Beware of Fragmented & Bundled Data Stores (Design Trap #4)

Just as we have old design traps for fragmented and bundled Business Events, we have the same design traps for Stored Data.

Beware of Fragmented Stores

It is typical to find fragmented "data sets" (i.e., data Entities) across old designer stores.

For example, information on a **Customer** could be scattered across **Sales Files**, **Purchased Products Files**, and **Service Agreement Files** as in Figure 9-7. If a **Customer** sends in a change of address, all three files would need to be updated. The analyst must bring these Data Elements together into a cohesive data grouping (i.e., an Entity) when repartitioning stores.

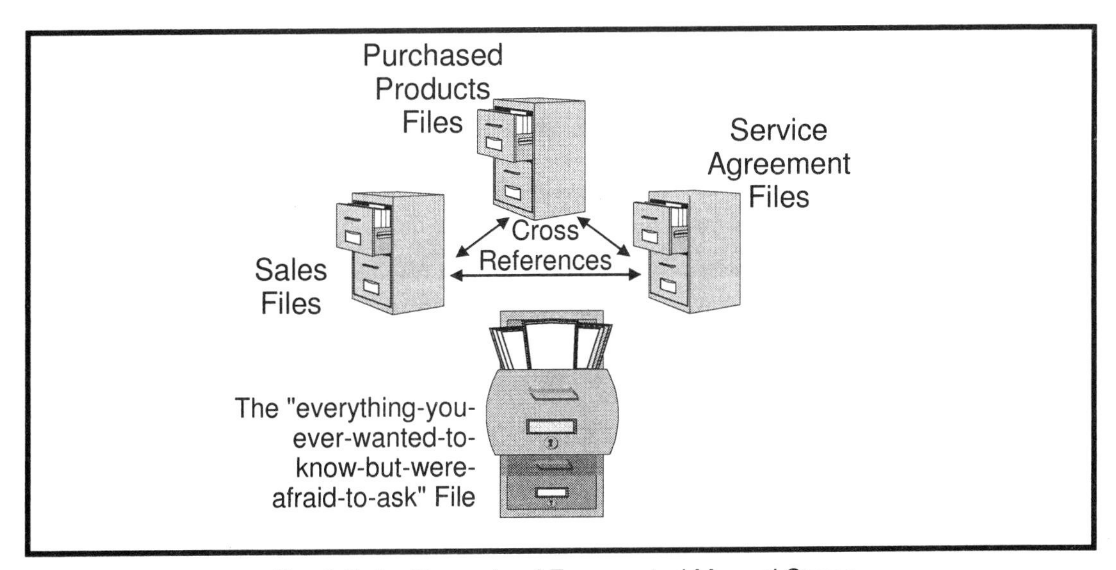

Fig. 9-7: An Example of Fragmented Manual Stores

Beware of Bundled Data

As we saw in the chapter ***System Archaeology***, bundled stores arise when the previous designer grouped separate sets of data together, possibly for the sake of efficiency. This is very common of the files used in "Edit-Update-Print" computer systems, because this data partitioning was based on process bundling for technology and efficiency. The analyst must break up these bundled stores (typified on the left side of Figure 9-8), and assign their Data Elements to cohesive data groupings (i.e., Entities) shown on the right side of Figure 9-8. We do this by observing the access key to any individual Data Element.

> For example, **Customer Delivery Address** may be retrieved via **Customer Number** in the **Purchased Products File**, **Customer Address** may retrieved via **Customer Date of Birth** in the **Sales File**, and **Customer Billing Address** via the **Customer ID** in the **Service Agreements File**.

This task typically is not as easy as in the above example because Customer Information may be stored and accessed via Customer Name and Address, Social Security Number, Customer Phone Number, or Driver's License Number, Passport Number, etc. The task will be made easier if we already have formed a Business Event Memory store as defined in the ***Partitioning by Business Events*** chapter and used a specific naming standard for each Data Element in those stores.

We need to take care of this trap by unfragmenting and unbundling the data to ensure data conservation as described in the ***Achieving Organizational Process and Data Integrity*** chapter.

A Typical Bundled Store	Partitioned Cohesive Stores
Customer Name	**Customer Personal Data**
Customer Address	Customer Name
Customer City/State/Zip	Customer Address
Account Number	Customer City/State/Zip
Account Balance	Customer Old Address
Account Creation Date	Customer Old City/State/Zip
Product Number	**Products Data**
Product Cost	Product Number
Product Lease Number	Product Cost
Credit Account Number	Product Lease Number
Credit Account Limit	Product Service Agreement
Customer Old Address	**Customer Account Data**
Product Service Agreement	Customer Account Number
	Customer Account Balance
	Customer Account Creation Date
	Customer Account Credit Number
	Customer Account Credit Limit

Fig. 9-8: Bundled vs. Cohesive Stores

What we are aiming at is a set of stores that are cohesive and whose names reflect the contents of those stores. Figure 9-9 shows two Business Events with their required, cohesive essential stores.

In this example, the **Customers** store contains only customer information about a Customer; the **Materials** store contains only information about the materials; and the **Customer Invoices** store contains only information about **Customer Invoices**.

Please note that the two Business Events shown may be merged if we found them to be two fragments of the same Business Event (Design Trap #1). In which case we would go on to question the need for the **Customer Invoices** store between them as it may only be a Convenience Store as discussed in the ***Partitioning by Business Events*** chapter (unless it is used by some other Business Event Partition). If we did combine these two processes into one Business Event, the **Customer Invoices** store would become a transient data flow. Also note the stimulating data flow **Customer Payment** would become a response data flow (a "pull"). The Event Context Diagram (and any lower-level subprocesses) for this Business Event would obviously also reflect this combination.

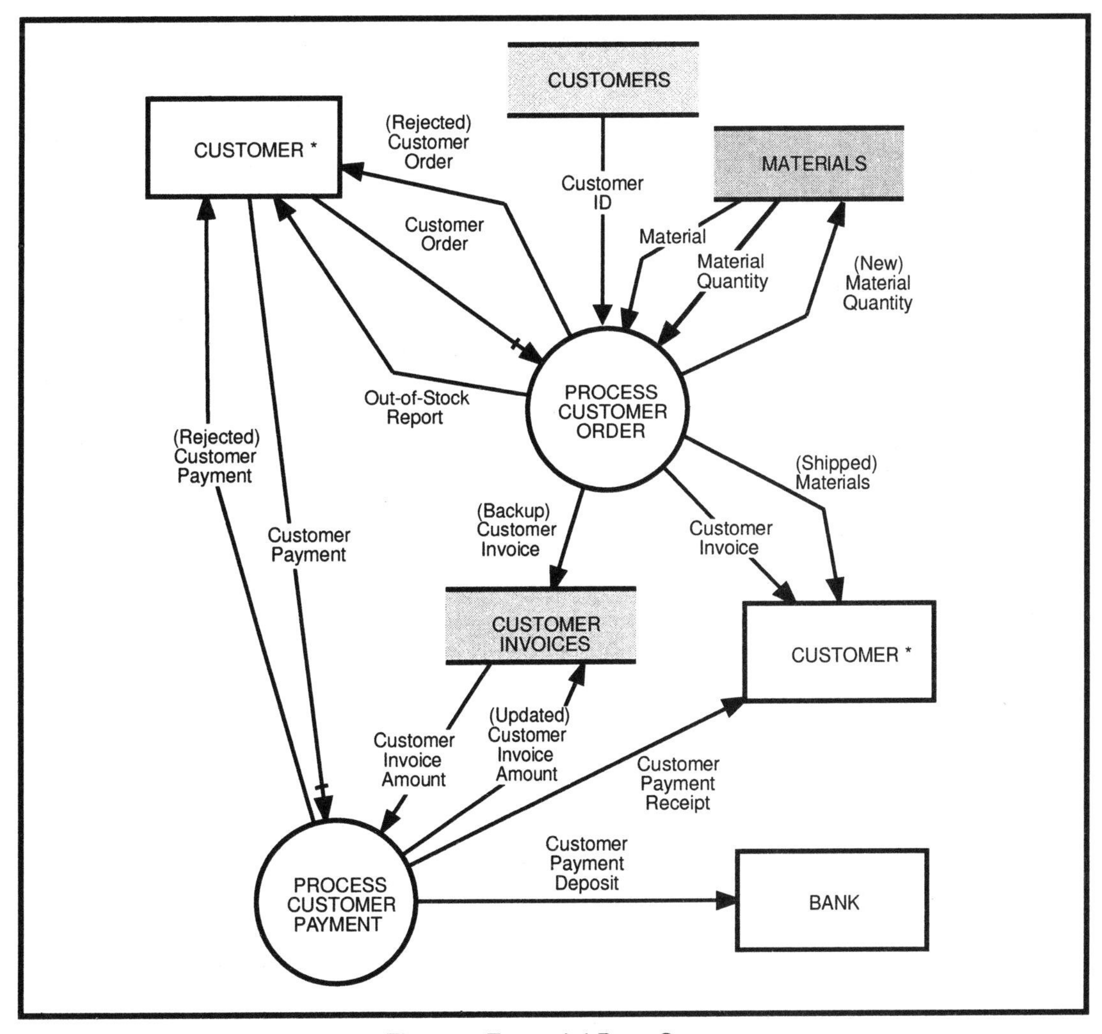

Fig. 9-9: Essential Data Stores

Using Business Events to Drive Associated Regulatory and Dependent Events

If we create a partition and model the essential processes and data needed to satisfy a Business Event, this partition would only contain necessary business data and processing. However, even after we've removed all the system issues from a Business Event Partition there will still be processing and data needed to support the Business Event. Figure 9-10 shows these supporting Events with dashed areas representing these issues. The figure shows the four essential processes and their data needed to satisfy our Business Event. It also shows the four support processes and data that aren't directly associated with our Business Event but that are required to satisfy the Dependent and Regulatory Event issues to support our Business Event.

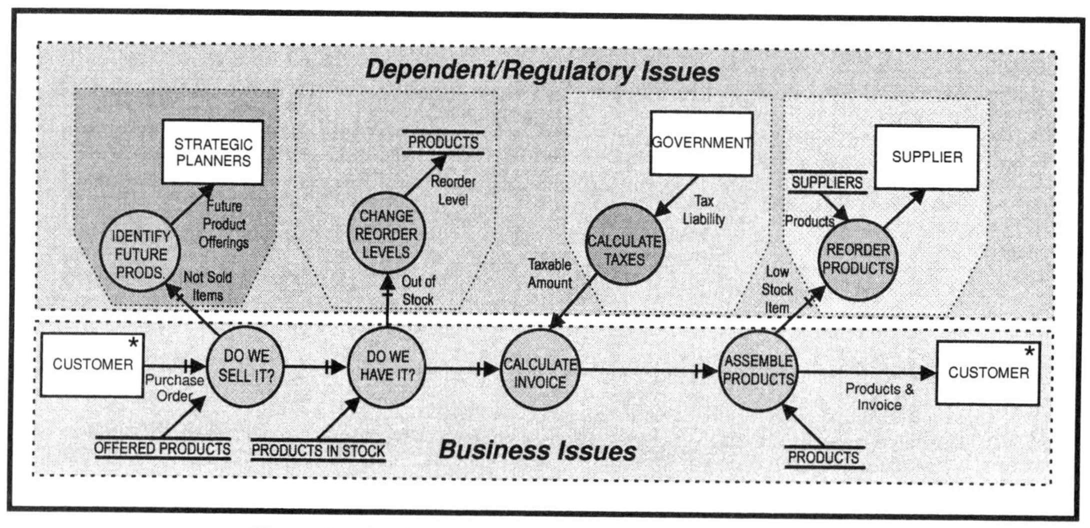

Fig. 9-10: Business Event with Its Supporting Events

If we have the ability to suggest changes to regulatory agencies or make an alliance with our supplier/vendor, then our Business Events can subsume disjointed Regulatory and Dependent Events and/or their aspects.

Dynamically Defined Business Events

Now that we've talked about the potential problem of historical Events, this leads us to talk about dynamic Events. The designers of old systems typically have not taken a customer "external" view of system partitioning; instead, they took the "internal" system view. These designers tended to look at a customer's need from the point of view of many small requests fragmented across internal systems. Rather than satisfy the customer's one request in full, they expected the customer to interact with the organization's systems to piece together fragmented requests to satisfy their needs.

Sometimes we will find that systems builders have tried to merge two or more of these small, fragmented requests into what the customer may actually want to do in one interaction with the organization.

> A bank deposit with the "less cash" option is an example of this merging: to get a significant complete task done, the bank's customer usually has to perform an interactive dialogue (with either a human or an automatic teller) and the dialogue is often interrupted by a number of returns to a menu screen, or there are new forms to fill out before his or her need is met.

The ultimate in customer satisfaction would be for the bank to find out exactly what the customer needed to do and then to perform the entire transaction seamlessly i.e., to create customized Business Events. In fact, the term, "customized Business Event" is redundant. If a Business Event isn't fulfilling one complete need of the customer, it's not a properly partitioned Business Event (see Figure 9-11).

For quality customer satisfaction, we need to identify each specific need of each customer. This might seem ridiculous to some veteran computer people who are used to generic transactions in a system. However, for ultimate customer satisfaction, we truly want to have something at the beginnings of our systems that can form a unique Business Event Stimulus each time and bring together whatever data and processing is necessary in our organization to accomplish that Business Event.

We typically can see this implemented in a manual system because human beings can dynamically adjust to what a customer wants (and probably even recognize and remember that customer's past requests). They can bring together the necessary processes and materials to satisfy a particular customer's specific need; we can do this in a new system without requiring the customer to go to other departments, or to fill out many different forms.

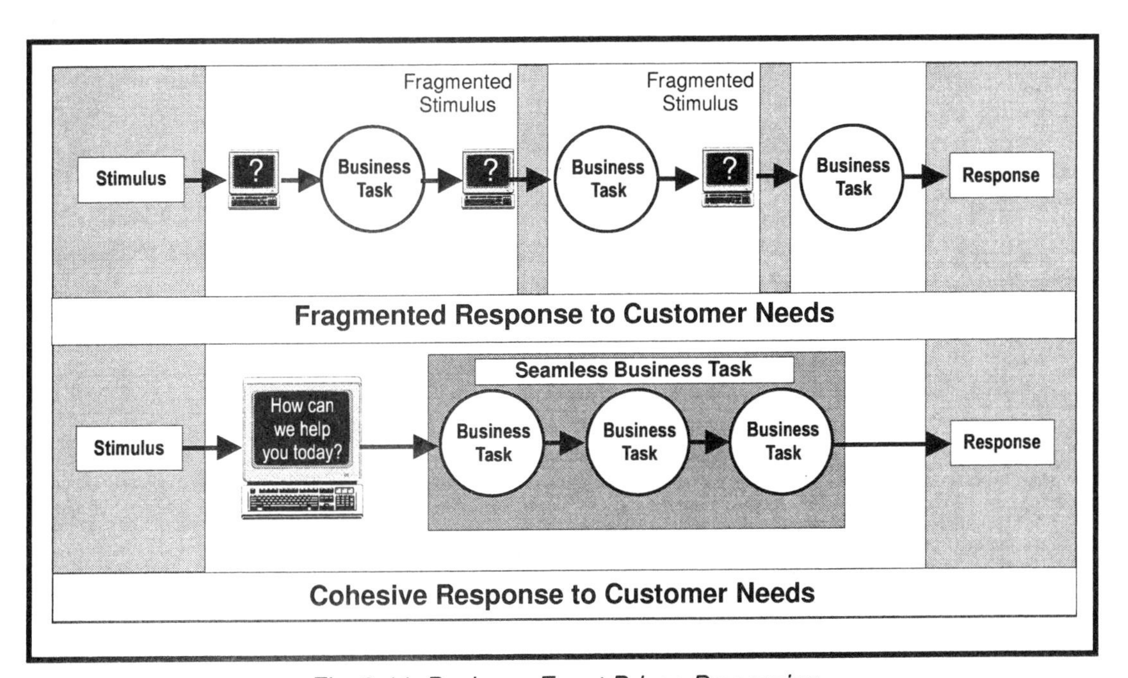

Fig. 9-11: Business Event Driven Processing

With today's technology we can be far more dynamic and interactive. In fact there is a demand to empower employees with multiple skills, to computer assist employees, and to satisfy the customers' needs with total on-line, real-time computer processing. In order to provide custom Business Events, we first need to discover what Business Events are typical from the customers' point of view. Once we've identified the processing and the data required to satisfy each of these typical Business Events, then we can satisfy dynamically changing Events by bringing together reusable parts of that processing and data on the fly.

Of course, if the customer requests that we embark on a new Business Event such as new processing and new data, then this involves a Strategic Event and requires us to conduct an analysis.

> In our banking example, the customer may request to buy theater tickets with seat selection. If the Strategic Planners say we wish to satisfy this as-yet undefined Business Event, we analyze our Business Policy for this new Business Event. We can satisfy that customer's need with the data and processing within our organization's context or we can create them.

This is no different than you having the set of skills and knowledge (processing and data) in a particular subject matter, and having someone request you to do something you've never done before. If you understand the request, you can satisfy that need by pulling together the procedures required to satisfy the demand.

The analysis of a Business Event becomes the act of identifying what the customer wants. The definition of the Business Event Partition becomes an issue of the collection from our organization's knowledge base of the processing and data required to satisfy the request.

> To beat this to death, let me give an example of a valid bank customer request where: we find out that every other Friday a customer wants to deposit a paycheck; the same time he wants to pay a standard mortgage payment, take out $200.00 in "less cash," put $100.00 in a checking account, and put the rest into a savings account. If we make our customer perform four separate transactions (either in a manual or an automated system), we have not satisfied the customer's needs seamlessly. Now, the customer may accept that that's how it's supposed to be, and he may accept multiple returns to the menu screen, or fill out multiple slips of paper to make his transactions. The sum of these separated transactions are really that customer's Business Event. We need to satisfy this specific need seamlessly before another bank offers this type of new service.

Now this dynamic forming of Business Events may not be easy to implement. We may decide to go back to a menu screen that gives a wide range of customer requests. However, for our Customer Focused Engineering analysis effort we are chartered to produce a customer focused Business Model and not worry about how we are going to ultimately implement it.

In our banking example we may decide in the implementation to use interactive television technology and put together whatever transaction the customer wants and store it in the television memory "add on" supplied by our organization. So, in this scenario, there's no need for the form, there's no need for the customer service desk, and there's no need for the menu screen on the terminal as we know it today.

For ultimate customer satisfaction, we need to accommodate dynamic Business Events and put together custom Business Events using our system knowledge base to satisfy our customer's needs. Today we can do this bringing together of dynamic Events. For those of you familiar with databases and high-level computer languages (those based on human languages such as English), this dynamic integration is currently available to us. However, this is usually on an ad-hoc basis, recreated each time and requires all the data be available before executing the dynamic request).

When defining a Business Event and its resulting partition our task is to put together that collection of nouns and verbs that satisfy our customer's true needs and this can then be used for new systems design and implementation.

Summary

Unfragmenting Business Events from old design traps is where a Customer Focused Engineer earns his or her keep. Bringing together Business Events is the most important step in producing a true Business Model.

Keep in mind there will be even more traps than I have identified here that are specific to your area of business. I hope the ones we've gone through give you the framework needed to recognize other design traps.

The next chapter discusses the remainder of the analysis portion of Customer Focused Engineering. This task is to flesh out the Business Library in greater detail.

Chapter 10

The Detailed Business Event Specification

I would never have succeeded in life if I hadn't given the same care and attention to the little things as I did to the big.

Charles Dickens

Now that we've talked about models and Business Events with Business Event Partitioning (and how they're the ultimate view of the business), we can talk about how we actually document a detailed Business Event Specification. This is where the "tire meets the road" within each Business Event Partition.

Even though I focus on a Business Event in this chapter, the concepts of subpartitioning also apply to Regulatory Events and Dependent Events. If we create a separate Meta Model, these concepts also apply to Strategic Events.

As stated in the title of a previous chapter, I believe the Business Model IS the business (or the only way we're going to see the true business). In this chapter we will bring together a set of models to specify each Business Event Partition and the complete library of these Business Event Specifications will form a detailed view of our business.

In the upcoming ***Designing and Implementing Business Event Systems*** Chapter we'll extend this specification to include the design aspects of a Business Event Partition. With these two views we can see what the business is and how it is/will be implemented.

Up to this point in the book I have recommended we show one process (for Business Logic) and one data store (for any memory) within the Business Event Partition. Before creating a fully-partitioned Business Event Specification, we need to talk about how we partition complex processing and memory within the organization's response to a Business Event. This chapter explains the reasons for subpartitioning and how we conduct this additional process partitioning and data (memory) partitioning.

Our task here is to model our organization's total response to each Business Event at a detailed level. Remember, a Business Event Partition is the most natural business structure for satisfying one specific need of a customer.

A Business Event Partition consists of a Business Event Stimulus plus all associated Processing, stored Memory, and outgoing Responses (along with the Recipient) that constitute the organization's complete reaction to a Business Event.

Given this definition, and the constituent parts of a partition, we need to decompose further any of the components of our Business Event Partition until we reach a level of specificity where our data and processing are at a primitive level (i.e., understandable to whomever will need to read our specification). We can then use this business analysis model to invent the design of our organization.

I've already been using the tried and tested models and symbology of process and information modeling as our means of depicting Business Event Partitions. In this chapter I'll use the extended symbology to accommodate Business Event issues. Remember, it's not the type of model that's important, but rather its ability to capture the business aspects of our organization in an understandable way for us to obtain validation of the model.

Subpartitioning of Processing

Up to this point, we've replaced old partitioning based on systems, departments, divisions, etc. with one based on a truly functional Business Event Partitioning. Another milestone of our progress is we have each cohesive Business Event Partition represented as a single process and data model. If the single-process Business Event Partitions are small and understandable by anyone needing to read our specification, then any further subpartitioning is not necessary. We can immediately specify the details of our business logic. If the single-process Business Event is complex, voluminous, or unwieldy, then we need to partition it further.

Just as it's beneficial to sub-partition the old design into sub-procedures within a task and fields within a file, it's helpful to do a further subpartitioning within a major Business Event Partition. These further decompositions may be of help, especially where they would identify reusable processes and groupings of stored data and, hence, save repetition of effort and redundancy in future procedures, programs, and files. It also helps to handle the complexity of a Business Event Specification to progress from a high-level abstract to a low-level detailed specification. Figure 10-1 shows an example of decomposition and leveling of processes on a Business Process Model until we reach what we call a Functional Primitive[1] Process.

---

1 A term introduced by Tom DeMarco (see Bibliography) that I feel is very descriptive of a detailed process (i.e., the process accomplishes a function at a primitive level of detail.

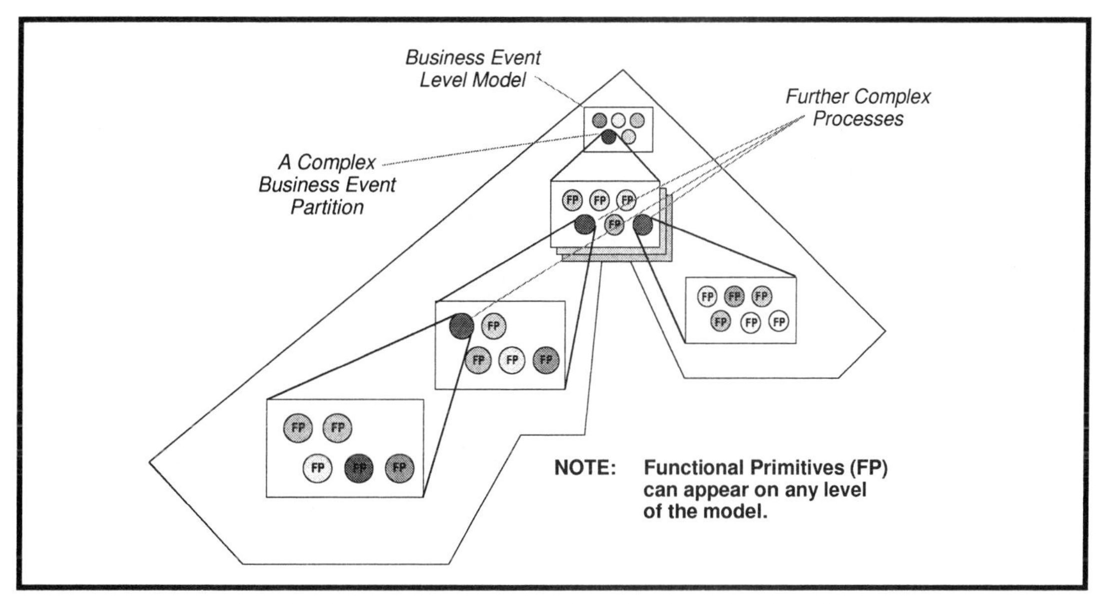

Fig. 10-1: Decomposition of Complex Bus. Ev. Partitions

Individual Process Specifications are used to declare the Business Policy (i.e., business rules as opposed to human or computer solution procedures) governing the transformation of Input Data Flow(s) to Output Data Flow(s) for each lowest level process (functional primitive) on an Information/Data Model such as a Data Flow Diagram.

These functional primitive process specifications should be written to enable *someone else* to carry out the Process' policy without our assistance; the specification should be a stand-alone maintainable product. When subpartitioning our Business Event Level Model we must be cautious not to be misled by the old partitioning. We don't want to depict any of the partitions shown in Figure 10-2.

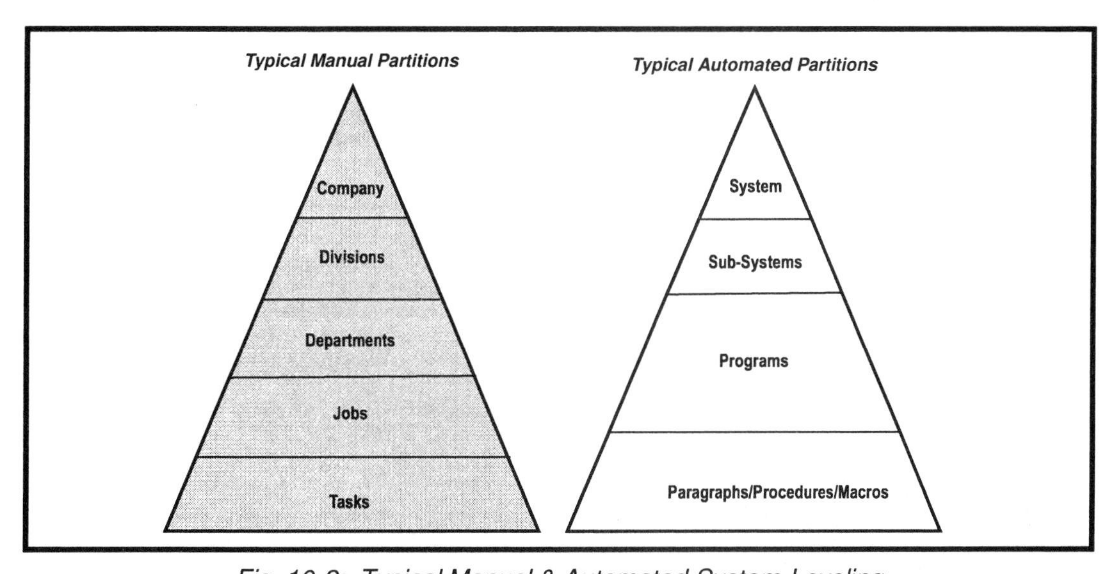

Fig. 10-2: Typical Manual & Automated System Leveling

A valid Business Event level logical process model would show one process stimulated by at least one data or control flow for one Business Event.

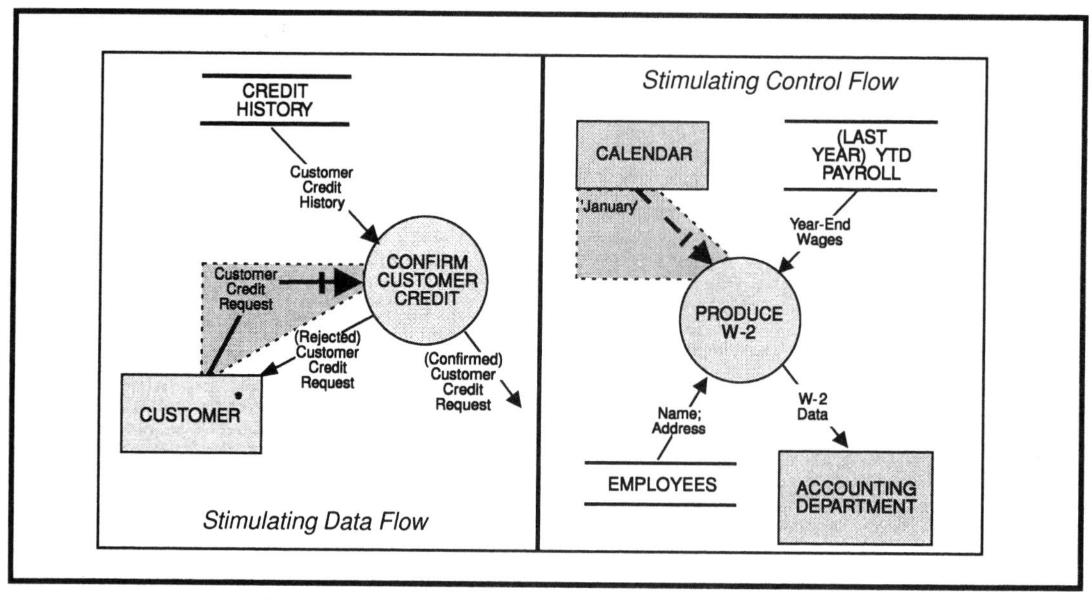

Fig. 10-3: Sample Stimulating Data & Control Flows

Remember, stimulating flows can be data- or control-oriented as indicated in the modeling chapter. Figure 10-3 shows the symbology used to indicate a Stimulating Data Flow and a Stimulating Control Flow.

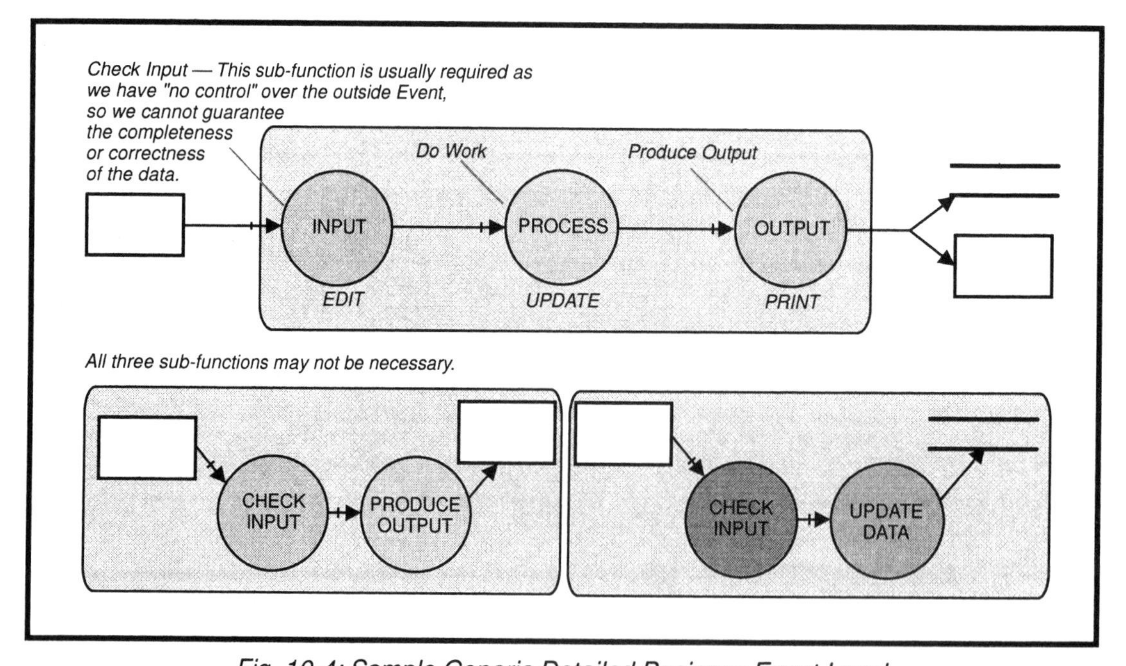

Fig. 10-4: Sample Generic Detailed Business Event Level

Even though "Edit-Update-Print" and "Input-Process-Output" are poor partitions for a system or application, they are helpful as partitions within an Event. Therefore, partitioning within an Event can take on this initial structure, using it as a basis if further partitioning of functions is required beyond the Event Context Level (see Figure 10-4). Further partitioning is helpful when the Business User/Policy Creator identifies significant sub-functions within an Event or where the reusability of detailed data or processes can be found.

When we decompose one process into sub-processes, we should continue to show the stimulating flow to each and every process.

When a Business Event Stimuli arrives from the outside world, it triggers **all** the processing and data required to respond to that Event.

Figure 10-5 shows the partition for Customer Wants to Order Our Materials in which the arrival of a stimulating **Customer Order** from the **Customer** triggers three consecutive processes.

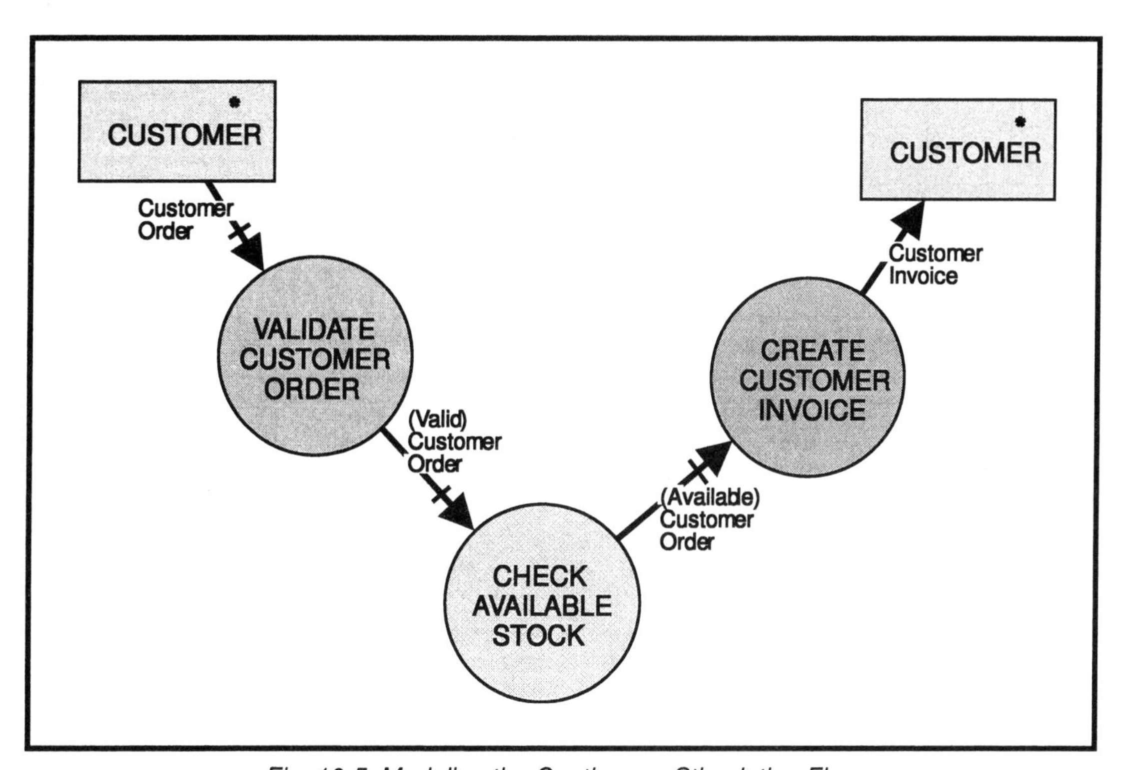

Fig. 10-5: Modeling the Continuous Stimulating Flow

For each Business Event, we identify one Business Event Stimulus from the outside world, but that doesn't mean we can't have more than one stimulus for the sub-processes for that Business Event.

In a business environment, it is obviously possible to have two concurrent sub-processes feed another. *(As the real world is not sequential, our modeling tool must not be limited to sequential views.)*[2]

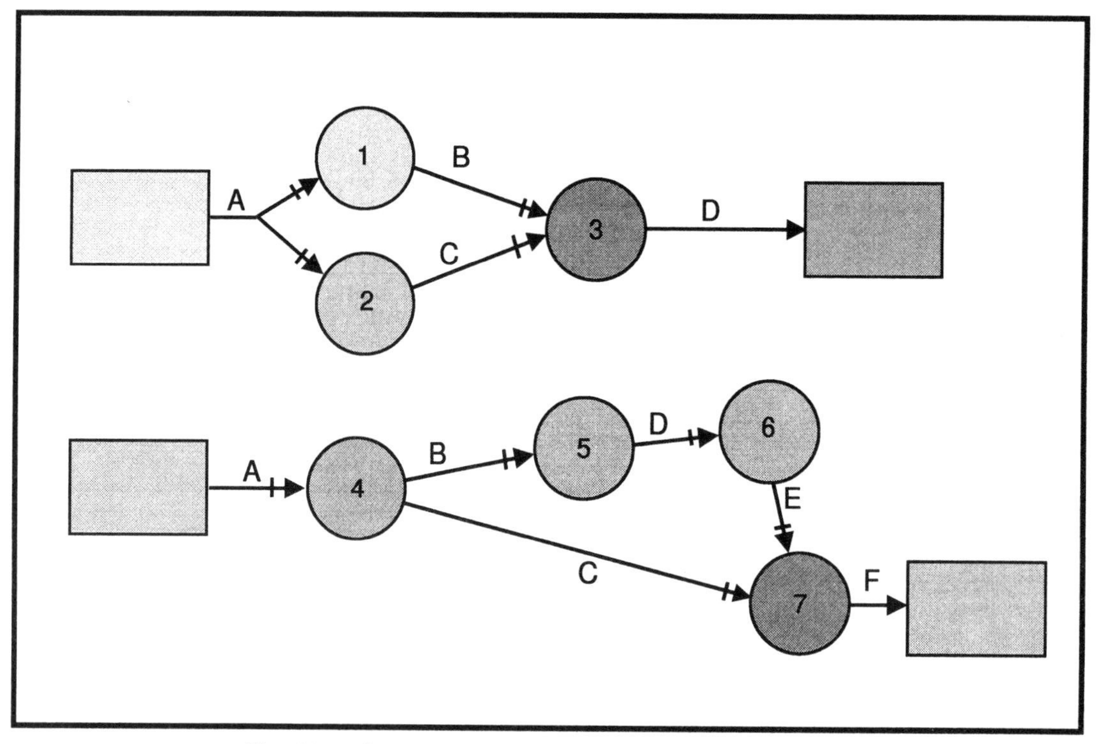

Fig. 10-6: Concurrent/Delayed Stimulating Data Flows

In the upper model of Figure 10-6 we see that process **3** cannot begin until stimulating Data Flows **B** and **C** arrive. Note that the business policy for process **3** will dictate whether **B** *and* **C** are both needed before it can begin its work, or whether only **B** *or* **C** is needed. In the lower example in Figure 10-6 we see that process **7** is stimulated by two Data Flows, **C** and **E**. If process **7** requires both Data Flows **C** and **E** to start up, then either Data Flow **C** or **E** will have to wait in their data-flow pipeline until the other one is available. Note, we do not make Data Flow **C** or **E** into a data store on our Business Model because they are transient within one Business Event.

The Reasons for Subpartitioning Processing

When we conduct process subpartitioning, we could say that the ultimate, lowest level partitioning is a single instruction (i.e., a verb-object statement). But this isn't a realistic business partitioning since all business functions don't partition down to one instruction. Obviously a better partition would be a process that does one complete, specific task.

2 Anyone who has used the Critical Path Method (CPM) or Program Evaluation and Review Technique (PERT) will recognize the need for a non-linear model for depicting processing.

We sub-partition, and hence stop partitioning, for reasons such as:

- Reusability (that is, a process can be re-used across Business Events). The reasons for partitioning for reusability tend to be internally driven within an organization.
- The Business Policy Creator sees the process as a single function. The reasons for the Business Policy Creator determining partitioning also may be internally driven, but should be based on how the customer is seen interacting with the organization's systems
- The process needs to be partitioned down into manageable units for human complexity reasons. This last rationale for partitioning is obviously a design limitation for system development because people have to build and maintain complex systems.

How we decompose processing within a Business Event Partition should be based on what are "natural," understandable reasons to our customers, as well as influenced by the reasons above.

For example, processing the Business Event Customer Wants to Purchase Our Materials resulting in the stimulus "Regular Order" can include three distinct activities:

Confirm Materials Available
Obtain Payment
Ship Materials

Or, six distinct tasks:

Validate Materials Requested
Confirm Materials Available
Validate Customer Identification
Calculate Invoice
Validate Payment
Ship Materials

Note that we may encounter a problem when subpartitioning based on the current Business Policy. Depending on who we speak to, we may also find different policy being stated by different detailed policy representatives within our organization.

For example, we may interview people in our organization responsible for the same Business Event: Customer Wants to Purchase Our Materials.

Figure 10-7 shows four different models based on our interviews with four policy representatives.

View 1 appears to be high-level, but the business person may not see the need to decompose it. It could be a perfectly valid detailed process from a business point of view and it can be immediately supported by a textual process specification. This view indicates that the **Customer Order** includes the **Payment**.

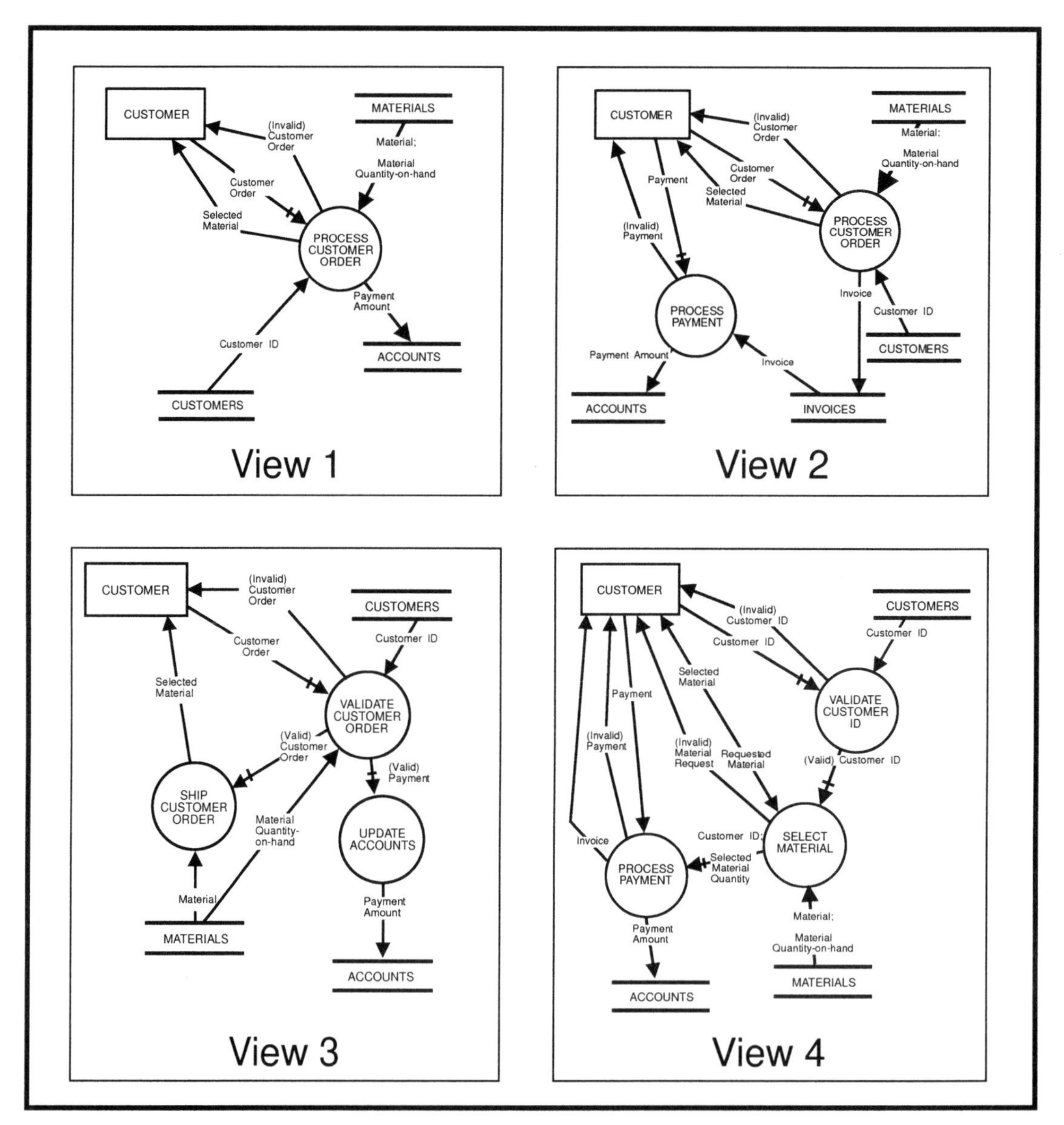

Fig. 10-7: Four Different Views of a Business Event

View 2 is from a business person who sees the same Business Event as two distinct Business Events: the first is the processing of the **Customer Order** information itself and the second is the processing of the **Payment** for that order.

View 3 comes from a business person who sees one Business Event decomposed into three distinct processes (the second and third processes triggered by transient data resulting from the first process).

View 4 comes from a business person who sees one Business Event decomposed into more detail from a data point of view where processes get only as much data as they need to do their job. Notice that in this view the second and third processes are not stimulated from the outside world, but by transient data even though new data are "pulled" in from the outside world (the **Customer**).

View 2 resulted in forming two Business Events from a single Event; this may not be a good thing to do if the customer wants to accomplish everything at once.

Views 1, 3, and 4 are valid for our Business Event. The difference between these views is what I call the distinction between *data cohesion* and *data conservation*. In views (1) and (3) all the **Customer Order** information comes in at once, so our Business Event Stimulus would be termed a *cohesive* set of data.

View 4 partitions the Business Event Stimulus (i.e., it's *fully data conserved*) and produces a potentially more reusable set of processes that are easier to develop and modify. Data conserving the Stimulus and Responses produces a better Business Model. In my experience, building fully data conserved models can get quite complex and time-consuming. In this case, an organizational goal for system reusability and maintainability should be in place before we invest in fully data-conserved models.

Subpartitioning Business Processes

A genuine business process (one that is allowed on the Business Model) must transform data in one of three ways:

- It creates new data (e.g., **Invoice Total** given the **Item Quantity** and **Price**).
- It changes the value of data (e.g., **Old Customer Balance** to **New Customer Balance** updated with **Deposit Amount**).
- It changes the status of data (e.g., **(Approved) Loan Amount** given **(Requested) Loan Amount** and **Credit Rating**).

Of course in a control-oriented process we can also produce additional control output (e.g., **'Room Temperature Too Hot'** results in **'Start Air Conditioner'**). We can use the above reasons for partitioning processes.

Figure 10-8 shows two *functional* Business Event Partitions:

- Customer Wants to Order Our Materials and
- Customer Wants to Pay for Previously Ordered Products

These two Business Events are separated by a dashed line. Each Business Event Partition consists only of business policy responses. The business policy responses consist of processing, any intermediate data, interactions with data stores, and issuance's of outbound data. In these Events we can see an example of subpartitioning. The Business Event, Customer Wants to Order Our Materials, results in the Stimulus, **Customer Order**, which stimulates processing that is shown sub-partitioned into four sub-processes. The first of these is to **Validate Customer Order**. This process performs a data status change from *Raw* **Customer Order** to **(Valid) Customer Order** or **(Rejected) Customer Order**. The **(Valid) Customer Order** stimulates another process **Check Material Availability** that changes the data status from **(Valid) Customer Order** to **(Available) Customer Order**. The **(Available) Customer Order** can then stimulate two processes — **Create Customer Invoice** which creates new data (this will probably be reusable in other Business Events) and **Ship Materials** which changes the status of **Materials** from *Stored* to *Shipped*.

Remember that this is a Business Model, so we are not interested in whether **Ship Materials** is manual or automated — the model just shows the processing and data that are triggered by a **Customer Order** coming into the business. (This process may have been separated out because the Business Policy Creator sees it as a significant separate task.)

Note that not all processes relate to stored data. Some may simply perform calculations to generate data that is transient to the system.

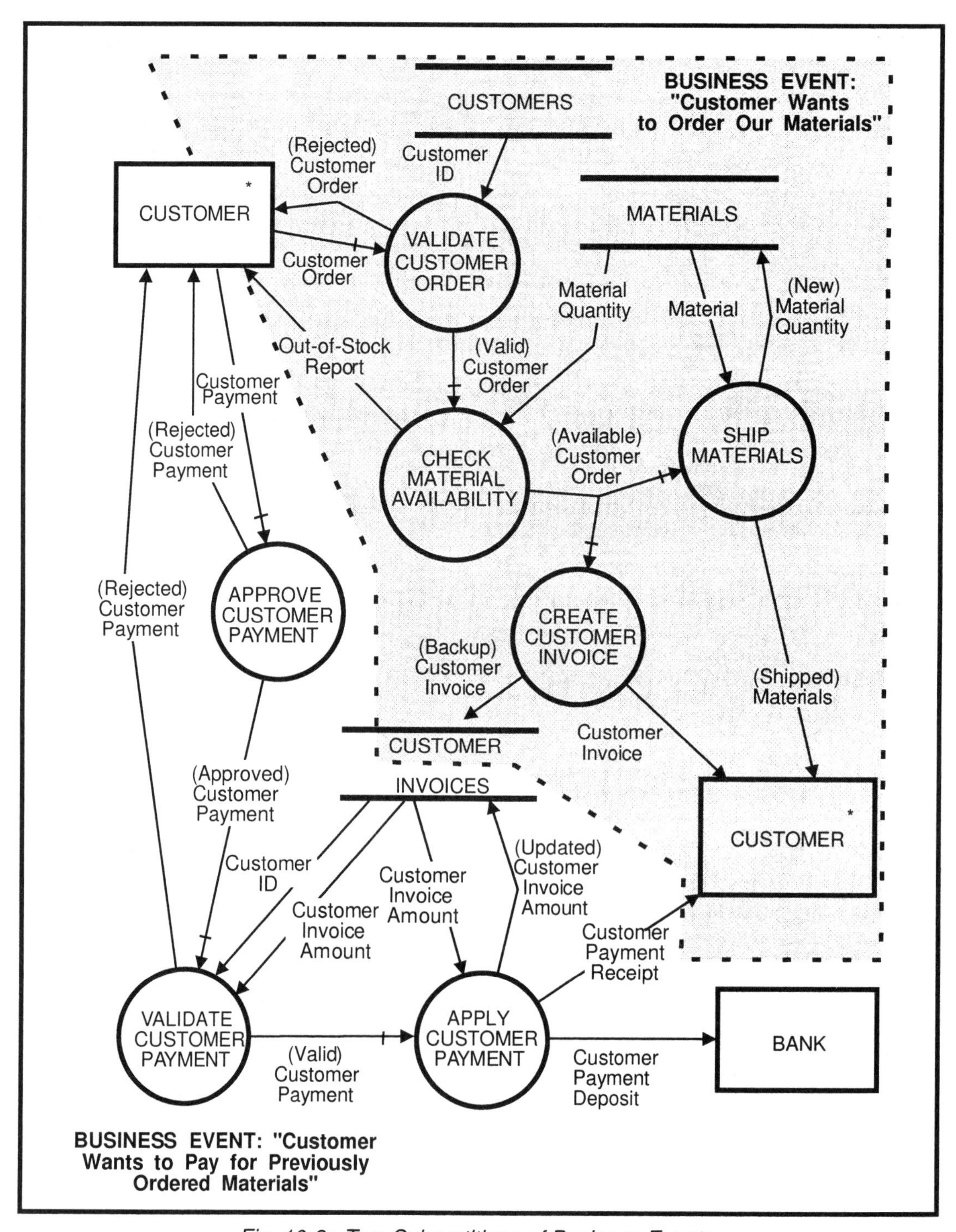

Fig. 10-8: Two Subpartitions of Business Events

The Business Event Partition stimulated by **Customer Payment** is also sub-partitioned into three lower-level processes.

In this case the payment is transformed in the first two processes from **Customer Payment** to **(Approved) Customer Payment**, and then to **(Valid) Customer Payment**. These are again *status changes* because they have not changed any data content, just the status of **Customer Payment** from *Raw* to *Approved* to *Valid*. However, the third process **Apply Customer Payment** does change the data content. The **Valid Customer Payment** and the existing **Customer Invoice Amount** are used to form an **Updated Customer Invoice Amount**, a **Bank Deposit**, and a **Customer Payment Receipt**.

We follow one Business Event Stimulus through any and all business processing until it terminates in a data store or an external interface. These are the only two terminations of data that can occur in a Business Model.

If a decomposition of any Business Event Process is still complex, we may want to further partition to another level. We would do this to help us manage complexity.

> For example, we could decompose a process called **Calculate Taxes** into **Calculate Federal Taxes**, **Calculate State Taxes**, and **Calculate Local Taxes** if we see the need for this further decomposition.

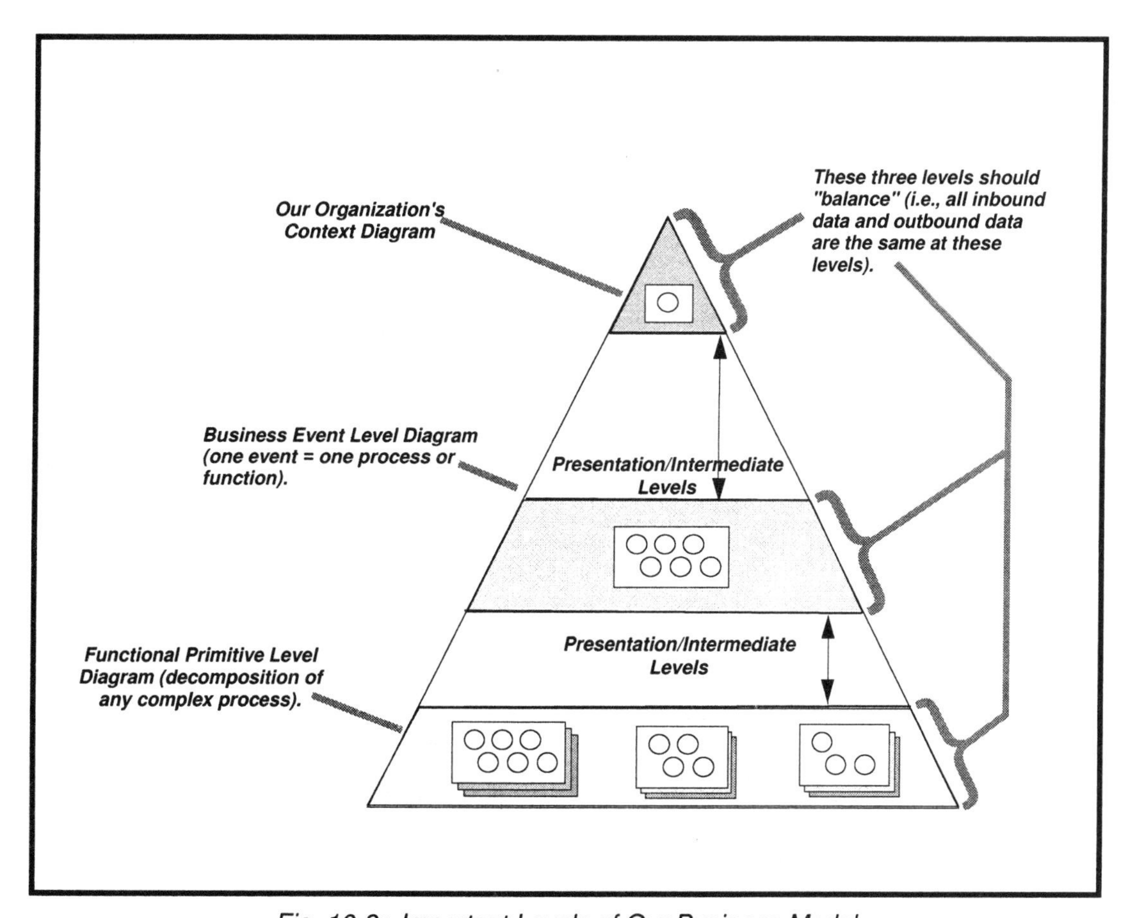

Fig. 10-9: Important Levels of Our Business Model

We don't want to get too concerned with any intermediate level models or spend too much time on their production. We do, however, want to be quite concerned with the correctness and completeness of the lowest-level model.

Figure 10-9 reflects the three important levels of our Business Model. They are the Business Context Diagram Level, the Business Event Context Level, and the Detailed (Functional Primitive) Level. Any other levels are only presentation levels to help overcome human complexity problems and do not need to be validated rigorously because they are never implemented.

Our Business Model will be supported with:

- Data Dictionary entries for all Data and Control Flows.
- Memory Specifications for stored data entities.
- Process Specifications for the functional primitive (lowest-level) processes.
- Optional process notes for any intermediate-level processes to help clarify these processes when we return to document them in detail.

The techniques of Business Process Analysis and Business Information Analysis needed to create these supporting specifications are covered in greater detail in Book III of the series.

Subpartitioning of Memory

One of the issues that must be taken care of by the Customer Focused Engineering analyst is that of extracting the data from old files and bringing them together into business entities. For this the old stores of data can form one of our inputs. A better source though would be the Business Event Memory stores as introduced in a previous chapter. The Business Event Memory stores only contain the *essential* stored data needed for the Business Event Partition (as opposed to using what happens to be in the existing stores).

Based on our rule that stores are only essential when data are shared between two or more Business Event Partitions the Business Event Level of our Business Model will be the first level at which genuine Business Event Memory (stores) need to be shown.

The Reasons for Subpartitioning Memory

When we conduct stored data partitioning we could say the ultimate partition is a single field (Data Element); in most cases this isn't a realistic business partitioning since we don't just keep one data field about something important in our business. A better partitioning would be a grouping together of only a strongly related *cohesive set* of stored data. Similar to process subpartitioning, we would like this subpartitioning of data to be because of:

- Reusability (i.e., the subpartitioned memory store can be re-used across many Business Events).
- The Business Policy Creator may see the set of stored data as being cohesive.

And also because:

- All the Data Elements within a cohesive unit are accessed by the same business key (i.e., **Product ID** to access **Product Price, Product Quantity on Hand, Product Weight, Product Quantity Discount**, etc.).

Repartitioning Business Event Memory

Since "processing is typically local and data is typically global" we may see little or no duplication of logic or transient data across Business Events. This is not true for stored memory. Each Data Element of memory will be used in the processing of at least two Business Events, otherwise we wouldn't be storing the data.

Figure 10-10 shows the seamless views that would result from a truly functional partitioning. From a customer's point of view their Business Event Partition appears to be seamless. The dashed lines in the figure separate the Business Events. Note that stores only occur at the intersection of the Business Events. While this is true, also note that to avoid complexity, I have not shown the potential that any Business Event can use data from any other Business Event Memory.

We will have one Business Event Memory store for each Business Event Partition, and each Business Event Memory store can contain a diverse set of data. During the development of our Business Model we must create and maintain an adequate Data Dictionary that defines each Data Element with a meaningful naming convention.

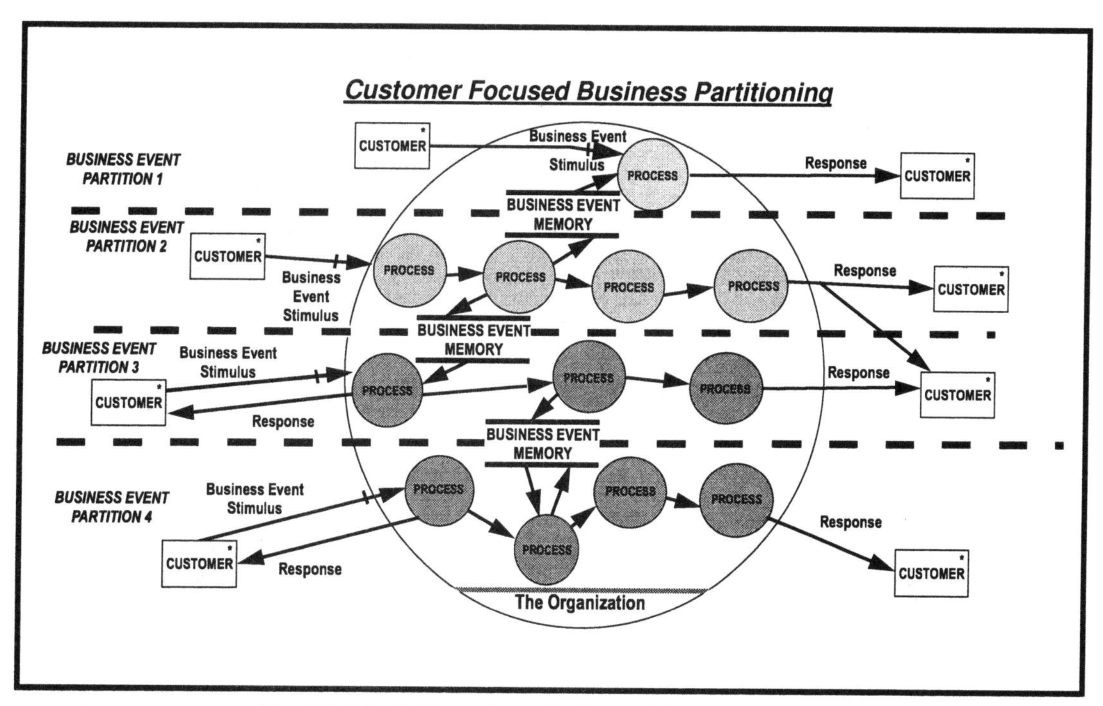

Fig. 10-10: A Seamless Business Event Partitioning

Any Data Element in one Business Event Memory store must be duplicated in at least one other Business Event Memory store. If not, it indicates what we call a non-conserved data store (i.e., either a "black hole" where data goes in, but never comes out, or "magic"store where data comes out, but never went in). The recommended naming standard will help us bring all of the data names (Data Elements), with the same qualifier, that are duplicated across Business Event Memory stores, into a collection of reusable, non-redundant, cohesive data sets (see Figure 10-11).

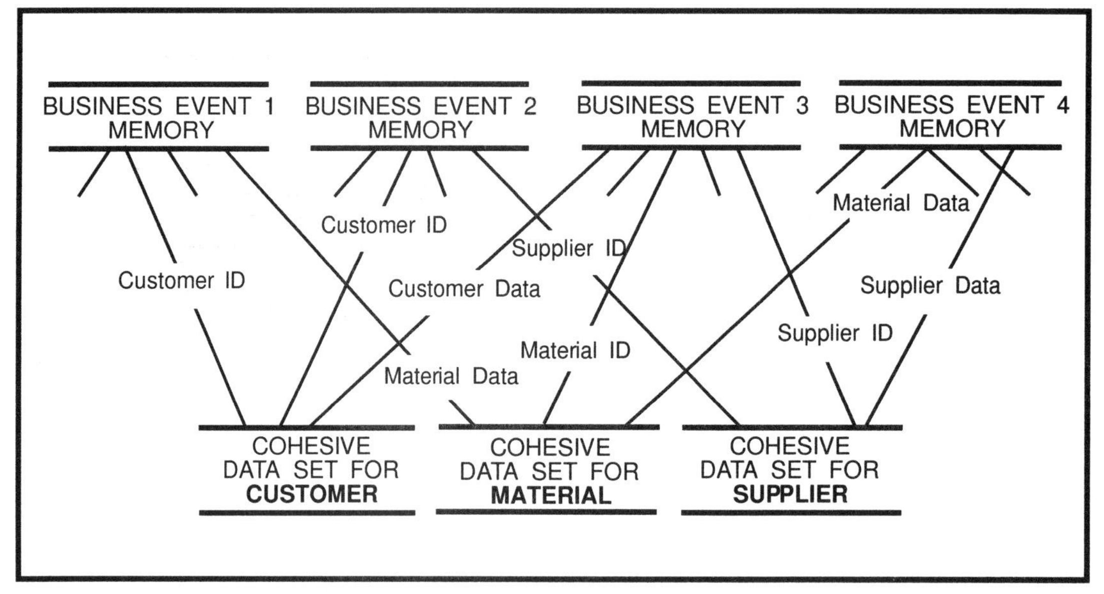

Fig. 10-11: Collecting Cohesive Data Sets from BE Memory

We must not lose the relationships between these sets of stored data; in other words, we must not lose how the processing relates one set of data to another (examples are: **Customer** *Orders Many* **Materials**, **Supplier** *Supplies Many* **Materials**, and a **Product** is *Supplied by Only One* **Supplier**).

From a data analysis point of view we have some techniques from the D.P. world that allow us to arrive at a reusable, cohesive set of stored data. One of these techniques is Data Normalization, which involves producing sets of data where every Data Element in the set can only be dependent upon a specific identifying key for that set of data. This key can be a "compound key." For example, we may need Product ID and Warehouse ID to determine the **Product Quantity on Hand** because each warehouse may have a different product quantity for the same product depending on the size of the warehouse.

This is where we can use an Information/Data Model such as an Entity Relationship Diagram to model cohesive sets of stored business data (Entities) and their business associations (Relationships) in our Business Event Partition definition. We don't need any modifications to the symbology already recommended for ERDs. We will, however, make a modification to how we partition the ERD. Whereas Entity Relationship Diagrams have been proposed as a means of representing a complete enterprise (organization), we can use the Business Event to provide an additional business-oriented partitioning for this model so we can produce a partitioned Information/Data Model such as an Entity Relationship Diagram for each Business Event Partition.

If we take the Business Event Partition, Customer Wants to Order Our Materials, shown on the top part of the Process Model in Figure 10-12, we see that by using a meaningful data flow naming convention it uses stored data about **Customers**, **Materials**, and **Customer Invoices**.

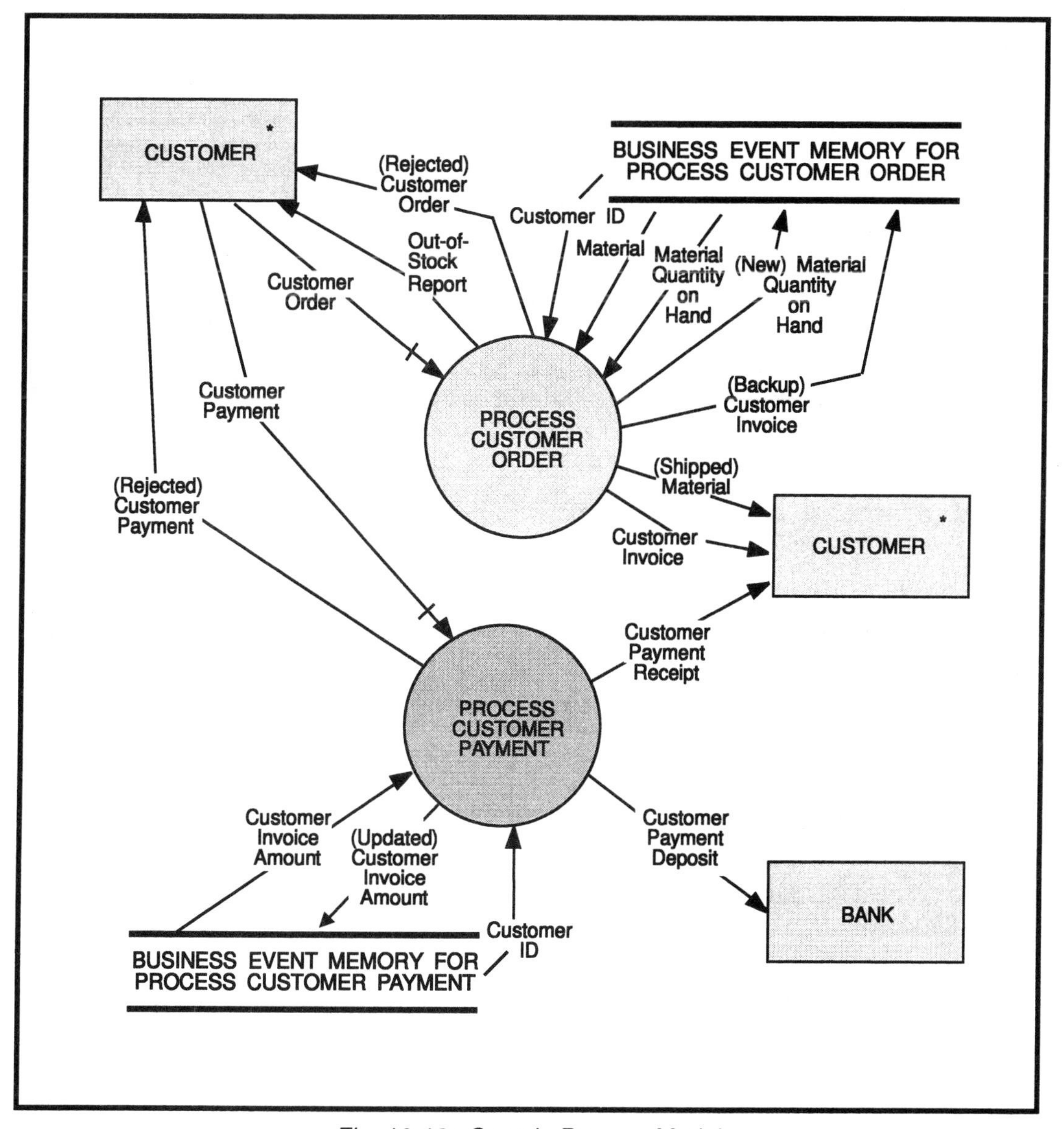

Fig. 10-12: Sample Process Model

If we look at the logic inside of **Process Customer Order**, we may discover that *Many* **Materials** are sold to *One* **Customer** via *One* **Customer Invoice**. So, we will produce a partitioned Information/Data Model for this Business Event Partition (as shown in Figure 10-13, using a Chen-style Entity Relationship Diagram).

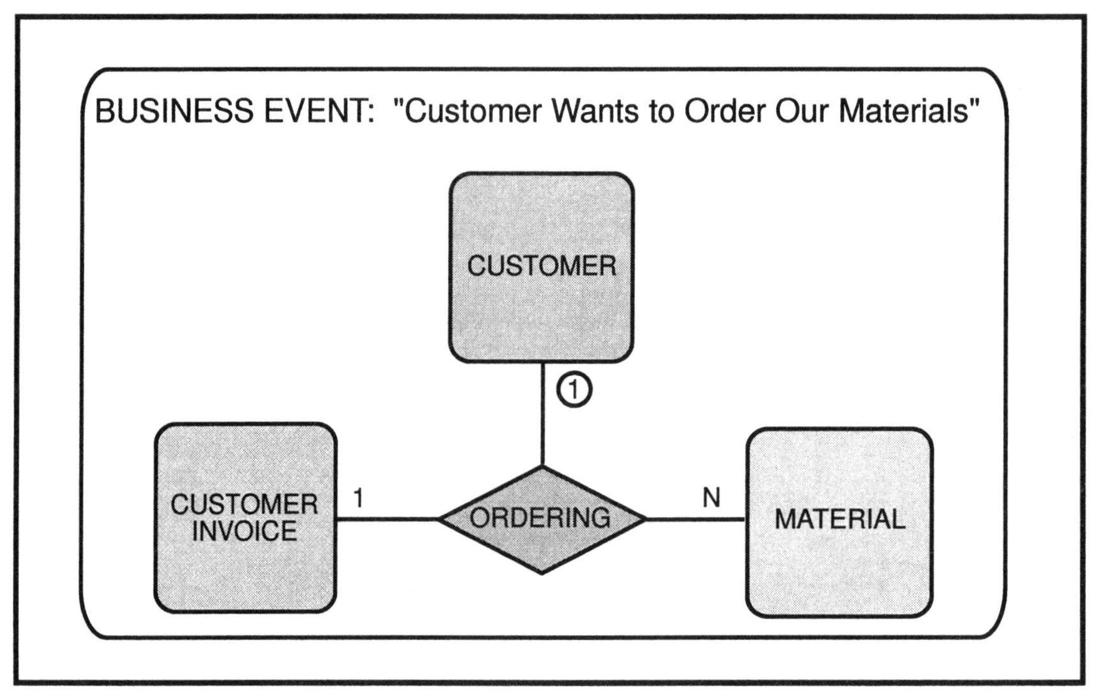

Fig. 10-13: Sample Partitioned Information Model

Applying the same discovery process on the Customer Wants to Pay for Previously Ordered Materials Business Event results in the model shown in Figure 10-14. This Information/Data Model states that *One* **Customer** is involved in a *Paying* Relationship with *Many* **Invoices**.

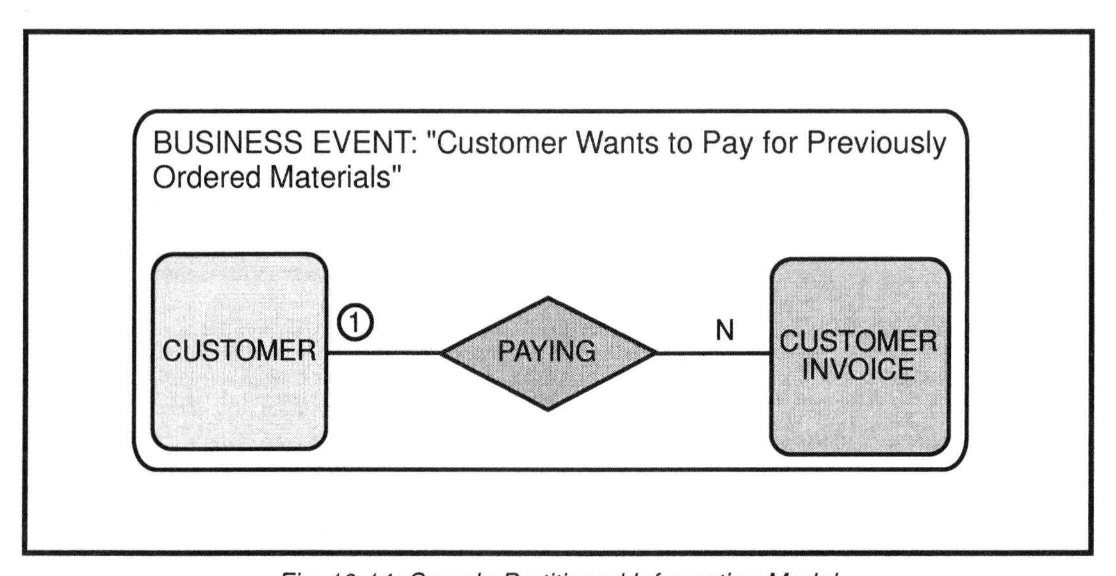

Fig. 10-14: Sample Partitioned Information Model

For the Information/Data Model to show this 1-to-N relationship, the Logical Process Specifications for this Business Event will have indicated *One* **Customer** can pay off *Many* **Invoices**. Notice the **Customer** and **Customer Invoice** Entities are repeated between the two Business Event Partitioned Information/Data Models. Again, this is only a modeling redundancy and the **Customer** and **Customer Invoice** Entities are the same on each model.

If we wish to bring all of the resultant sets of Business Event Partitioned Information/Data Models together, we can form the Organization's Information/Data Model where these Entities would be shown only once (see Figure 10-15).

Once we have our Cohesive Entities partitioned on our Business Event Partitioned Information/Data Model, we may want to return to our Process Model and replace our Business Event Memory store with these Cohesive Entities.

We need to support the Information/Data Model with Data Dictionary entries for:

- Entity Specifications — one for each Entity on our Information/Data Model
- Relationship Specification (optional) — one for each Relationship on our Information/Data Model
- Data Element Specifications — one for each Data Element in each Entity on our Information/Data Model

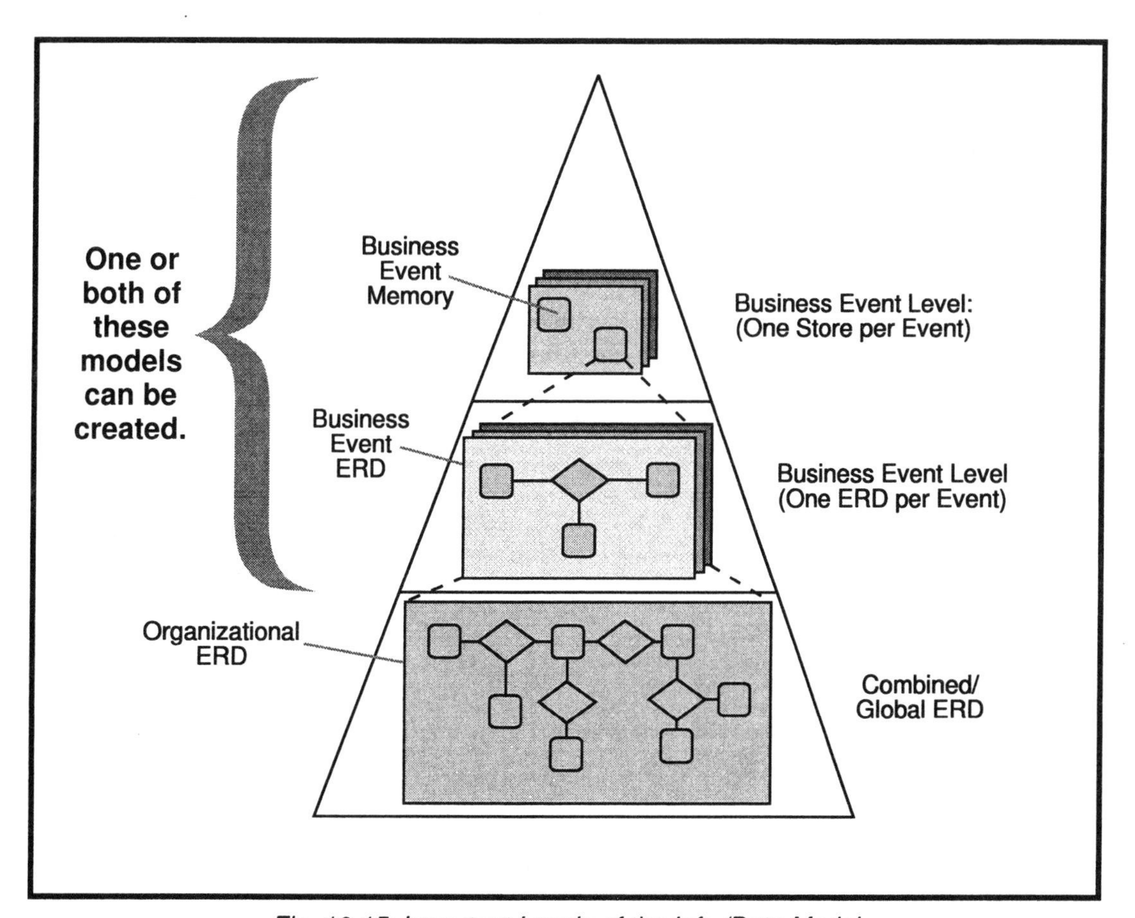

Fig. 10-15: Important Levels of the Info./Data Model

The technique of Business Information Analysis is covered in greater detail in Book III in this series.

The Progression of Analysis Models and Their Levels

In this chapter we have talked about producing a detailed business specification for a Business Event (in addition to specifications for Dependent and Regulatory Events). However, during analysis we may want to produce different types of models to ensure that we understand the business and to verify with our business users that there are no omissions of Business Policy.Figure 10-16 is based on ideas in Tom DeMarco's book [3] and shows the analysis portion of the development life cycle. As Customer Focused Engineers we can use this portion of the life cycle to assist in developing our Business Model.

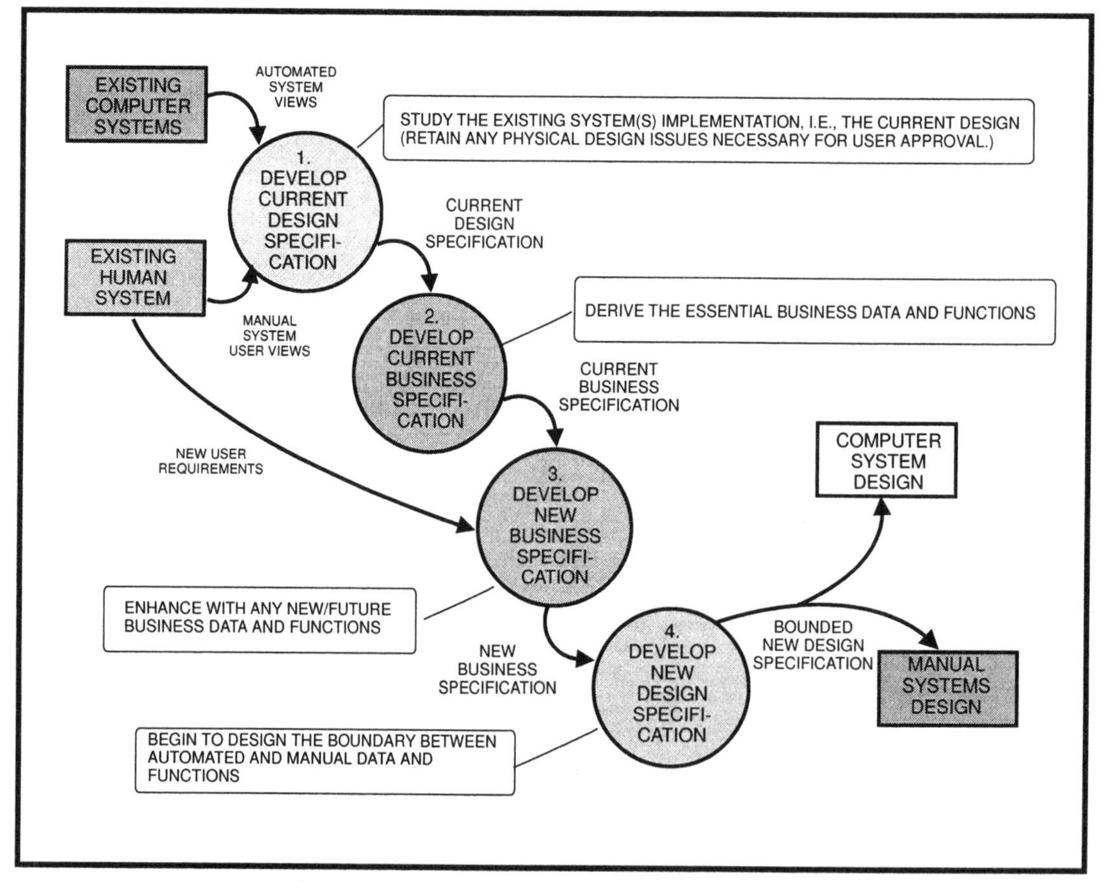

Fig. 10-16: The Progressive Stages of Analysis

3 *Structured Analysis and System Specification* by Tom DeMarco — See Bibliography

1. This figure indicates in Step 1 that we can produce a specification of the existing design (if such a specification doesn't exist) for the purpose of validation and to ensure that the context of study is comprehensive for our project.
2. In Step 2 we can then use this model to derive the true business specification based on Business Model Events. This would be where we would extract any existing design characteristics, remove System and Strategic Events, and partition by Business Events.
3. The resulting specification which shows all existing functionality of the business (after approval), is then used in Step 3 to add or remove any Business Events or Regulatory Events and/or Dependent Events.
4. The new business specification will be used to go on to develop the new design (shown in Step 4). This is where we would add design or implementation characteristics such as an "automation boundary" and any new System Events to accommodate the new design.

As Customer Focused Engineering analysts we're most interested in Steps 2 and 3, **Develop Current Business Specification** and **Develop New Business Specification**.

Figure 10-17 builds on the progression shown in Figure 10-16 adding layers of detail. This shows that as a Customer Focused Engineer, we may want to take each Business Event in our organization and produce a set of separate models of varying detail.

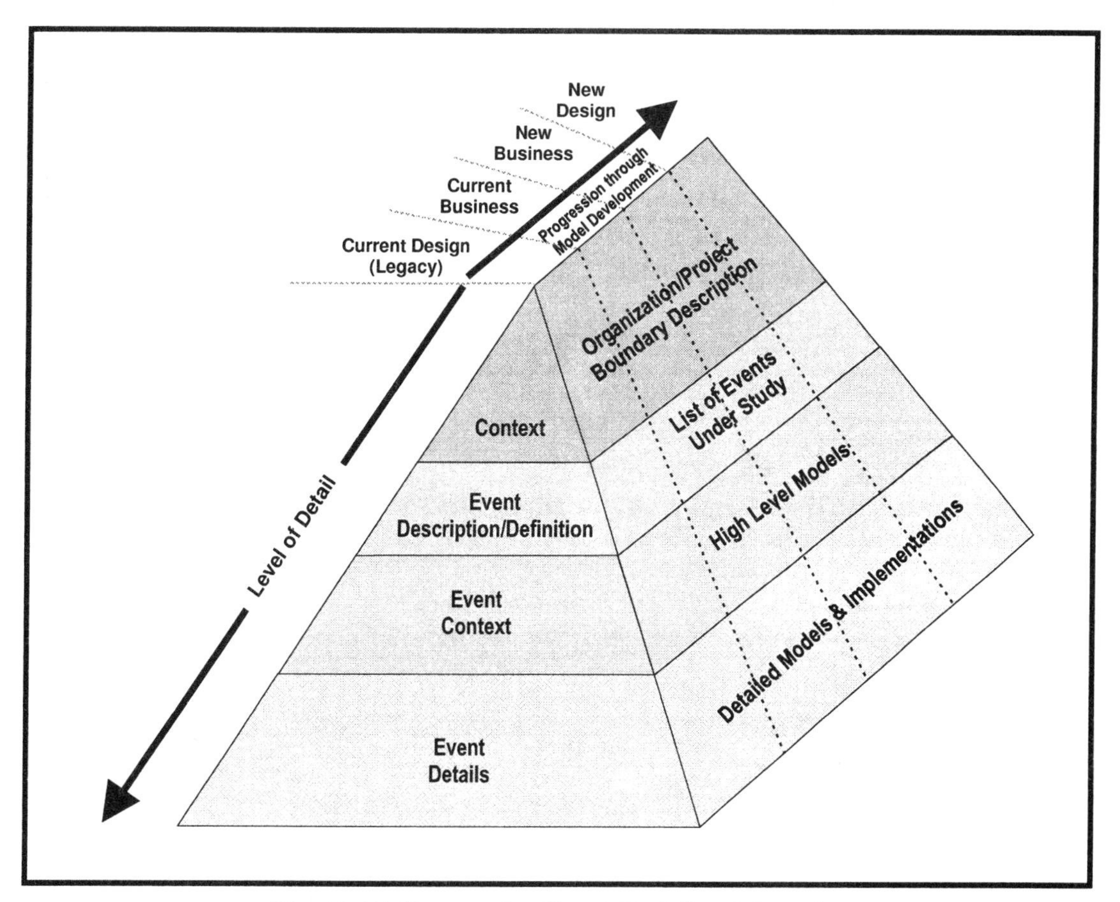

Fig. 10-17: Progressive Event Modeling & Layering

The number of different models we produce and their degree of detail needs to be evaluated on a project-by-project basis depending on the environment under study. This is the realm of Business Process Analysis which is covered in greater detail in Book III of this series.

The Business Library — Documenting the Analysis Business Event Specification

Now we can bring together the data and process models for each Business Event.[4] I recommend that we produce a set of Business Models for each Business Event. Together, these sets constitute what I call the Business Library. Figure 10-18 shows a sample of an entry in our Business Library. It shows the analysis documentation for our previous Business Event, Customer Wants to Order Our Materials." In this figure I have specified our Business Event using two analysis models, a Process Model and an Information/Data Model (although other analysis models are equally valid). These Business Models are also supported with specifications in the Business Library Entry.

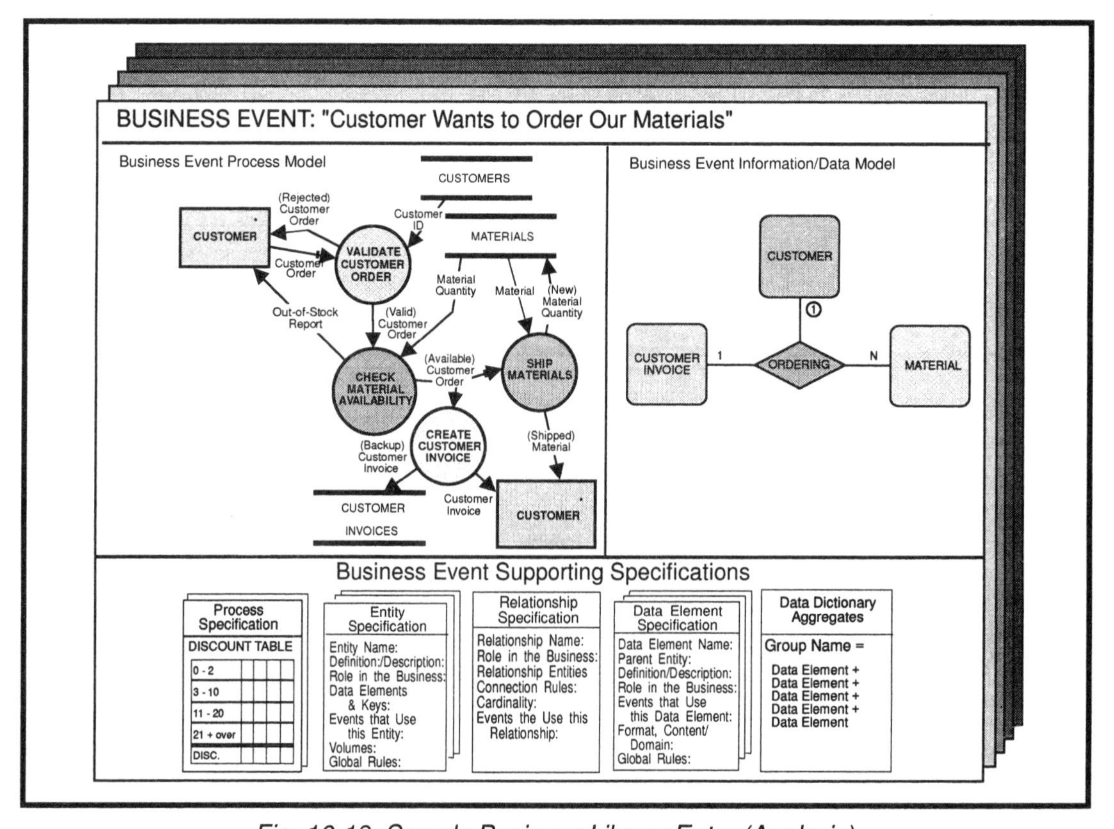

Fig. 10-18: Sample Business Library Entry (Analysis)

4 I believe these should never have been split apart in the first place. This is where an Object Oriented Model is superior because Process and Data are not separated. If you feel your business folks can relate to an Object Oriented Model (with some minimum training), then this can be used here.

Managing the Library via the Business Library Conservator

If our business responds to a hundred different Business Event Stimuli, then we will have one hundred of these Business Event Specifications. I propose each organization have one or more "Business Library Conservators" to maintain the consistency and the integrity of the Organization's Business Library.

> For example, if one project team is developing a model for a Business Event that requires Retrieval of **Material** data, the Business Library Conservator ensures that some other Event has already Created the **Material** data and ensures that a project doesn't reinvent the wheel and write duplicate Data Definitions or Process Specifications (and hence, procedures or software) that already exist and can be re-used.

This library could be part of what's known as the Organizational Repository (or encyclopedia) where we keep all data about our organization. This repository is perhaps the most important asset of an organization and should be treated as such. No haphazard modifications should be allowed to this repository.

The Business Library Conservator (also known as "Conan the Librarian") protects the Business Library's contents just as we would protect our financial assets. This Business Library should also be the responsibility of the Chief Information Officer (CIO), a title that is becoming popular at many organizations.

Summary

Now we can see the detailed documentation needed for the analysis stage of Customer Focused Engineering. The most important thing here is Business Events can be treated as individual portions of the business (i.e., each entry in the Business Library can be treated as a mini-project for Strategic Planning and implementation purposes.

The data stores across Business Events are the only items with which we'll have to be concerned if we decide to implement Business Events separately. Then we may have to create intermediate steps in our Implementation Plan to make sure that separate temporary stores are not "out of sync."

To return to my house-building analogy, when a client's requirements completely match the specifications of another house (verified by looking at such models as blueprints), then the client can buy the already existing models, or even buy the pre-built house from the factory.[5]

---

5 I wouldn't dream of buying a new house without seeing its engineered blue-print, and similarly, I would recommend not buying a new system package to run an aspect of my business without seeing its engineered specifications.

By applying engineering principles to business systems, there is every reason to exercise the same option, that is, to buy existing models and systems from an organization's Business Library. This assumes that our specifications are engineered and worth buying. [6]

---

6 In the late 1980s I met with some CEOs of software/consulting companies and I recommended that they undercut their competition's rates by half for their consultants at a client's site. This was providing they could keep the Business Models from the assignment. I foresee a library of skeleton Business Models that any organization can use to modify for their own Business Model. You already know I'm not talking about accounting models, word processors, or databases, etc., but true Business Models (for example, the model of a training company or an airline company).

Chapter

11

Achieving Organizational Process and Data Integrity

Science is organized knowledge.

Herbert Spencer
English philosopher, 1820 - 1903

One of the biggest problems that an organization faces is to organize its business knowledge and to maintain the integrity in and across that knowledge.

Some of the most costly corporate errors I've seen have been due to maintenance "fixes" that caused some other unforeseen change in another part of the organization's systems. Many of these multi-million dollar errors are not heard of outside the organizations. The ones that make it into the news media are the most devastating to an organization.

Without an engineering discipline applied to its systems development an organization should expect these errors to occur. Even with an engineering discipline applied to individual systems development, we still need to maintain engineering concepts across all the organization's systems. Using the Business Event Methodology we can obtain this extra important level of organizational integrity.

We achieve this integrity by maintaining a set of Business Model Event Matrices. I propose we maintain up to five of the following matrices dependent upon the modeling tools we use:

- Business Model Event/Reusable Process Matrix
- Business Model Event/Data Element Matrix
- Business Model Event/Object and Methods Matrix
- Business Model Event/Data Entity Matrix
- Business Model Event/Relationship Matrix
- Business Model Event/Engineered System Matrix

With these matrices we can realize the significant benefits of Customer Focused Engineering across the whole organization. By using the Business Event Methodology, we gain a better implementation of a Business Event plus:

- Process Integrity (Reusability)
- Data Integrity (Conservation)
- Substantial Savings of System Development Time/Investment
- Ease of Maintenance

As I mentioned in chapter on ***The Detailed Business Event Specification***, I believe we need to have one or more Business Library Conservators to manage our Business Library. The task of the Business Library Conservator is to take care of the first two items in the list above. Each system's Development Manager would take care of the third and fourth items. The Business Library Conservator maintains and conserves our organization's data and processing:

- Across Business Model Events in our organization, whether implemented as human-based or computer-based systems.
- Within all computer databases and manual files.
- During the development of new systems.

Using these Business Model Event Matrices will yield positive results in terms of managing business or system changes, re-using data and processing, and allowing our organization to easily identify and remove dead data and processing.

Before going further, it's important to realize that the main role of the Business Library Conservator is to be concerned with data about our organization's data. In other words, they maintain the information about the various Data Elements themselves, as opposed to being responsible for the contents of those Data Elements. So the statements in computer systems and manual procedures should protect the actual data contents of Data Elements (fields). The Business Library Conservator protects the Data Elements and Business Logic. The Business Library Conservator may also wear other hats in a small organization such as Data Administrator, but it's important to keep the conservator role distinct and separate.

Determining the Potential for Reusability

> The one example that we're probably most familiar with where we see reusability is in the automotive world. When a new automobile is introduced, it's never completely new from the point of view of its design. In this industry we can see large scale and small scale reusability. We see engines and maybe even chassis re-used across models, but the most reusability we see is when we look at the smallest components. It's very unlikely that the designers of a new vehicle invented their own nuts, bolts, washers, etc. and had them specially machined.

Notice that the reason for ceasing to subdivide components for reusability would be the functionality of the component.

Take for example, the case of an automobile battery. Even though it does have components, there's no point in breaking it down into smaller components, at least from a customer's point of view. Another example of this component functionality is in such items as spark plugs and wiper blades. The customer is interested in being able to replace a spark plug rather than an entire engine, or the wiper blade, as opposed to the complete wiper assembly. These items are typically reusable even across manufacturers.

The acid test of whether someone will use a component is how much extra work is necessary to re-use a unit of processing or how much extra baggage is associated with re-using a unit of data. The 10,000 line computer program or the three inch thick procedure manual is reusable in a sense, but usually with massive wastage. We are looking for process and data units that give a one-for-one match between the functionality of content needed, and the functionality of content provided.

The cornerstone of what makes any component reusable is the fact that it is a quality engineered product.

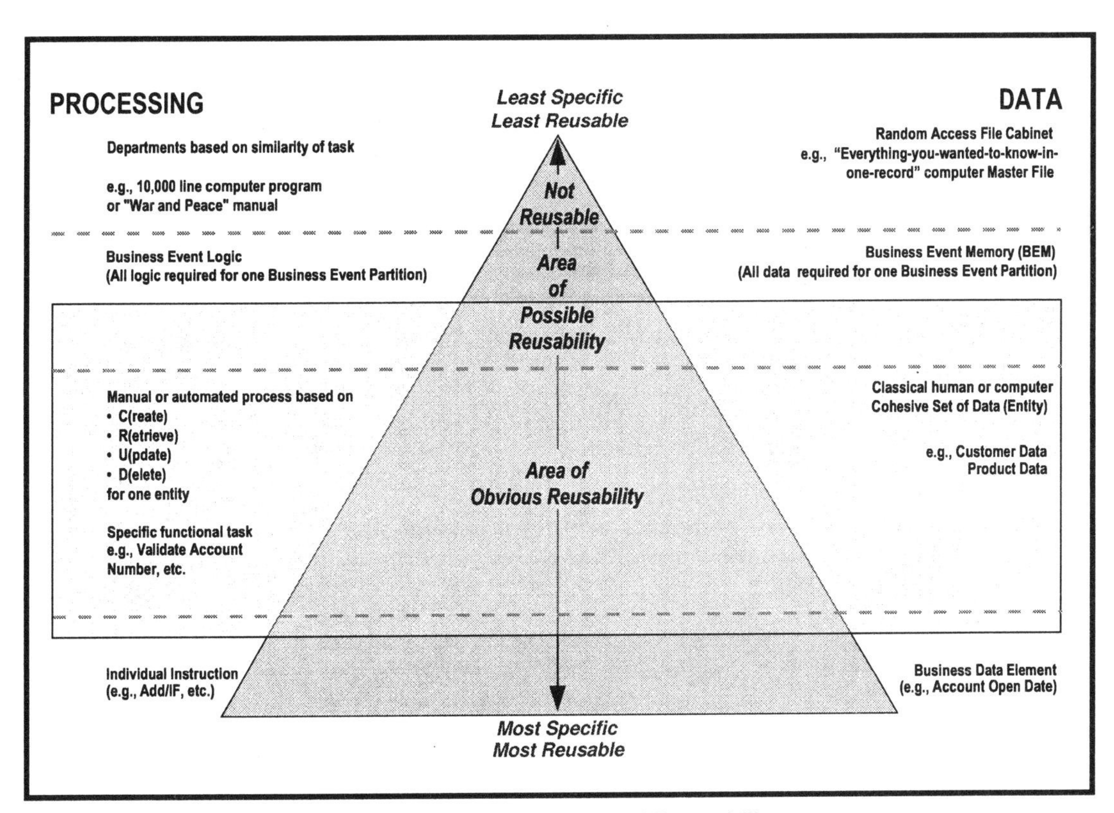

Fig. 11-1: The Pyramid of Reusability

Figure 11–1 shows what I see as a *Pyramid of Reusability*. The left side of Figure 11–1 refers to processing and the right to data description. At the bottom of the pyramid we see a single instruction and a single Data Element. These are reusable, but they are not very useful for our purposes. Just above this, we see the primitive process and the data aggregate level. This level is very reusable. Above that, we see the specific customer data (e.g., **Customer Account**) and the Create, Retrieve, Update, and Delete (C.R.U.D.) logic that is specific to the Entity. This level, too, is reusable, though an Entity may contain too much data for a single procedure or module. The lowest levels of reusability are at the top — the traditional 10,000 line program and "kitchen sink" master file.

This diagram also shows where the Business Event Level fits. Notice that it's possible to re-use a Business Event Partition's Specification as a whole.

> For example, the Business Event Partition for Customer Pays Our Invoice could also be subsumed within the Business Event Partition for Customer Wants to Purchase Materials and Pay for Them Now.
>
> Another example may be that our organization decides to introduce A Special Order Business Event that is similar to a Regular Order except for needing a longer lead-time, an extra stocking charge, or a smaller discount. In this case we may be able to completely re-use Regular Order processing within Special Order processing. Of course, we would still document the new Business Event in the Business Library with its additional components for Special Order processing or data.

Reusability Incentives

I urge organizations to consider introducing some or all of the following measures to encourage reusability:

- *Introduce Reusability Standards* — To promote reusability and discourage the "Not-Invented-Here" syndrome, organizational standards should be set for: procedure writing, consistent human interfaces to computer users and between departments, interfaces for data and control flow, internal module structuring aimed at easy modification and the separation of business issues and technology issues, technical environments aimed at portability, and the level and quality of all documentation.
- *Introduce Centralized Naming* — A central function should assign names for business terms, files (Entities), processes (functional modules), and technology interfaces.
- *Set up a "Reusable Library"*— A well-indexed and cross-referenced on-line automated repository of off-the-shelf components and procedures will do much to alleviate the complaints that finding a reusable procedure or software component is more trouble than starting from scratch.
- *Launch an Incentive Program* — Introduce some form of recognition for the producers of reusable software, such as posting reusability figures, such as the number of reusable modules created by project and even by employee. Offer an incentive for people to use reusable components, for example, a percentage of time saved on a project can be rewarded as time off.

Obtaining Process Integrity (Reusability)

Process Integrity is a feature that applies to every individual process indicating that all data input into it is utilized and that all data output from it is derived only from its input. Process Integrity is an important concept when we want to re-use processing across Business Model Events and hence across systems. Each process that is re-used must not be corrupted by haphazard changes and must be used in a consistent manner.

If a proposed modification to a process will affect its reusability across Business Model Events or applications, the Business Policy Creator must make the decision to either ensure that the proposed change is acceptable in every Business Model Event that uses the process, create another reusable process and leave the original process unchanged, or sacrifice the advantage of having the process re-used and create other non-reusable processes.

Re-using processing and data is one way we can increase productivity during system development as well as during global maintenance/modification changes. We need an organizational objective and incentives in place for reusability to gain the full benefits of reusability.

As I said in a previous chapter, we should emphasize the reusability of both data and processing during the analysis and design modeling stages of Customer Focused Engineering. We can spot reusability during the analysis and design stages by having a consistent, organization-wide naming standard, a Data Dictionary, and by requiring every project to interface with the Business Library Conservator.

Again, reusability is one of the major reasons for performing sub-partitioning of Business Event Partitions in the first place. Without it, we re-invent the wheel and may end up with monstrous programs and procedures.

The Business Model Event/Reusable Process Matrix

In our role as Business Library Conservator, we can use a Business Model Event/Reusable Process Matrix to obtain Process Integrity across our entire organization by keeping track of reusable processes as they are created, modified, or deleted.

The Business Model Event/Reusable Process Matrix would list on one axis all the Business Events, Regulatory Events, and Dependent Events to which our organization must respond. So the size of this axis obviously depends on the scope of the business we're modeling. The other axis would list every reusable process available in the organization.

> For example, an organization that specializes in shipping packages would need to respond to fewer Business Model Events than a conglomerate in the airline, pharmaceutical, and restaurant markets.

This matrix would include all potentially reusable processes, even if they are currently used only once. It shows all of the places where a reusable process is used. If the last Business Model Event where a reusable process is used gets canceled, the process itself can be eliminated or flagged as inactive.

The Business Library Conservator ensures that reusable processes are used and not corrupted, catalogs new processes, and enforces a naming standard for all process and their implementations. This is done via the Business Model Event/Reusable Process Matrix as shown in Figure 11–2.

Just as we have a dependency of data for something such as **Customer Address** (street, city, state, and zip), we can also have a dependency of reusable processing. The Business Model Event/Reusable Process Matrix diagram can be used to show this relationship between the subordinate and superordinate processes.

For example, the reusable process, **Calculate Invoice**, may rely on two reusable subordinate processes: **Calculate Commission** and **Calculate Sales Tax**. Changes to any of these three processes will affect how our organization creates its invoices.

This integrity issue can be taken care of using the third dimension of the Business Model Event/Reusable Process Matrix. That is, we may have notes or pointers to dependent processes. These pointers could be bi-directional between subordinate and super-ordinate processes.

Business, Dependent, & Regulatory Event Names	Reusable Process Name	Calculate Order Qty.	Create Customer Invoice	. . .	Create New Customer	Delete Customer	Calc.Federal Taxes	Validate Payment Type
B.E.1 -- Customer ceases business.					X			
B.E. 2 -- Customer pays our invoice.		X					X	
B.E. 3 -- ...	X		X					
B.E. 4 -- New customer wants to sign up with us.				X				
B.E. 5 -- ...			X			X		
D.E. 1 -- Supplier sends back-ordered goods.	X		X					
D.E. 2 -- ...						X		
R.E. 1 -- Time to make quarterly tax payment.			X			X		X

Reusable Process Z

Pointer to implemented process specification

Pointer to any dependent process

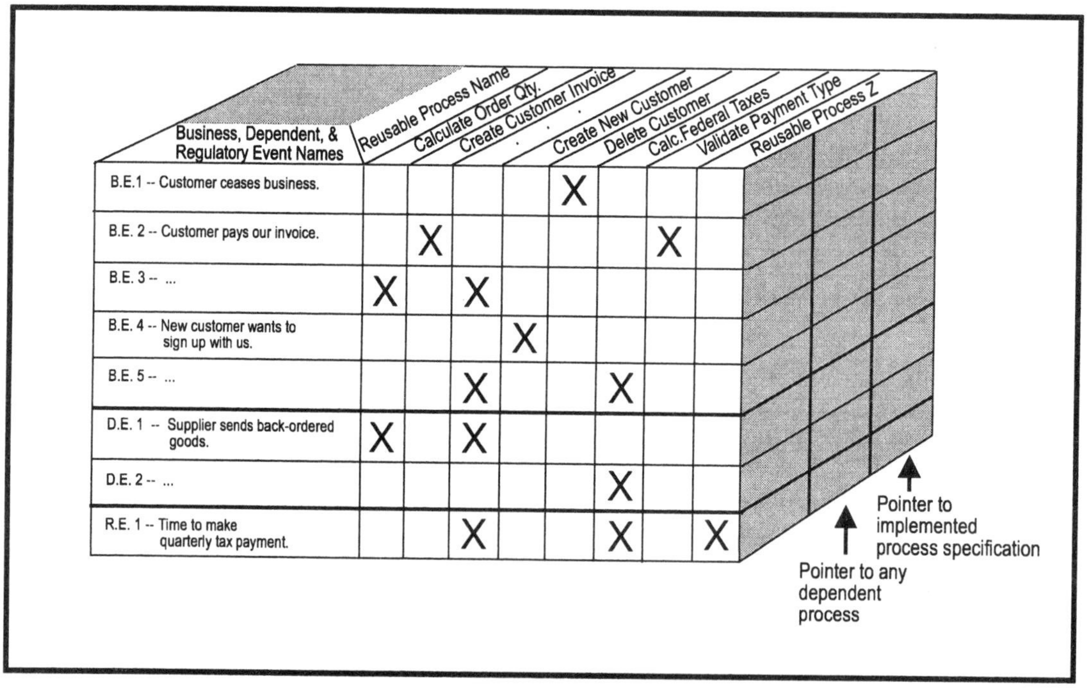

Fig. 11-2: Business Model Event/Reusable Process Matrix

It's largely from the Business Model Event/Reusable Process Matrix Library that we determine system development productivity. One way we can gauge our development productivity is by how many Business Model Events use each reusable process. I feel that an organization should be averaging 60 – 70 percent reusability and about two-thirds of a Business Event Partition should be made up of reusable processes (or processes that can be re-used in the future). (This percentage is an aggregate of reusable components discovered through analysis, design, and implementation.) This means that if your organization is not using a Reusable Library, then each new system development is wasting up to two thirds of its budget. Preparing an engineered component for reusability should cost no more than one which will not be re-used. The payback to our organization for keeping a Business Library Conservator is many fold based on this level of reusability.

Process Reusability Is All in the Name

All business processes must be meaningfully named. Ensuring unique and standardized names is one of the responsibilities of the Business Library Conservator. Standardized naming is essential for reusability. Without it, the function and purpose of a piece of software, or a human-based procedure may be so obscure and difficult to locate that it defies any hope of reusability.

> ***We should avoid using generic names in favor of specific names which will allow us to find a reusable component in the future.***

Limiting the number of reusable verbs that are used in a verb-object process name will help searching; the object naming in the process can be the unique identifier.

In Table 11–1 I offer examples of names for processes and procedures using process verbs (and the synonymous terms they absorb) for use in our specification. Add to this list verbs that are applicable to your business.

At the detail level, we should avoid high-level verbs such as Do, Analyze, Match, and Execute. When naming business processes, we should also avoid verbs that suggest getting and putting of data, such as Collect, Read, Gather, Assign, Issue, Package, Communicate, Respond, Distribute, Write, Access, Formalize, Record, and Submit. These may well be used as verbs for reusable components such as modules identified during design.

Sample Verb	Replaces ...
Create	Define, Build, Propose, Generate, Develop, Establish, Derive, or any other verb that involves producing something that did not exist before.
Validate	Calibrate, Check, Verify, Inspect, Edit, or any other verb to do with detecting potential errors.
Confirm	Any verb to do with asking whether data are there or not.
Compare	Any verb to do with the association of two or more pieces of data or processing.
Update	Any verb that uses existing data, modifies it, and replaces it.
Delete	Destroy, Eliminate, Cancel, or any verb implying removal of data.

Table 11–1: A Sample List of Possible Process Name Verbs

Obtaining Data Integrity (Data Conservation)

Data Integrity is a feature that applies mainly to stored Data Elements but also somewhat to transient Data Elements. Data Elements, stored data sets (Entities), and their relationships must have meaningful names. The Business Library Conservator ensures a naming standard is adhered to and eliminates the use of alias names across systems and the whole organization.

When an organization's customers change an address, that change should be reflected across the whole organization. We should not be required to know the number of internal "systems" within that organization that use our address and send in multiple address change requests. The ensuring of Data Integrity is another responsibility of the Business Library Conservator.

The business analyst's responsibility for developing the process models should ensure that within their area of study, the stores are data conserved. However, the Business Library Conservator needs to ensure the conservation of data across the whole organization. Figure 11-3 shows a hopefully obvious example of an analysis model where the conservation of data can be readily observed before design and implementation. This figure shows a model where stores **Ws**, **Ys**, and **Zs** are not fully data conserved. In the case of store **Ys**, what goes in doesn't come out (it's a "Black Hole" store) and in the cases of stores **Ws** and **Zs**, what comes out doesn't go in (they're "Magic" stores).

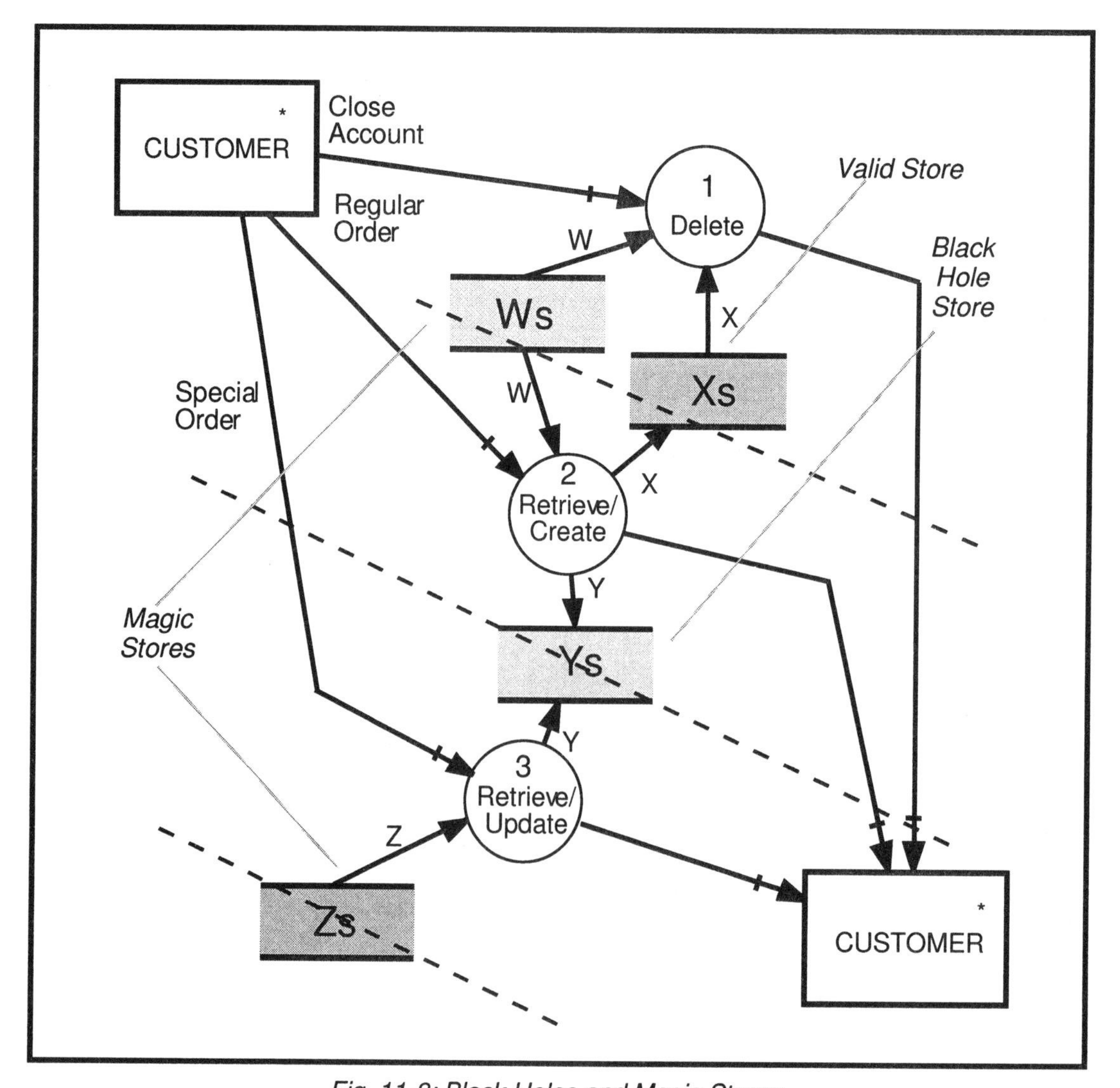

Fig. 11-3: Black Holes and Magic Stores

Of course spotting data conservation via such a model is simplistic because we are more concerned with specific Data Elements that are usually within groupings (records) and may not be easily observed on data flows as shown in Figure 11-3. This chapter offers a far more rigorous and comprehensive way of maintaining data conservation than simply observing the input and output of stored Data Elements.

Notice that this Data Integrity checking across Business Event Partitions is not intended to guard against the imperfections of the outside world — we cannot prevent a customer from trying to pay an invoice which does not exist, or from entering bad data. Such imperfections require us to build the usual edits into our Business Event Responses. The Business Policy Creator ensures that these edits, etc. are included in the business.

The Business Library Conservator ensures integrity of the Data Elements across the total Business Model.

Maintaining "Meta Data"

At this point I need to re-introduce the concept and use of the term "Meta Data." We need this term to separate the issue of "data values" verses "data about data." I mention this here because the following matrices are data-oriented. When describing these matrices the Business Library Conservator will concern his or her self with identifying what cells need to go into the matrices as opposed to the actual data values that will populate the contents of the matrices' cells.

Anyone familiar with spreadsheets is already aware of the difference between setting up the rows and columns of a spreadsheet versus populating the cells of the spreadsheet with data. The Business Library Conservator maintains the Meta Data (the cells) while the Business Model Events themselves running in a production system in the real world perform the actions that Create, Retrieve, Update, and Delete the contents of the cells (see Figure 11–5).

So, when relating to one of the data matrices in this chapter, a Delete operation noted in the matrix indicates a deletion of the contents of the cell and NOT the cell itself. The deletion of the cell would be requested by the Business Policy Creator via a Strategic Event and would constitute a change to the Meta Model and therefore a change to the matrices. The actual value (along with the Data Dictionary definition) is the responsibility of the Business Model Events that use the Data Element, Entity, or other cell contents.

The Business Library Conservators are not concerned with "Data Value Conservation" but with "Data Element Conservation." Therefore, in this case the Business Library Conservator's job is to Create, Retrieve, Update, and Delete components of the matrix itself. In other words, they deal with the Meta C.R.U.D. *(excuse my French).*

Data Reusability Is All in the Name

A convention for giving unique names to all Data Elements (attributes) can be made up of three parts: a Qualifier, an Adjective, and a Reserved Word. In Table 11–2 I offer a taxonomy for data naming.

Sample Qualifier (Mandatory)	Sample Adjective (Optional)	Sample Reserved Word – Data Types (Mandatory)
Supplier	Daily	Name
Part	Weekly	Description
Customer	Monthly	Number
Order	Yearly	Amount
Seminar	Minimum	Quantity
Student	Maximum	Count
Product	Average	Price
Account	Federal	Identification
	State	Date
	Standard	Deduction
	Special	Item
		Indicator
		Code

Table 11–2: A Sample List of Possible Data Names

We should ensure that the reserved words (Data Types) are never used on their own for any Data Element. Let the qualifier and adjective(s) help form the unique identifier.

A Note about Data Ownership

I've read many technical books that talk about identifying "who owns" data. Business Event Partitions own data — people or job titles **don't** own data any more than do computer programs. Having said this, it's important to realize that the organization is the ultimate owner of its data.[1]

> For example, using the government's tax tables an organization needs to Retrieve tax data to conduct business, but the government is the only organization allowed to Create the data. Notice how the concept of stores needed only between two Business Events also works between organizations, not just within a single organization.

1 I say "its" data because where two organizations need to share data, the data now has dual ownership — still across two Business Events.

Figure 11-4 shows that at the Business Event Level Business Event Partitions own the data. At the system level it may be helpful to identify the systems as the users of data via Business Model Events that have had a design placed on them. It is important to recognize that Business Event Partitions own data because, when a business change occurs, we need to go back to the Business Model Events that own the data before looking at the implemented systems. This realization that only Business Event Partitions really own data helps eliminate "private" personal/departmental data problems within an organization.

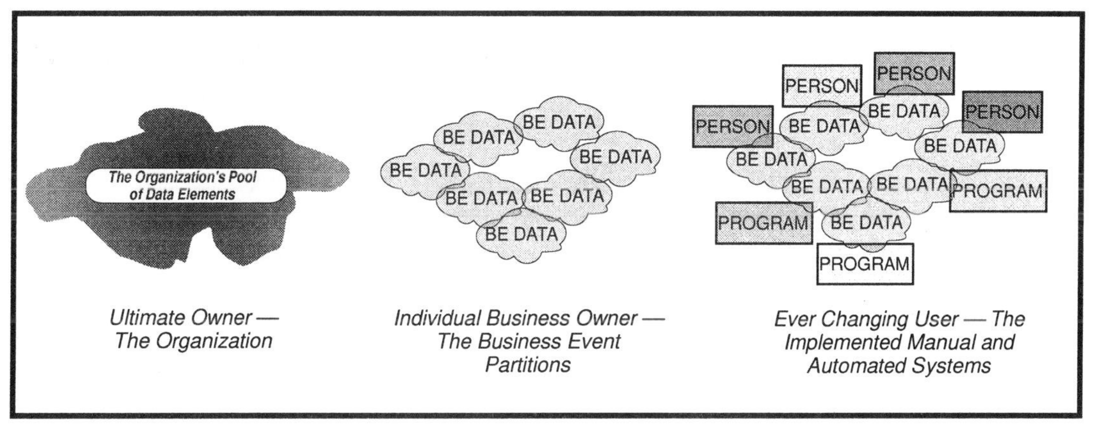

Fig. 11-4: The Progression of Data Ownership

The Business Model Event/Data Element Matrix

The most important matrix to be maintained by the Business Library Conservator is the Business Model Event/Data Element Matrix. Creating and maintaining this matrix is vital to support the Customer Focused Engineering effort and to maintain the on-going quality of our organization's systems. An important by-product of using this matrix is it allows the Business Library Conservator to keep track of the use of data and it ensures Data Conservation throughout our organization. Data conservation is the practice of ensuring:

- Data that is Retrieved, Updated, or Deleted by a Business Model Event must have been Created in another Business Model Event.
- Data that is Created by a Business Model Event must be Retrieved in another Business Model Event.

So even though we as the Business Library Conservator are not interested in Data Element values, we are interested in where data values are Created, Retrieved, Updated, and Deleted.

The Business Model Event/Data Element Matrix (see Figure 11–5) lists on one axis all the Business Events, Regulatory Events, and Dependent Events to which our organization must respond. On the other axis, the matrix lists all the Data Elements identified in our organization. This matrix may have several thousand columns and its organization and storage would probably require computer assistance. At each intersection of a Business Model Event and a Data Element, we would indicate whether the Business Event Partition **Creates, Retrieves, Updates** and/or **Deletes (C.R.U.D.s)** the data in this Data Element. The third dimension of the matrix could show anything associated with the Data Element that has to do with its data conservation or integrity. These could be such things as pointers to Data Element Specifications, to dependent Data Elements, and to individual processes that use a particular Data Element.

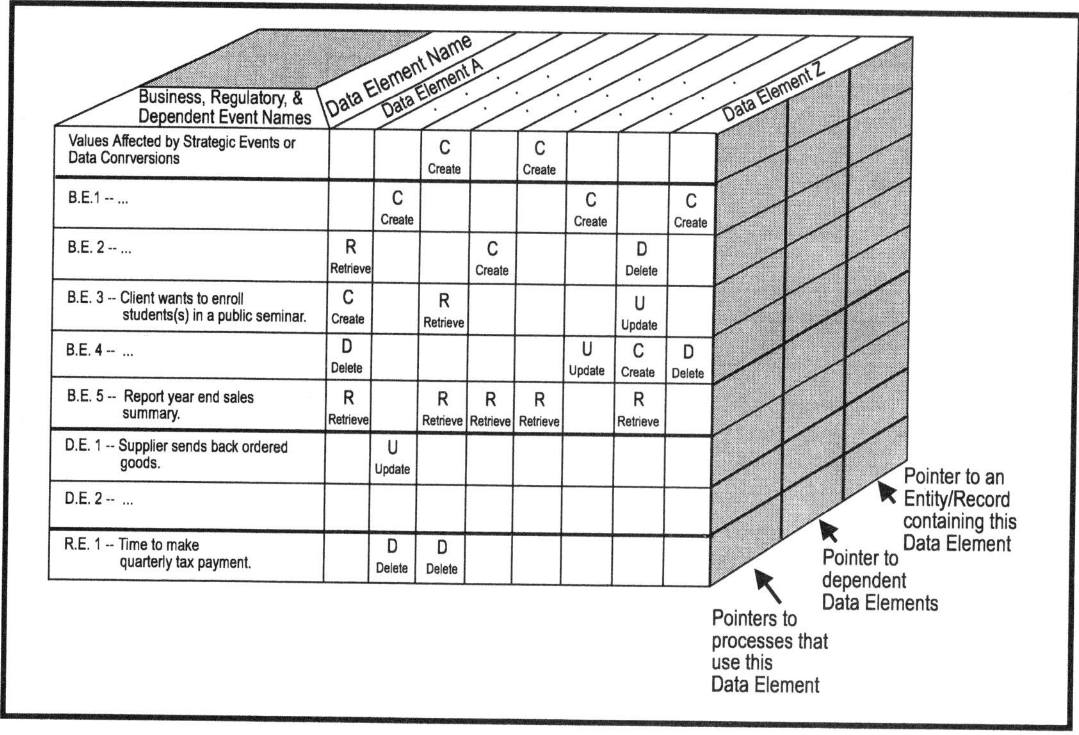

Fig. 11-5: Business Model Event/Data Element Matrix

For this matrix the Business Library Conservator would:

- Perform the initial set up of the Business Model Event/Data Element Matrix. (This task includes identifying all of the Data Elements and all of our organization's Business Events.) This is also known as setting up a data Meta Model.
- Identify where the Data Elements are Created, Retrieved, Updated, and Deleted by each and every Business Model Event in the matrix.
- Maintain the integrity of the Business Model Event/Data Element Matrix during production changes or new systems development.

The Business Library Conservator's role in maintaining the integrity of the organization's processing and data involves ensuring that the column for each Data Element contains at least one Create, Retrieve, and Delete.

The Business Model Event/Data Element Matrix is used to track how all of the data in your organization are used to support all of the defined Business Model Events.

Now I can explain why I talked about separating Strategic Events from other Events when they are combined at our Event Horizon. We need to make sure we have data conservation in this matrix.

When we affect the value of a Data Element (a noun) other than via a Business, Regulatory, or Dependent Event these changes need to be indicated in some way in either the Business Model Event/Data Element Matrix, the Business Model Event/Entity Matrix, or the Business Model Event/Relationship Matrix. This type of non-Business Model Event driven change can happen in two scenarios: one is via a data conversion effort and the other occurs when a Strategic Event places an initial value in a Data Element, changes the value of a Data Element, or sets up a relationship between data Entities.

For example, the Strategic Planners may identify a new **Book** to be handed out with a specific **Seminar**, therefore forming the initial relationship between these two Entities.

We should indicate the dualistic nature of these Events in the Business Model Event/Data Element, Business Model Event/Entity, and Business Model Event/Relationship Matrices. I recommend that we show how these Data Elements, Entities, and Relationships are data conserved with an initial line in the matrices to ensure the integrity and conservation of the total organization's data as I've shown in the matrices in this chapter.

In the ***Recognizing the Five Types of Events that Stimulate Our Organization*** Chapter we used an example of the government requiring our organization to add a new excise tax and specifying that the value of the tax must be 4 percent. If we treated this as a Strategic Event and missed its associated Regulatory Event, our Business Model and the matrix would not have a Create for the Data Element and therefore would not be data conserved. It would have a Retrieve of this field in the course of responding to an Event, but the matrix would indicate we never Created it.

In this example we should not mistake the adding of a new field as a Regulatory Event because it can't modify (add nouns and verbs to) the Business Model even though the Event came from a regulatory agency.

In those cases where the creation of a Data Element's value is accomplished in a one-time conversion procedure there may not be any Business Event that Creates these Data Elements in the course of our organization's normal operations. In these cases a Data Element may show only a Retrieve and a Delete or a Retrieve, Update, and Delete.

Some examples of this are the creation of Data Elements such as Fiscal Year End Date, the Organization's Name, or the Incorporation Date.

Ideally, each column in the matrix would have a Delete to ensure that implemented data files don't grow forever. Even a seven-year tax file should have a Business Model Event that Deletes (archives) the data after the statute of limitations has expired. However, we may find some data that is only deleted at the demise of our organization such as the **Organization Address** or its **Fiscal Year End Date**.

When aspects of the business data change, the Business Library Conservator is responsible for ensuring there are no violations of Data Element Integrity within and across Business Model Events.

This conservation process involves tracking any additional Business Model Events and their implementations that will be affected when a change to a Data Element is made. A requester of the business change such as the Strategic Planner can then be told the ramifications of a requested change. If the impacts of the change are acceptable, we can make changes to the computer code, the procedure manuals, forms, screens, etc. knowing that we have covered all of the bases. This procedure eliminates dead data and its associated dead processing in our manual and/or automated systems. It also eliminates any potential system down time that would result from inadvertently deleting something that was used in other Business Model Events.

In this way the first use of this matrix will be to identify a "Where Used" List of Business Model Events when a Strategic Event affects the Data Element itself (e.g., Advance Order Quantity Data Element is no longer required) all other Business Model Events that use that Data Element will need modification.

If the Strategic Planners eliminate a whole Business Model Event, and this Event is the only place a particular Data Element is Retrieved because the Business Event uses the Data Element for derivation of other data or for reporting outside the Business Event, then all other Business Model Events where the Data Element is Created, Updated, or Deleted will have to be modified. In other words, to obtain data conservation with no "dead" data and no "magic" data, each column of the Business Model Event/Data Element Matrix should have at least one Create, one Retrieve, and one Delete.

The "Where Used" List of Business Model Events will also be of help when a design change occurs (e.g., when the format of a Data Element changes).

This indicates the Business Library Conservator's responsibility to bring to the attention of the Business Policy Creator (the person requesting the change) any Business Model Event Integrity problems caused by a Business Model Event's deletion or modification.

> For example, if in the second to last column in Figure 11–5 Business Event 5 is removed, then we must ask the Strategic Planners what to do with the logic associated with the Data Elements in Business Events 2, 3, and 4 as we have just deleted the last place where this Data Element is Retrieved.

Transient data (data between processes) within a Business Model Event do not need integrity as they "burn up" after a Business Model Event has finished doing its job. This is similar to what happens to a manual "scratch pad" or working storage within a program. But we should still note that transient data need to be defined in our organization's Data Dictionary (with adherence as much as possible to our naming standard) for format and content reasons. Stored data, on the other hand, require integrity because they are being used across Business Model Events and are therefore capable of perpetuating their errors across other processing. For these reasons, the Business Model Event/Data Element Matrix is for stored data only.

The Business Model Event/Data Entity Matrix

We can also generate a Business Model Event/Data Entity Matrix using our Business Model Event/Data Element Matrix.

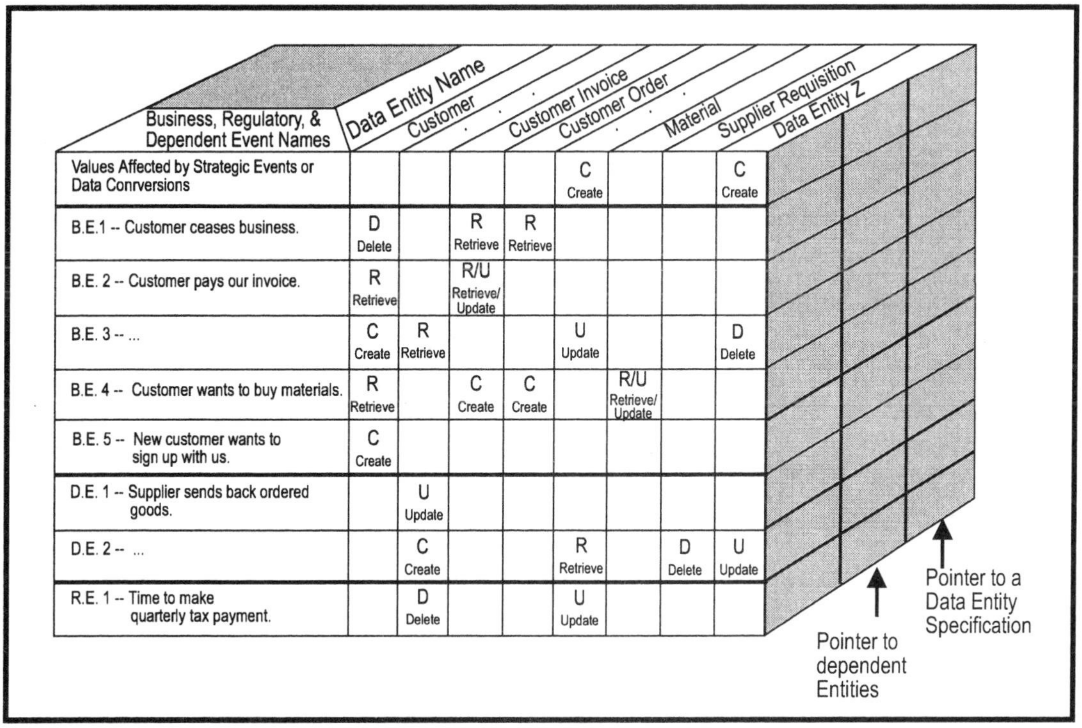

Business, Regulatory, & Dependent Event Names \ Data Entity Name	Customer	.	Customer Invoice	Customer Order	.	Material	Supplier Requisition	Data Entity Z
Values Affected by Strategic Events or Data Conrversions					C Create			C Create
B.E.1 -- Customer ceases business.	D Delete		R Retrieve	R Retrieve				
B.E. 2 -- Customer pays our invoice.	R Retrieve		R/U Retrieve/ Update					
B.E. 3 -- ...	C Create	R Retrieve			U Update			D Delete
B.E. 4 -- Customer wants to buy materials.	R Retrieve		C Create	C Create		R/U Retrieve/ Update		
B.E. 5 -- New customer wants to sign up with us.	C Create							
D.E. 1 -- Supplier sends back ordered goods.		U Update						
D.E. 2 -- ...		C Create			R Retrieve		D Delete	U Update
R.E. 1 -- Time to make quarterly tax payment.		D Delete			U Update			

Pointer to a Data Entity Specification

Pointer to dependent Entities

Fig. 11-6: Business Model Event/Data Entity Matrix

The Business Model Event/Data Entity Matrix (see Figure 11–6) would list on one axis all the Business Events, Regulatory Events, and Dependent Events to which our organization must respond. On the other axis we would list the entire set of the Cohesive Data Entities (groups of Data Elements). Again, these are related to stored data; not a transient record of data. The third dimension of the matrix could show anything associated with the Data Entity that has to do with its data conservation or integrity. These could be such things as pointers to Data Entity Specifications or pointers to dependent Entities (Entities that have some related integrity to this Entity).

In some Business Model Events we will Create and Delete complete Data Entities as opposed to individual Data Elements (also, most databases work at the Entity level), so this matrix will be more helpful in these situations.

For example, if a strategic decision was made to become a "Cash and Carry" operation, we may no longer store Customer Invoices or Customer Orders. We will then use this matrix to see all places where these Entities are used.

Please note that this Business Model Event/Data Entity Matrix is somewhat redundant with the Business Model Event/Data Element Matrix. The Business Model Event/Data Element Matrix is more important and is where we show the actual Create, Retrieve, Update, and Delete of each Data Element within an Entity. We may Create and Delete whole Entities and possibly use all of an Entity's Data Elements in a Retrieve. However, we don't really Update whole Entities, we Update the Data Elements within Entities. In Figure 11-6 we can see that we must be missing some Dependent Events.

For example, how do we create a new Supplier Requisition?

The Business Model Event/Relationship Matrix

The Business Model Event/Relationship Matrix (see Figure 11-7) is used to track and conserve the Relationships between the Entities listed on the Business Model Event/Data Entity Matrix. When a Relationship is established between two Entities (or two occurrences of the same Entity), this is reflected with a Connect in this matrix. When a Relationship is severed, this is indicated by a Disconnect in this matrix. Of course, this could also be related to the creation or deletion on an Entity. The third dimension of the matrix could show anything associated with the Relationship that has to do with its data conservation or integrity. These could be such things as pointers to cardinality or pointers that indicate whether the Relationship is mandatory or optional.

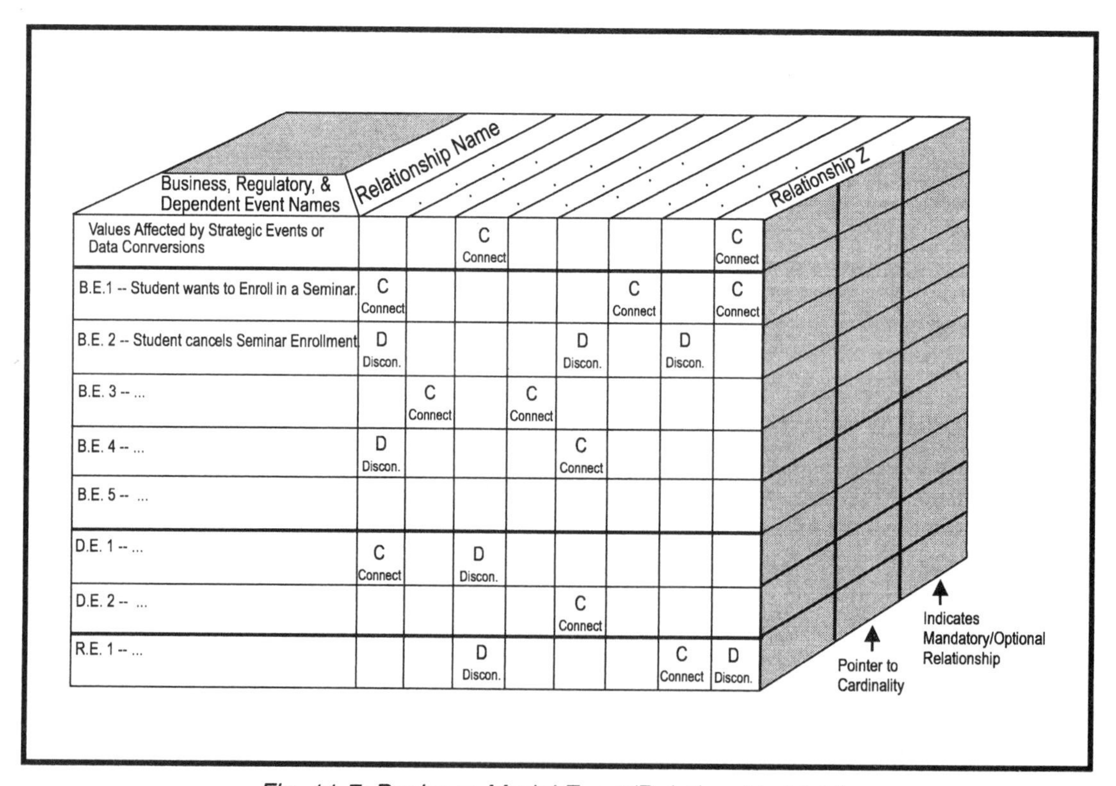

Fig. 11-7: Business Model Event/Relationship Matrix

This matrix allows us to ensure the conservation of Relationships (i.e., a Disconnect must previously have had a Connection associated with it). Also, when we see a Connect we must ask if there should be a Disconnect in some Event to sever this Relationship.

> For example, in my organization, an existing individual in my **Clients** file may ask to be "signed up" for a **Public Seminar**. This results in no new data, just a connection (Relationship) between the existing **Client** and the existing **Public Seminar**. I would also need to investigate a Business Model Event that allows a **Client** to cancel out of a **Public Seminar** (i.e., sever the Relationship between that **Client** and that **Public Seminar**).

The Business Model Event/Object and Method Matrix

If we are using an Object Oriented Methodology, the preceding matrices may be combined and we can keep track of Objects and their Methods via a Business Model Event/Object and Method Matrix (see Figure 11-8). In the Object Oriented Method data and its processing are encapsulated within an Object, so our matrix can ensure the reusability and integrity of Objects across Business Model Events by tracking which methods within an Object are being initiated by the Event.

Again, the Business Model Events are listed on one axis of this matrix and the Objects are listed on the other. The cells would indicate the individual methods within an Object that are used for any given Business Model Event. The third dimension of the matrix could show anything associated with the Object that has to do with its Data Conservation or Integrity. These could be such things as pointers to libraries of reusable objects, pointers to message sources (senders), or pointers to message recipients.

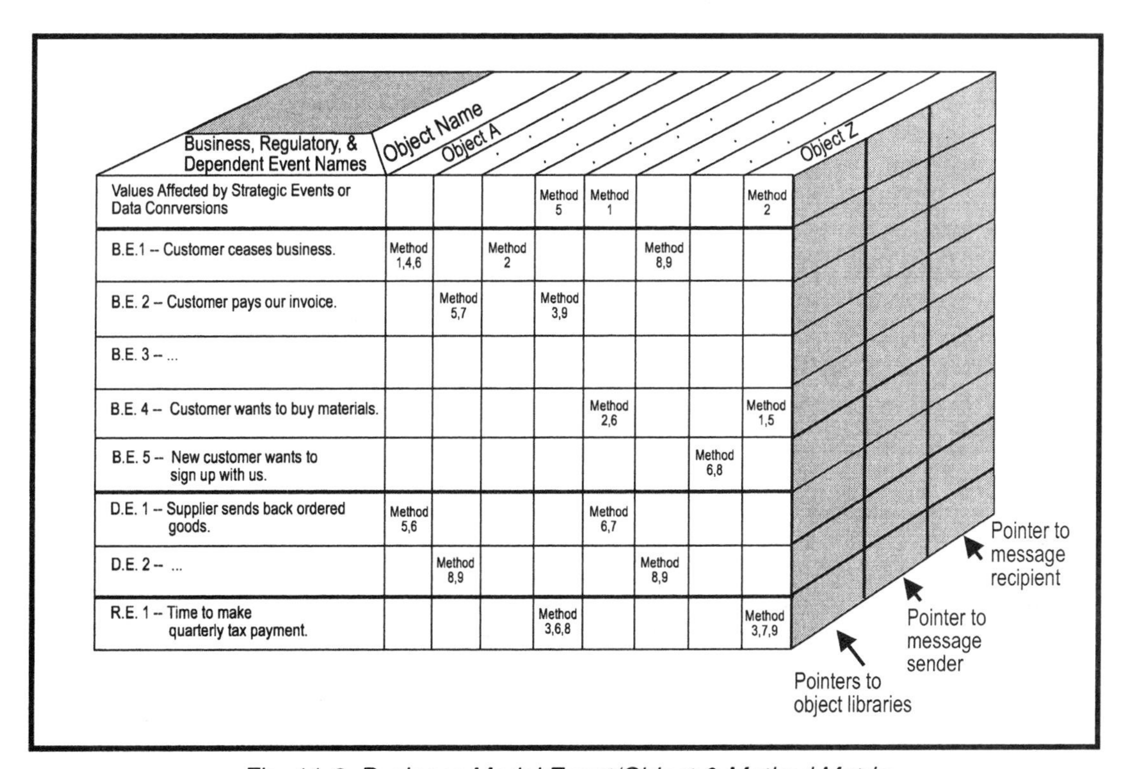

Business, Regulatory, & Dependent Event Names	Object A							Object Z
Values Affected by Strategic Events or Data Conrversions				Method 5	Method 1			Method 2
B.E.1 -- Customer ceases business.	Method 1,4,6		Method 2			Method 8,9		
B.E. 2 -- Customer pays our invoice.		Method 5,7		Method 3,9				
B.E. 3 -- ...								
B.E. 4 -- Customer wants to buy materials.					Method 2,6			Method 1,5
B.E. 5 -- New customer wants to sign up with us.							Method 6,8	
D.E. 1 -- Supplier sends back ordered goods.	Method 5,6				Method 6,7			
D.E. 2 -- ...		Method 8,9				Method 8,9		
R.E. 1 -- Time to make quarterly tax payment.				Method 3,6,8				Method 3,7,9

Fig. 11-8: Business Model Event/Object & Method Matrix

The Business Model Event/Engineered System Matrix

When we come to the implementation phase of a Customer Focused Engineering project, we can produce a Business Model Event/Engineered System Matrix.

Business, Dependent, & Regulatory Event Names	System Name	Customer Service System	.	Supplier Service System	.	.	Regulatory Conf. System	. System Z
B.E.1 -- Customer ceases business.	X						X	
B.E. 2 -- Customer pays our invoice.	X			X				
B.E. 3 -- ...					X			
B.E. 4 -- New customer wants to sign up with us.	X			X			X	
B.E. 5 -- ...		X						
D.E. 1 -- Supplier sends back-ordered goods.			X					X
D.E. 2 -- ...							X	
R.E. 1 -- Time to make quarterly tax payment.		X				X		

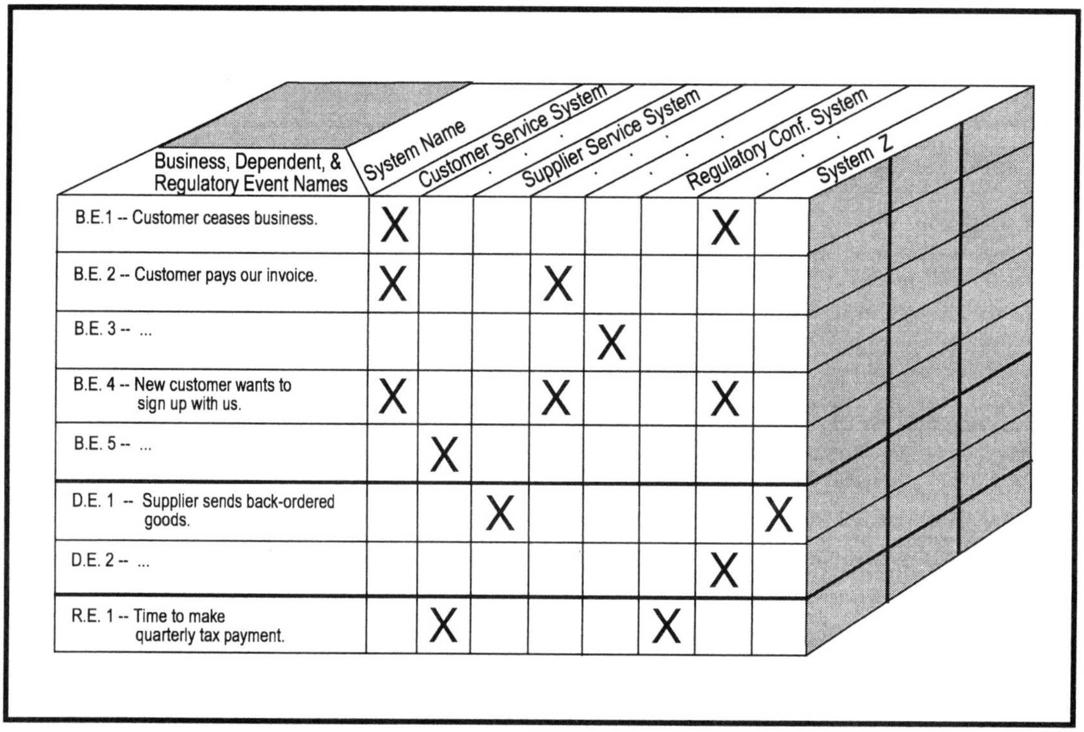

Fig 11-9: Business Model Event/Engineered System Matrix

As we will discuss in the ***Designing and Implementing Business Event Systems*** Chapter, we try not to fragment Business Model Events across systems. However, we may form cohesive collections of related, complete Business Model Events within one system. If we did this, we may want to keep track of these systems in a matrix (see Figure 11–9). In other words, we can use this matrix to keep track of the designs for how we implemented our Business Model Events.

The Business Model Event/Engineered System Matrix would list on one axis all the Business Events, Regulatory Events, and Dependent Events to which our organization must respond. On the other axis we would list the systems we have engineered to satisfy these Events. This matrix will be helpful in the modification of production systems because it points to the implementation of a Business Model Event.

Saving System Development Time/Investment

I mentioned at the beginning of this chapter that we can obtain substantial savings of system development time/investment. I hope you can see this is achieved using reusability via these matrices.

There is another benefit regarding system development that is worth mentioning here. Most system development projects have a deadline as their main constraint. If our project is using a linear dependency "Waterfall Methodology" of analysis, design, implementation, and installation, then all of these activities are on the "critical path." In other words, a slip in one phase automatically produces an equivalent slip in the succeeding phases. This is acceptable if we have no easy way of partitioning the analysis activity itself because the system cannot be partitioned and is considered a whole unit. But with Business Event Partitioning we can treat each Business Model Event as a "mini-project" because they naturally stand alone — connected only by stores. If we have multiple project members, we can assign each individual one or more Business Model Events. This won't cause any interface problems; especially if we're using a Reusable Library and a Data Dictionary with the recommended naming standards.

Figure 11–10 shows how we can use Business Model Events to significantly compress the development life cycle and meet a deadline without any disruption to the final system or sacrifices in quality. Of course we use more people resources at any one point in time.

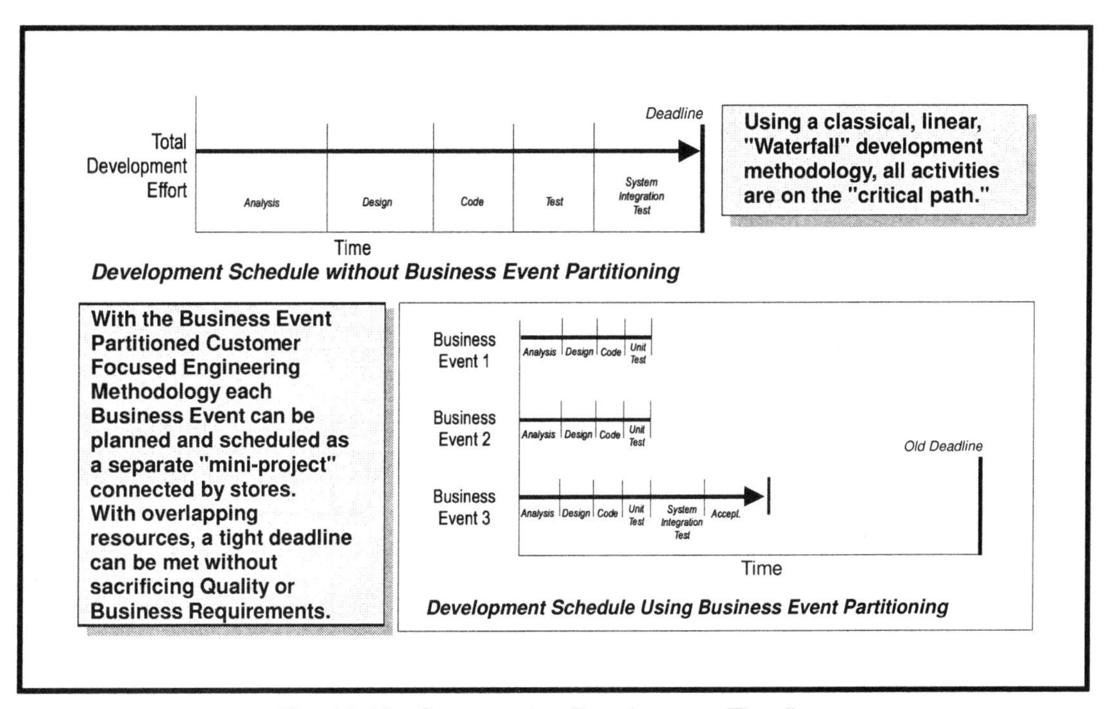

Fig. 11-10: Comparative Development Timelines

Summary

The matrices described in this chapter will allow the Business Library Conservator to maintain and verify that the Business Models partitioned by Business Model Events are completely reconciled and consistent across the whole enterprise.

With these matrices we can readily assess the impact on the whole business of a new Business Event Partition (i.e., a new line of business), or a change to an existing Business Event Partition. Such matrices would also readily show the availability of reusable data and processes, and so avoid, if not eliminate, many re-inventions of existing processing within the organization. This would enable us to respond more quickly to business changes.

I haven't found a single organization in over 30 years that would take me up on the bet that I can't find dead data and dead processing in their systems. I hope you can see that by using this Customer Focused Engineering Methodology and its matrices, this elimination of dead data and processing is not only possible, but quite attainable. [2]

---

2 At the time of the writing of this book, LCI is in the process of building a Customer Focused Engineering computer software tool that will take care of most of the systematic cross checking in the matrices. When using this tool the enforcement and maintenance of the matrices will be fully computer assisted to greatly ease the workload of the Business Library Conservator.

Chapter

12

Designing and Implementing Business Event Systems

In a time of drastic change it is the learners who inherit the future. The learned usually find themselves equipped to live in a world that no longer exists.

Hildebrant
The Counterfeiters

I find this design stage to be the easy/fun part of Customer Focused Engineering, providing I have a comprehensive Business Specification as input.

This is where we take our Business Model and turn it into one or more system models by adding design features. Now we put on our "designer hats" and concern ourselves with using the best appropriate technology to implement the business. We can apply the technical skills of one or more individuals as well as a technical set of hardware to implement each Business Event Partition; therefore, I class people (carbon based units) as a technology issue just the same as computers (silicon based units).

Using the Business Event Methodology we have the benefit that we do not need to concern ourselves with any issues to do with business correctness; these issues have been taken care of during the analysis stage. This is no different than when a new house is being constructed, the bricklayer or carpenter does not need to worry about whether the door is in the right place or if the room is the correct size.

By the end of this chapter, we will see a Business Event Specification containing models from analysis, through design, to implementation. We will also see a scenario showing the old design and the new design of a typical Business Event.

System Support Issues

It's during the new design stage that we introduce any new System Events and associated system issues necessary to make the new implementation work (i.e., we introduce System Events and associated issues to "cover" our Business Events).

Because our Business, Regulatory, and Dependent Events will most likely require the ancillary support provided by our organization's internal departments (such as power and climate control, etc. from the building's Physical Plant and services from the Human Resources, Accounts, Information Technology, and Legal Departments) we will need to acknowledge these tie-ins to our Business Model. We can accommodate these system support issues with two types of implementations. The one we choose will depend on our aggressiveness towards creating a Customer Focused Organization.

The first implementation is to keep the existing concept of separate departments for these support services and to have all Business, Regulatory, and Dependent Events call upon these departments as necessary (see Figure 12-1). This will work OK as long as these system issues do not interfere with the efficient accomplishment of our Business Model Event.

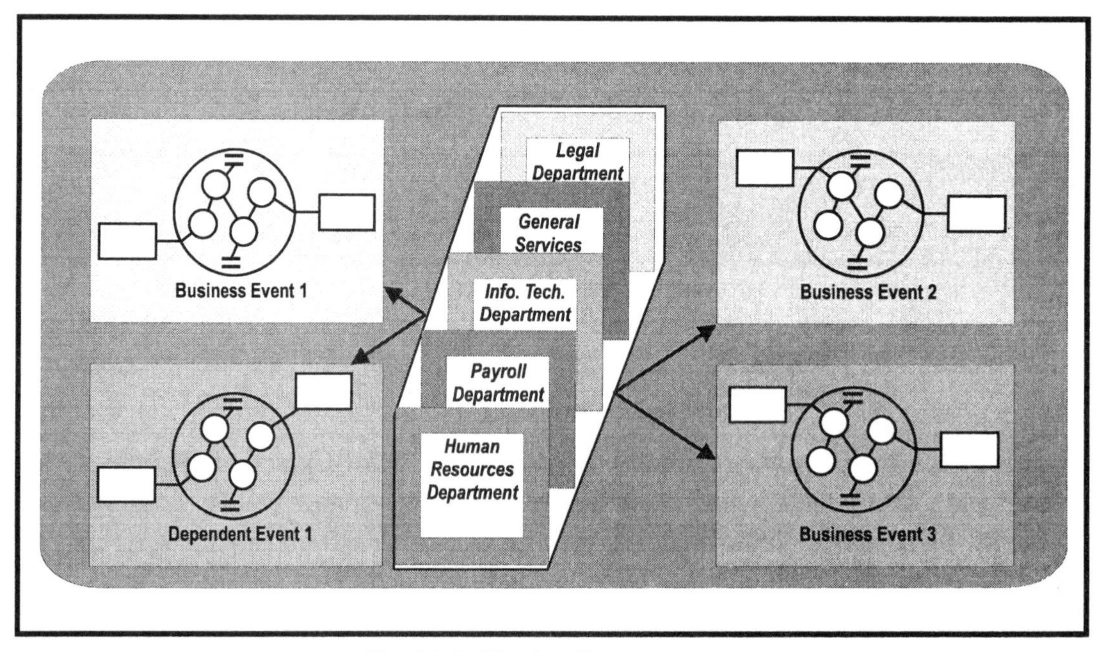

Fig. 12-1: Sharing System Issues

In the second type of implementation we apply the same concepts used for Business Events. Just as we found we could unfragment our Business Events and bring them together (even including their associated Regulatory and Dependent Events), we can now look at our System Events (both bundled and fragmented) to see where they can be subsumed within each Business Event (see Figure 12-2). We should do this if we find subsuming the system issues makes each Business Event more efficient and provides better control over its fulfillment.

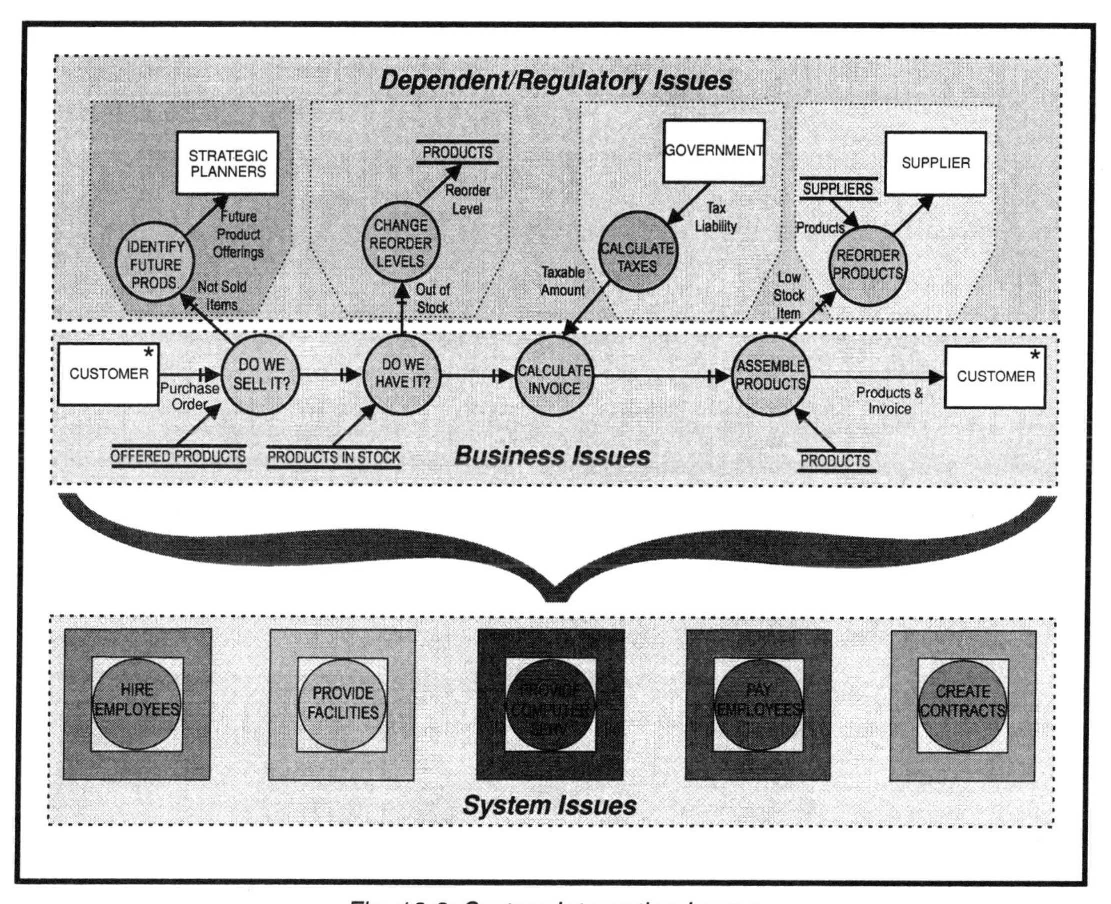

Fig. 12-2: System Integration Issues

When we introduce System Events and other system issues such as any associated System Event memory stores (e.g., computer **Employee** files or manual desk files to hold documents until the start of the work day), we can use the same concepts as those that underlie the Business Model Event/Data Element Matrix and other matrices to maintain the data and process conservation across our System Events during the time that those System Events are implemented.

Therefore, if we bring in **Employees** to implement a Business Event and need to capture Data Elements such as **Employee Name**, etc., we will need to Create, Retrieve, and hopefully Delete these fields to maintain data conservation. To do this we need to maintain a System Event/Data Element Matrix, a System Event/Entity Matrix, a System Event/Relationship Matrix, and/or a System Event/Object Matrix

Figure 12-3 shows samples of these System Event matrices.

Data Element Name / Data Element A . . . Data Element Z

System Event Names								
Values Affected by Strategic Events or Data Conrversions			C Create		C Create			
S.E.1 -- ...		C Create				C Create		C Create
S.E. 2 -- ...	R Retrieve			C Create			D Delete	
S.E. 3 -- ...	C Create		R Retrieve				U Update	
S.E. 4 -- ...	D Delete					U Update	C Create	D Delete
S.E. 5 -- ...	R Retrieve		R Retrieve	R Retrieve	R Retrieve		R Retrieve	
S.E. 6 -- ...		U Update						
S.E. 7 -- ...								
S.E. 8 -- ...		D Delete	D Delete					

Data Entity Name / Customer . Customer Invoice / Customer Order . Material / Supplier Requisition / Data Entity Z

System Event Names								
Values Affected by Strategic Events or Data Conrversions					C Create			C Create
S.E.1 -- ...	D Delete		R Retrieve	R Retrieve				
S.E. 2 -- ...	R Retrieve		R/U Retrieve/ Update					
S.E. 3 -- ...	C Create	R Retrieve			U Update			D Delete
S.E. 4 -- ...	R Retrieve		C Create	C Create		R/U Retrieve/ Update		
S.E. 5 -- ...	C Create							
S.E. 6 -- ...		U Update						
S.E. 7 -- ...		C Create			R Retrieve		D Delete	U Update
S.E. 8 -- ...		D Delete			U Update			

Relationship Name . . . Relationship Z

System Event Names								
Values Affected by Strategic Events or Data Conrversions			C Connect					C Connect
S.E.1 -- ...	C Connect					C Connect		C Connect
S.E. 2 -- ...	D Discon.				D Discon.		D Discon.	
S.E. 3 -- ...		C Connect		C Connect				
S.E. 4 -- ...	D Discon.				C Connect			
S.E. 5 -- ...								
S.E. 6 -- ...	C Connect		D Discon.					
S.E. 7 -- ...					C Connect			
S.E. 8 -- ...			D Discon				C Connect	D Discon

Object Name / Object A . . . Object Z

System Event Names								
Values Affected by Strategic Events or Data Conrversions				Method 5	Method 1			Method 2
S.E.1 -- ...	Method 1,4,6		Method 2			Method 8,9		
S.E. 2 -- ...		Method 5,7		Method 3,9				
S.E. 3 -- ...								
S.E. 4 -- ...					Method 2,6			Method 1,5
S.E. 5 -- ...							Method 6,8	
S.E. 6 -- ...	Method 5,6				Method 6,7			
S.E. 7 -- ...		Method 8,9				Method 8,9		
S.E. 8 -- ...				Method 3,6,8				Method 3,7,9

Fig. 12-3: The System Event Matrices

As far as possible, we must expand the business view of our Business Event Partitions into the design and implementation of systems. This may be a little more difficult in a manual implementation than in a computer implementation as humans are usually skilled in one subject but computers can mimic any systematic skill.

We obviously have many options to choose from when designing and implementing a Business Event Partition to obtain Customer Focused System such as:

- Empower teams of skilled individual employees with the responsibilities, authority, and the support systems needed to completely satisfy a customer's needs.
- Empower interactive teams of single-skill employees to bring together the resources needed to best meet the customer's needs.
- Create an environment that integrates the empowered employees with the computer support and other system technology to satisfy the customer's needs.
- Design and implement a fully automated system to meet the customer's needs.

With the above designs we obtain Customer Focused Systems. We will address these options in this chapter.

Business Event-Driven Systems Design

If we have the opportunity to implement our new system in accordance with the Business Model Events it supports, then each implemented Business Event Partition should endure as long as its associated business does (that is, as long as the organization is in that line of business), rather than for the duration of the hardware or support software used to implement it.

If we restructure manual areas, then Business Event Partitioning will also yield an ideal manual implementation partitioning.

> For example, we may find that having one employee call upon "subject area experts" when needed, putting together a team of people, or empowering one employee with the necessary skills to accomplish one Business Event Partition produces a higher quality product than a production line where each person sees only a small part of every end-product.

In the past, we have tended to design departments and programs according to similarity of tasks as described in the ***Systems Archaeology*** Chapter. This was partly for economy of scale and because of historical models of human efficiency reasons (based on Frederick Winslow Taylor's ideas of management from the 1920s).[1]

As you now know, Business Event Partitioning will yield an almost opposite view, with a Business Event Partition typically spanning many old design partitions.

Customer Focused Engineering via Human Systems

With human-based systems we can take advantage of the fact that humans can learn "on the fly" to dynamically adjust to the changing needs of a customer and modify the system design when it's not working well and still stay within the organization's business requirements since humans can adapt to unexpected conditions far more readily than computer systems. When customers interact with our systems, a human-being interface is usually (but not always) looked upon as a far more user friendly environment than a computer system.

As systems designers we must acknowledge our system users (customers) and we must be aware that computer technology can be intimidating to customers who are not computer literate. The word "empowerment" is important when it comes to effectively implementing Business Event Partitions with human beings.

In the past organizations that have had to immediately respond to a customer's need have tended to empower employees.

> For example, the emergency operator can take calls, allocate resources, dispatch rescue units, etc.

We can use this same empowerment idea even in systems that aren't time-critical. We obviously should still want to satisfy the customer's needs with an efficient design.

---

1 It was Frederick Winslow Taylor who said: "A worker should be no smarter than an ox."

To correctly empower people with the responsibility to take care of a complete Business Event Partition requires a break from the classical design structure of boss-subordinate relationships. This is difficult if you are a manager or an employee who has been used to this structure all of your working life. Maybe this structure was valid when we moved from the Agricultural Revolution to the Industrial Revolution (we couldn't have the free-form structure of a farm environment apply to a factory environment), but it is not valid in the Information Revolution.

When an organization does not give its employees the opportunity of being in charge of their own professional destinies and fails to provide adequate training and a supportive environment, it leads to the employees feeling a loss of control and even lowered self esteem. This is reflected in lower quality products and services. Therefore, quality is preempted by the managers not being true facilitators/leaders and employees not having an engineering vision.[2]

I realize that within an organization, human beings (employees), just like computer systems, are implementation devices for business systems. So, we should look on the human world of Customer Focused Engineering as the computer world of Customer Focused Engineering and expect "zero defect" systems.

How do we achieve zero defects in human-based systems? We may have to accept that a system implemented with human beings can have the potential for more errors than an engineered computer system because unlike in a human environment, a computer system (once in place) will perform exactly the same each time it's invoked. But this does not mean that we can't apply the same ideas of Customer Focused Engineering to the role of

2 This reminds me (for some strange reason) of a quote from Groucho Marx: " I don't care to belong to a club that would have me as a member." I see the "Groucho Syndrome" (as I call it) at some organizations (not all, thank goodness) reflected in their employees' attitudes as "If I were good, I wouldn't be working here."

human beings by empowering them with the correct skills, training, procedures, and equipment to do their jobs more effectively and by providing a work environment that promotes productivity.

By using ideas such as *Partitioning by Business Events*, we can simplify the forms and procedures for human beings to eliminate (or at least minimize) potential errors and omissions in their jobs. In addition, the introduction of technology, such as assisting the human being with powerful Personal Computers (PCs) and links to on-line databases can ensure minimum errors in data by eliminating redundancy, synchronizing data, and providing easily accessible data. There is no difference between partitioning software systems and partitioning the procedures in a human-based system.

Customer Focused Engineering via Computer Systems

If we look at the last 30 years of computer technological advances, we see orders of magnitude growth of capability and capacity. For example, today we get a lot more machine power for our dollar. Because of the advances from the early days of vacuum tube machines, something with the power of a "mainframe" machine that used to fill a large air-conditioned room now fits on my lap, and maybe in the palm of my hand. These kinds of improvements match what we strive for in a Customer Focused Engineering project, but without having the burden of purchasing new hardware.

The Benefits of Customer Focused Engineering Computer Systems

A benefit of implementing computer-based systems using an engineering discipline is any modifications are made everywhere instantly with no need for retraining. The capacity and speed of computers exceed that of human beings when implementing a systematic process; there is a growing market segment that prefers to interface with computer systems. Obviously, any external computer systems linking to our organization would be prime targets for linking to another computer system within our organization. Another benefit of using computers in design is a software product contains no material resources; it can be replicated at practically no cost once it has been created.

> A building, for example may cost millions of dollars and take months or years to build. Another identical building will cost almost the same and take equally as long. By contrast, an air traffic control system or an operating system for a personal computer may cost millions and take many person years to develop. However, once built the system can be replicated for the cost of the magnetic medium and of copying the procedures documentation. The implications of this are that the computer software industry can benefit a large market quickly.

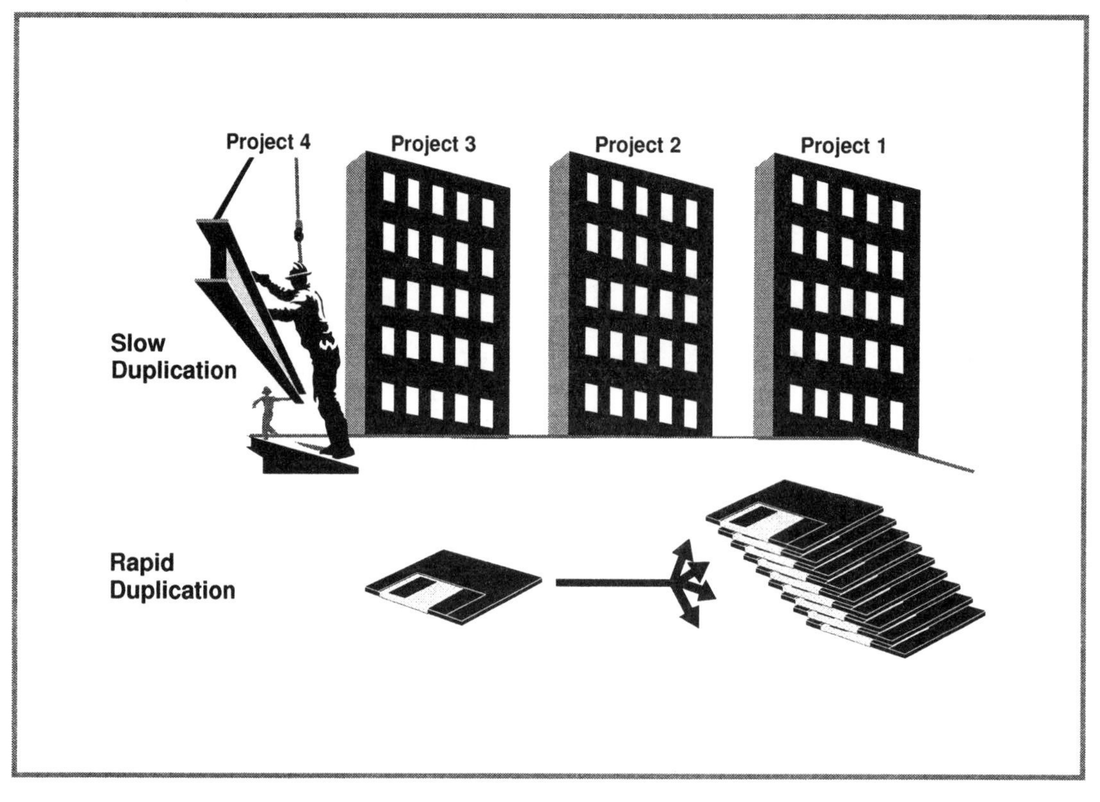

When we come to implement a new computer oriented design from our Business Event Partitioned model, we should still "package" as much as possible by Business Event Partitions, and not replicate the old system design partitions that we saw in the ***Systems Archaeology*** Chapter.

In Figure 12–4 we see five Business Event Partitions packaged into five computer partitions. The top two on-line computer partitions mirror their Business Event Partitions. The middle part of the diagram shows one Business Event partitioned by time into batch daily and batch weekly packages for a specific *design* reason. These two design partitions interface with a design tape store. If this whole Business Event Partition later became implemented totally on-line or totally as a daily batch run, we would remove this connecting design store and the **'End-of-Week'** system stimulus. The lower portion of the diagram shows two Business Events packaged into one ad hoc reporting system. With the exception of the middle diagram, Business Event 3, all these design packages are functional and have not fragmented their Business Event Partitions.

During the design phase of computer systems, one problem in the past, even with modern methods, has been deciding what constitutes a program and subsystem boundary. We can see now that the best program or subsystem partitioning is by Business Event Partitions. From a computer system implementation point of view, we have many design methods to choose from.

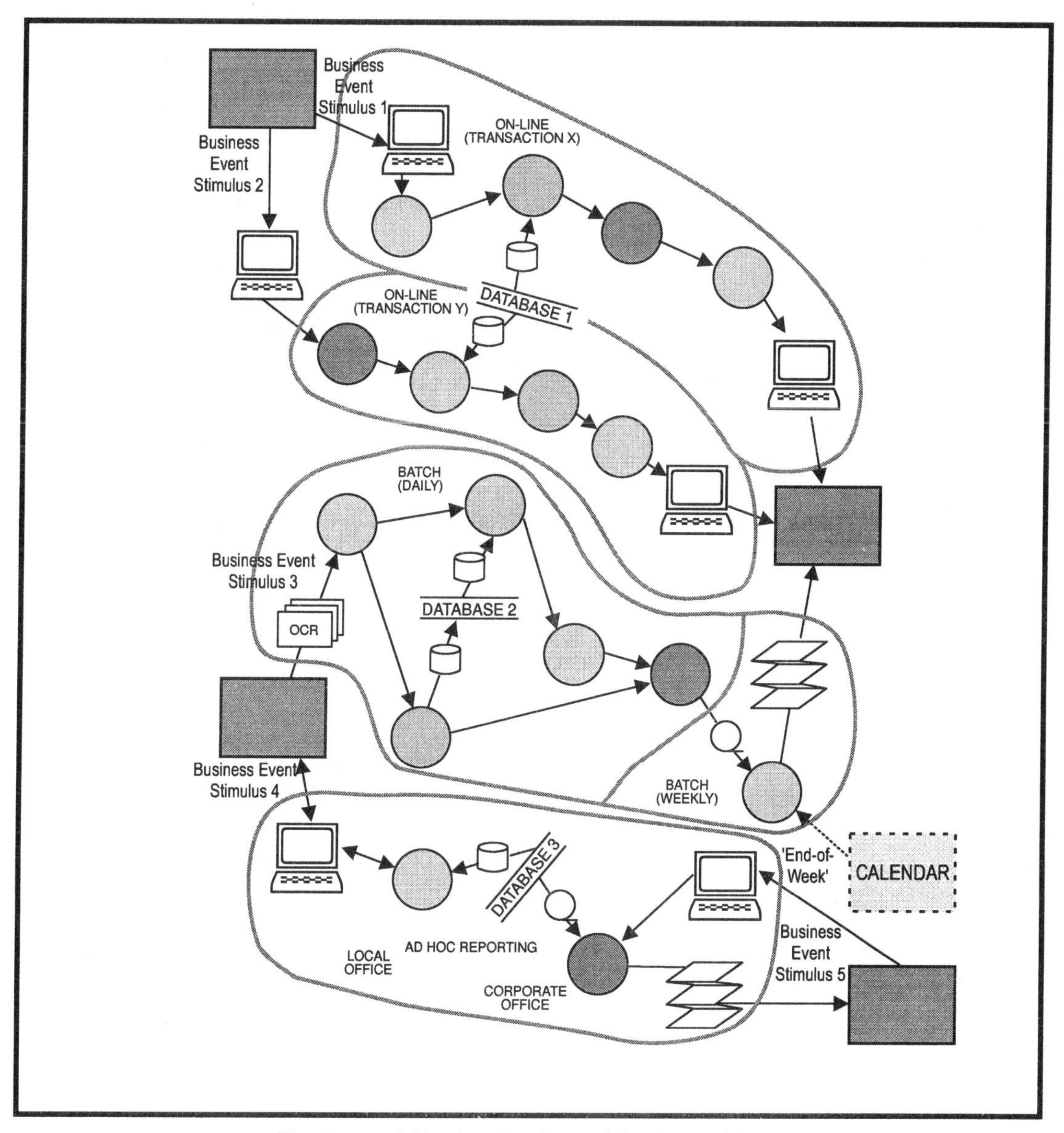

Fig. 12-4: A Design Packaged Business Model

Business Event Driven Design

We can implement a Business Event Partition with an empowered, multi-skilled person or a localized team with each member possessing a complex skill. A hybrid of computer assistance and manual operations may also be effective. Figure 12–4 shows a hybrid implementation using our Business Event example from the previous chapters, Customer Wants to Order Our Materials.

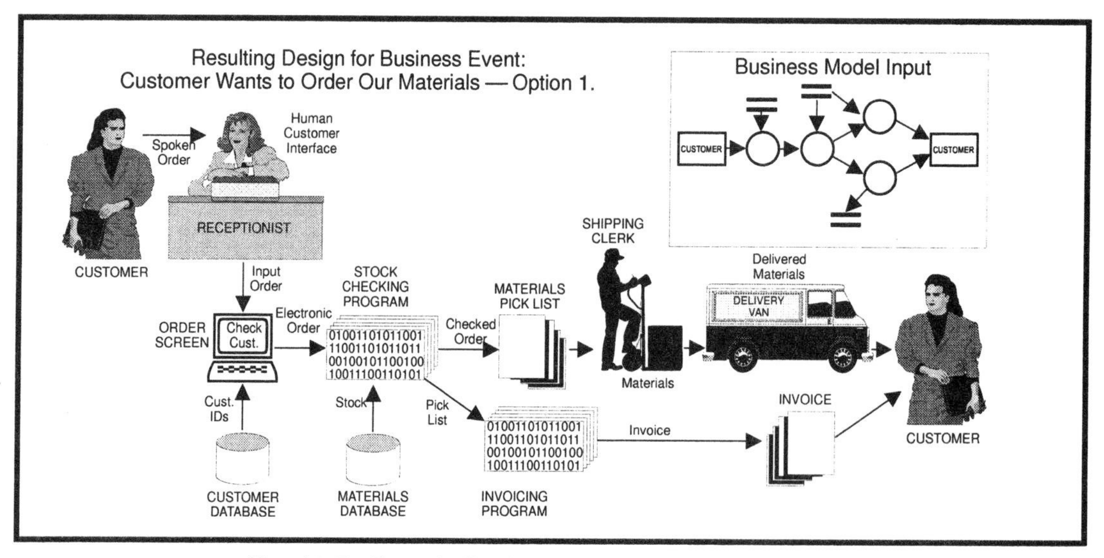

Fig. 12-5: Sample Business Event Partitioned Design

Figure 12-5 shows the design characteristics of how our Business Model's data and processes are implemented. Notice in the design stage there are many designs that can be used to implement our Business Model. Figure 12–6 shows another design for the same Business Model.

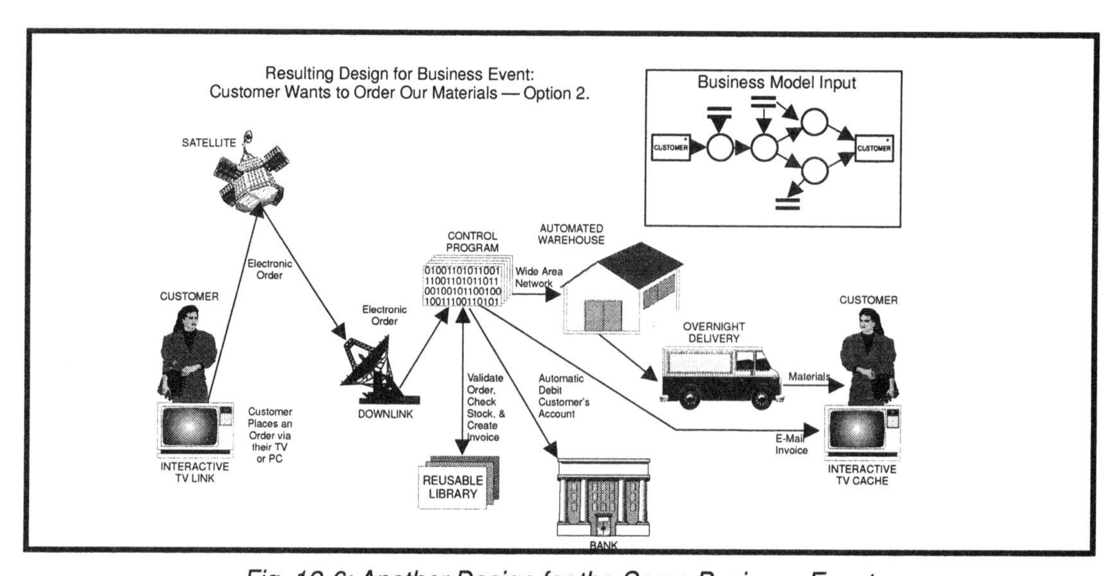

Fig. 12-6: Another Design for the Same Business Event

This second design shows an interactive TV or PC link from the **Customer** to our organization. Our organization processes the **Customer's** electronic **Order** immediately and interactively on-line; simultaneously, our implemented system triggers three things:

- It causes the nearest or most appropriate **Automated Warehouse** (which may be our vendor's warehouse) to ship the **Order** directly to the **Customer**.

- We send an electronic **Invoice** to the **Customer's** PC or Interactive **TV Cache** (internal storage).
- We obtain an automatic transfer of funds from the customer's bank **Account** into our organization's account.

Which design we use will depend on such things as our customer and our budget. The important issue here is to pick a design that is customer focused and that seamlessly implements the Business Event Partition.

Three Design Scenarios for a Business Event

The following sections show an air travel industry worst case, current case, and future case design examples of customer service. I have used the airline industry as the basis for the scenarios because the joys and frustrations of air travel are something to which many people can relate. As mentioned previously, we tend to believe that an existing design is the basis for future designs and as a result, we get stuck in perpetuating this same design.

The worst case scenario shows how intolerable air travel would be if the airlines followed what many organizations mistakenly call a "functional partitioning." This worst case scenario may have resulted from a designer focusing only on satisfying their organization's internal departmental managers' goals.

The current case scenario briefly outlines the processes involved in air travel today. The designer of this scenario obviously focused on satisfying both internal and external (customer) goals.

The future case scenario shows one new design for air travel, if we were allowed to apply a customer focused view.

The Worst Case (or I'd Rather Walk)

If the airline operations designer had come from an all too typical old mentality environment, their check-in operations design could require that all passengers allow one full day to take care of customer administration tasks before taking a flight. Figure 12–7 shows the airline operation with step numbers indicating a customer's step-by-step procedure in order to take a flight.

1. The customer sets up an account with the airline at their New Accounts Department, gets a credit check and, on approval, has an Account Number allocated with a pre-approved Credit Amount.
2. They find the Help Desk and research the flight's equipment Catalog Number for their particular flight.
3. Next they go to the Order Department and set up and document a Flight Order Form.
4. The customer must be prepared to get put on back order if the flight is full. If so, they go back to the Help Desk to see when the next available flight leaves.
5. As financial matters can only be assigned to an accountant in the Accounting Department, they go to the Accounts Receivable Department to pay for the flight.
6. As they want to be credited for flight miles, they must go to the Special Services Department to register their frequent-flier ID. Of course, if they don't have a frequent-flier ID or they forgot their number, they need to return to the Order Department to get one.

7. Now obviously their next stop will be at the Catering Department where they'll need to make their meal request. If they have no specific meal request, they only need to fill out the "short form."
8. If there is still time before the airline closes for the day at 5:00 PM, they can go to the Expedite Desk to turn over all of the assorted paperwork given to them by all the other departments. They take a number and, after the clerk gets around to inspecting all of their paperwork, the clerk types all of the same information on the forms into a computer terminal ready for their batch-process computer system. The computer batch system runs overnight. (This is because in the 1960s we were limited on computer time and since then the D.P. Department has done a line for line conversion of the batch system with the introduction of important new computer hardware.) The airline is currently making a significant investment via a modernization initiative in a new system to capture data at each department so that your wait time at each counter will be reduced!

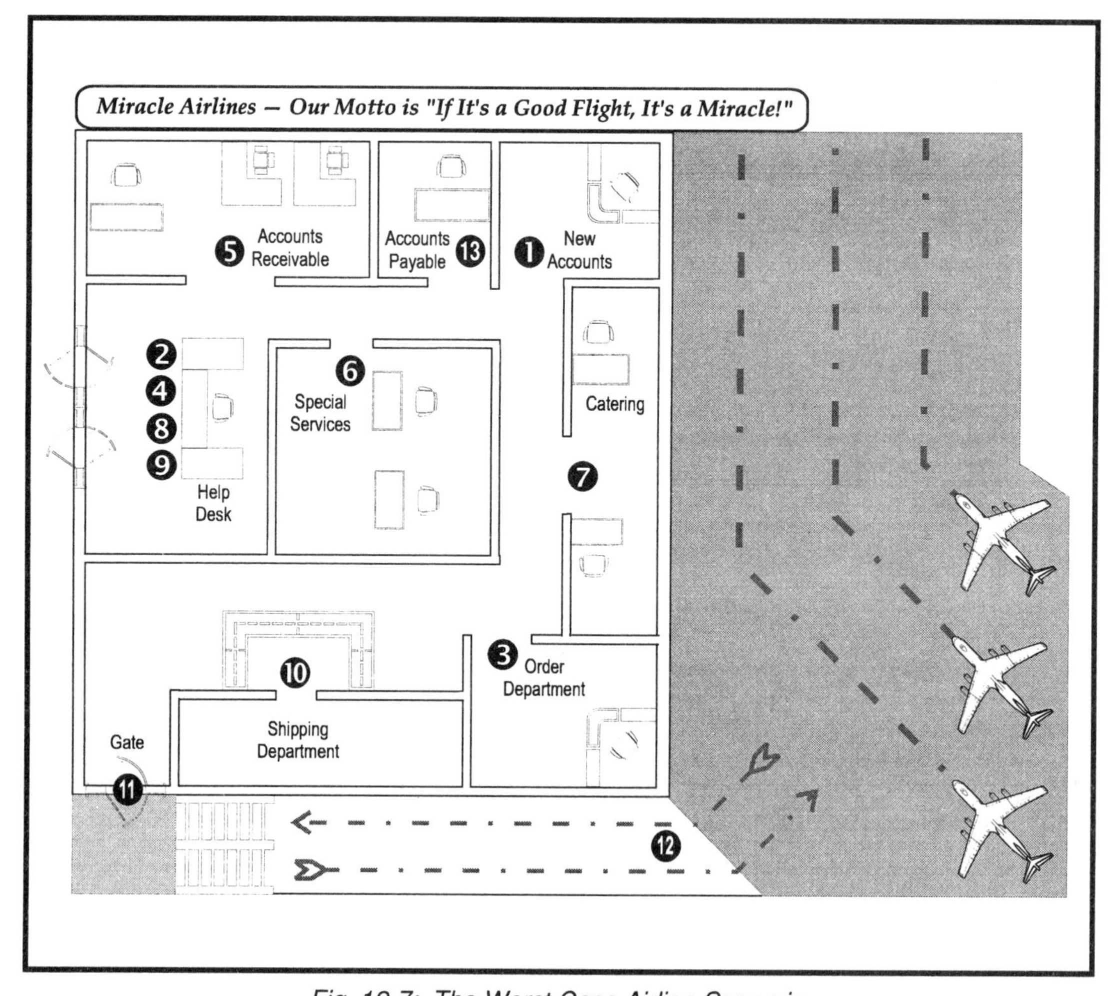

Fig. 12-7: The Worst Case Airline Scenario

9. They then show up the following morning at the Pick Up Desk to get their Flight Authorization Printout. Unfortunately, this may contain an error message for their Account/Reservation Number. If so, they start the entire procedure again but with a special form which enables them to jump each line in each department. If no errors are found or typed in wrong on their mountain of paperwork, they get a printed invoice that they use for boarding.
10. Before going to the Gate, they must stop at the Shipping Department where they leave their luggage. Of course, this is exchanged for more paperwork in the form of baggage claim tags.
11. Here another customer service representative helps them get their Departure Gate Number and directs them to the bus to get to the departure lounge and to where there are restaurants (providing the customer service representative is not too busy with the administrative procedures of the day). The departure lounge Gate is where they turn over their boarding documents to the clerk who promptly tears them in half and gives the customer back one half for important audit purposes. (The customer must not lose these bits of paper.)
12. The boarding Gate in the departure lounge leads to a path across the complex to the tarmac where the planes sit in a row in ascending order by Catalog Number. Unfortunately, when they finally board the plane, due to one or two minor "bugs" in the existing computer program, the customer might end up sitting on someone's lap. That's just part of traveling — they have to expect some bugs because the airline business is a complex operation.
13. If a customer misses a flight, the airline's modern daily batch computer system feeds a monthly batch system to take care of credits, refunds, etc. In this case customers can go to Accounts Payable to file a refund claim. Then return to the scene of the crime at the end of the month and go to the Refunds Department to receive a refund check. It's that simple!

The Current Case (or, Why's my luggage in Rangoon when I'm in LA?)

Looking at the current airline industry as a contrasting example, we see that even with its imperfections, it is somewhat externally (customer) oriented in a way that puts most other industries to shame. If I show up at the airport wanting to take a flight home as soon as possible, the check-in clerk will do whatever is necessary (within their Business Policy Rules) to get me on my flight.

This one person is empowered to take on the roles of Computer Operator, Baggage Handler, Cashier, Caterer, and Information Desk representative — all in order to get my business. They will find a flight or series of flights (even on another airline) on their computer terminal, tag and lift my luggage over to the baggage conveyor, take my food and seating preference, and direct me to a specific location (usually in a large, complex airport). They will even help me with questions about the location of airport lounges, concessions, restaurants, etc. All of this takes place in just a few minutes (see Figure 12–8).

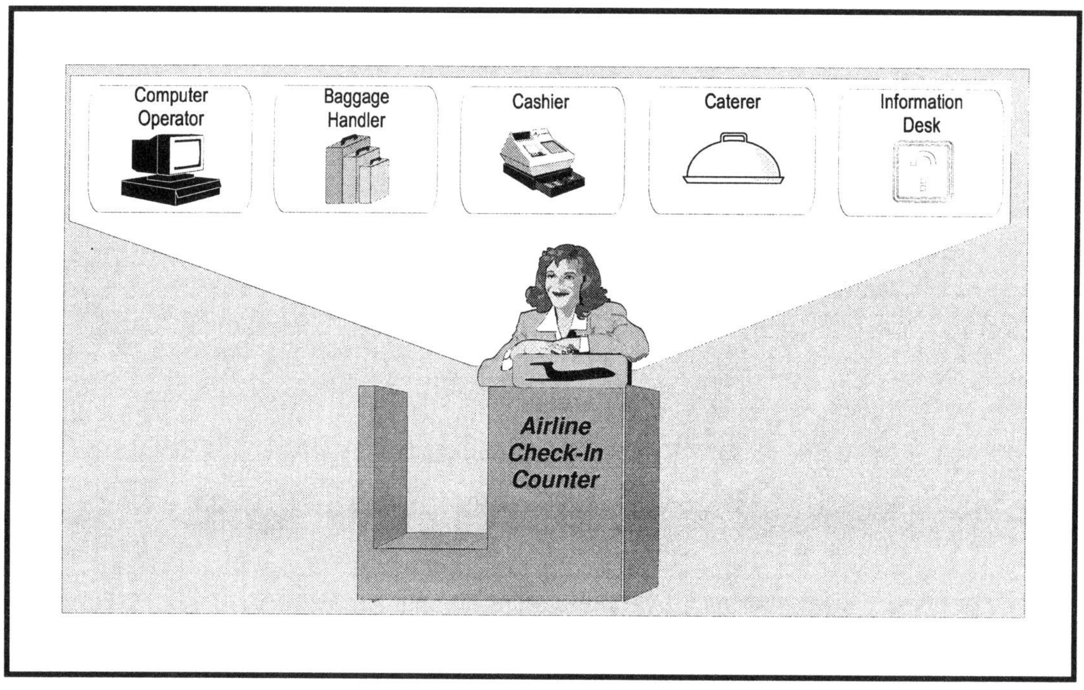

Fig. 12-8: An Empowered Employee

The customer-oriented design of this part of the airline operation is not reflected in all of the airlines' designs.

> For example, the baggage handling design leaves much to be desired. The problem with poor baggage handling is so endemic in fact (and unfortunately accepted), that almost every experienced air traveler takes on board with them some carry-on luggage with their toiletries, valuables, and important documents (this is even recommended by the airlines themselves).

In the current design the Business Event Partition for Customer Wants Us to Get Them to a Destination is fragmented, leading to bits of paper for separate systems and the separation of me and my luggage.[3]

The Future Case (or "You Want It? You Get It!" Airline)

As this is the design section of this book, I find myself concentrating on how to address the design of the baggage system mentioned in the preceding scenario. Baggage is part of my Business Event with the airline, so it should NOT be separated in design. My task as designer is to propose something for the Strategic Planner(s) to approve or deny and not to concern myself with costs.

---

3 This reminds me of the fact that I both dislike and like commuter flights. I dislike them because they are usually less comfortable "pond hoppers" which means I can expect multiple take offs and landings before I get to my destination. I like them because I get to take my precious luggage to the plane and my luggage is at the bottom of the plane's stairs when I get off at my destination. I don't have to cross my fingers and hope my luggage arrives with me. Also, I don't have to wait at a luggage carrousel in some distant part of the airport.

Remember that the initial setup of the banking ATM network was very costly, but after ATMs were in use, it cost banks less to use the ATM network than to use the manual teller-based design. Therefore, it may be that Economy Class Travelers eventually are the ones who pay for a new implemented airline system based on their sheer numbers.

Depending on our project and system scope, we could redesign the entire terminal to create a seamless system where our luggage is never separated from us. After all, we were capable enough to make it to the airport with our luggage, we can surely make it a few yards further.

Let's do an analysis of this Business Event first and then create a new design. The Business Process Model in Figure 12–9 (which I have already annotated with design features of one possible design) depicts what is needed to satisfy our requirements for air travel. The dashed line on the model represents the separation of manual and computer processing and data. A basic rule is wherever our automation boundary cuts across any data flow, we need to utilize some input/output technology.

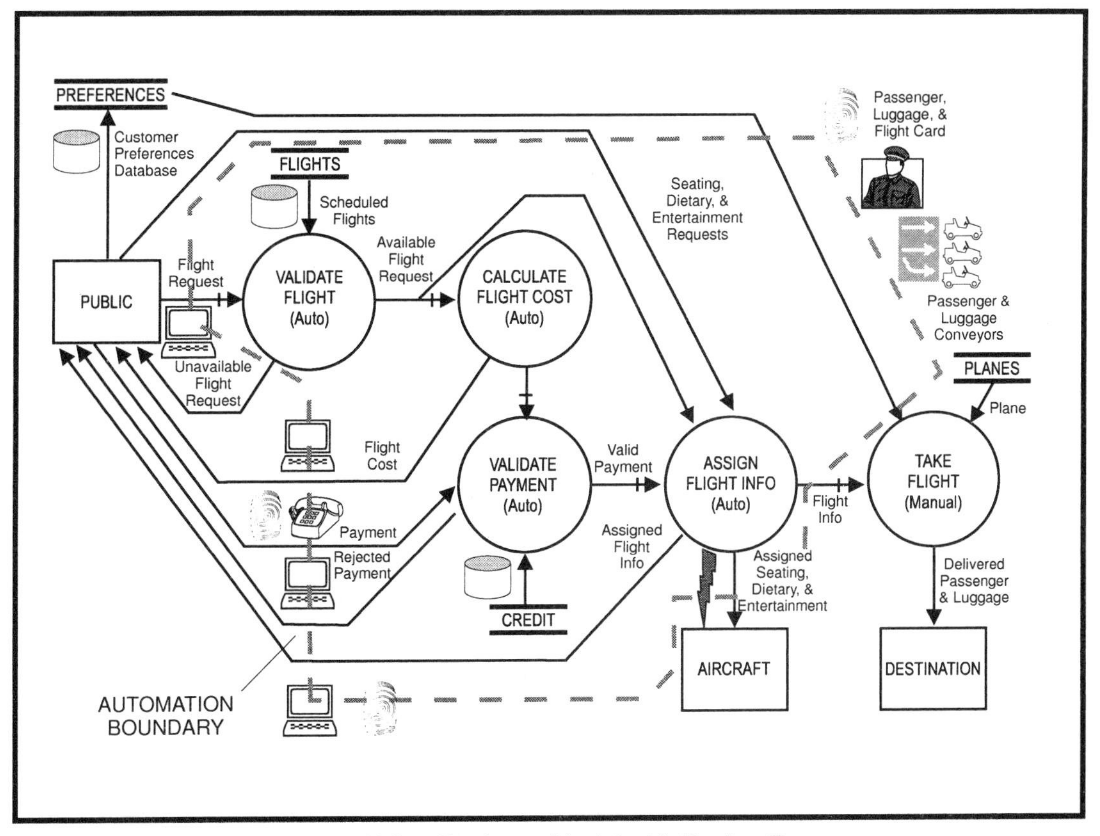

Fig. 12-9: Airline Business Model with Design Features

Given this Business Event Partitioned model, we can now put on our designer hats and come up with a design based upon this complete partition (see Figure 12-10).

Now let me describe this new design based on our Business Event Partition that was created from a Customer Focused point of view. Keep in mind you could create variations of this Business Model for Business Travelers, Occasional/Holiday Travelers, First Class Travelers, or Economy Class Travelers. Each type of customers' Business Event would result in variations on this same design.

In the new design you as a customer enter the airport and place your thumb on a thumbprint reader. If you have traveled before with the airline and have made a reservation in advance, the thumbprint reader retrieves a cohesive set of data from a database that includes your flight information (destination and gate), seating and meal preferences, frequent flier ID, music and movie preferences, etc. You don't even need to know a flight number in this design as this is the airline's internal flight ID and not something you care about. If you have not made previous arrangements, the thumbprint reader has a means for you to make your needs known to the system at the entry to the terminal.

At this point the system knows who you are, which flight you are taking, and everything else required to satisfy your needs in this Business Event. To simplify your travel the thumbprint reader has a direction indicator that points out which gondola-style conveyor to board to take you and your luggage to the correct gate.

The conveyor has a seat for you and a detachable baggage cart for your luggage. At this point an airline employee helps those passengers who want assistance to board the correct conveyor gondola with their luggage. This empowered employee is not a behind-the-scenes agent. They perform such tasks as assisting those needing special assistance, taking care of oversized luggage, and helping foreign travelers with the use of computer-assisted translators.

When the conveyor reaches the aircraft, the luggage carts split off and goes directly into the plane's cargo hold (if we're not allowed to redesign the airplane, that is). As you board the plane, your name is displayed above the seat you selected. In this system your portion of the luggage compartment in the airplane's hold and your seat assignment are directly linked, so that as you exit the plane, your luggage is shuttled off of the plane to meet your exit conveyor gondola (even on flights with transfers). You don't need any bits of paper such as luggage tags or boarding passes for other people to collect from you later.

When you arrive at your destination airport, the baggage compartment re-connects with the conveyor cart you board upon leaving the plane. Then, you and your baggage are driven to a selected destination in the terminal that you previously entered (e.g., hotel shuttle, taxi stand, car rental, etc.).

This is the end of the Business Event Partition as currently identified. If we think "out-of-the-box" the customer is interested in arriving at an airport, but they quite likely have another destination. We as an airline organization could extend our Business Model (and hence its design) to get the customer to his or her ultimate destination.

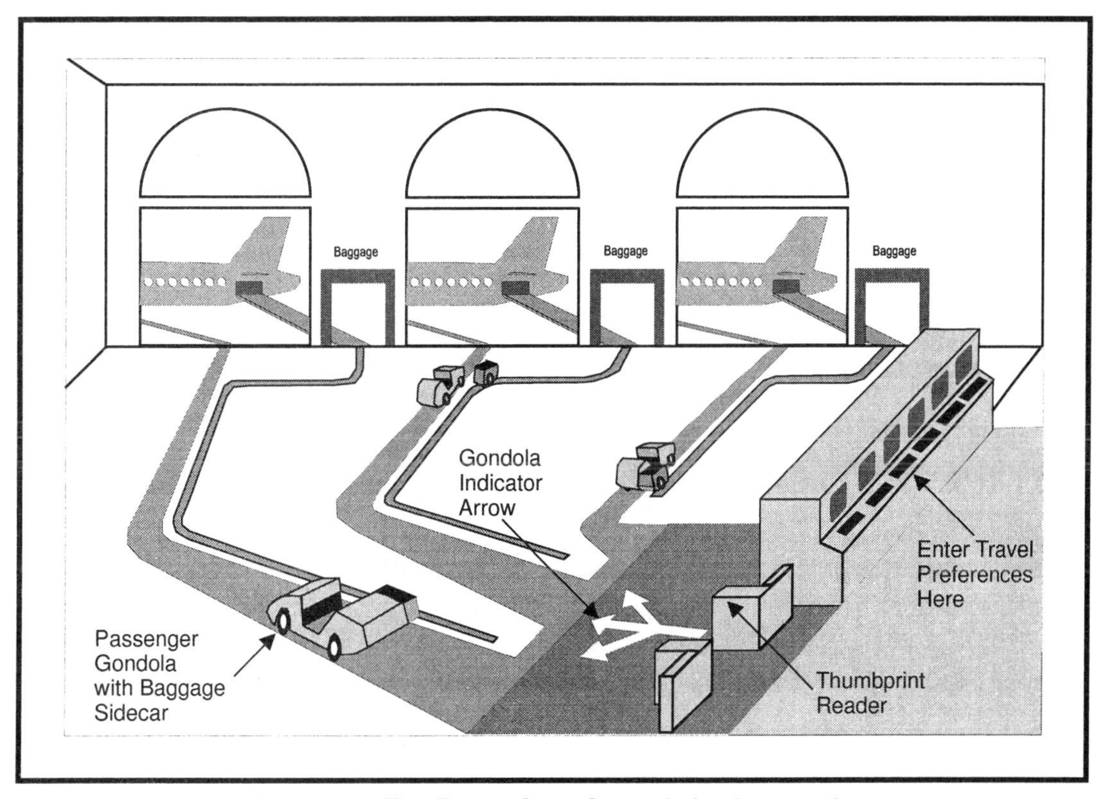

Fig. 12-10: The Future Case Scenario Implementation

From an implementation point of view, any subset of this seamless, ticketless, paperless journey could be phased in.

> For example, the first phase may be to create an Airline Travel Card instead of a thumbprint reader that debits your bank account for the flight's cost (minus your frequent-flier discount, of course) and has your seating and dietary preferences encoded. These advantages could be provided in stages where First Class Travelers get the benefits in the system's first release, the Business Class Travelers get included in the next release, and then Economy Class Travelers. A subsequent phase of the system could introduce the thumbprint reader to eliminate the card entirely.

The Complete Business Event Specification — An Example

Moving from generic examples to a more specific example; at Logical Conclusions one critical Business Event is when a client wants to enroll one or more students in a public seminar. Let's study the set of models through to the design stage that we can use to completely specify the Business Event Partition for Client Wants to Register Student(s) in a Public Seminar. I will concentrate only on the detailed level for this example.

Sample Business Event Analysis Specification: Data on the Move

First, we can produce a Business Process Model for this Business Event (see Figure 12-11). If LCI had an organizational Business Information/Data Model such as an Entity Relationship Diagram, we would probably already have identified the three Entities (**Clients**, **Seminars**, and **Enrollments**) shown as data stores on the Business Process Model. Otherwise we can produce a single Business Event Memory store and repartition it into the three stores after we produce the detailed Information/Data Model.

The Business Process Model in Figure 12–11 should speak for itself, but would obviously have to be supported with an associated Data Dictionary and Process Specifications.

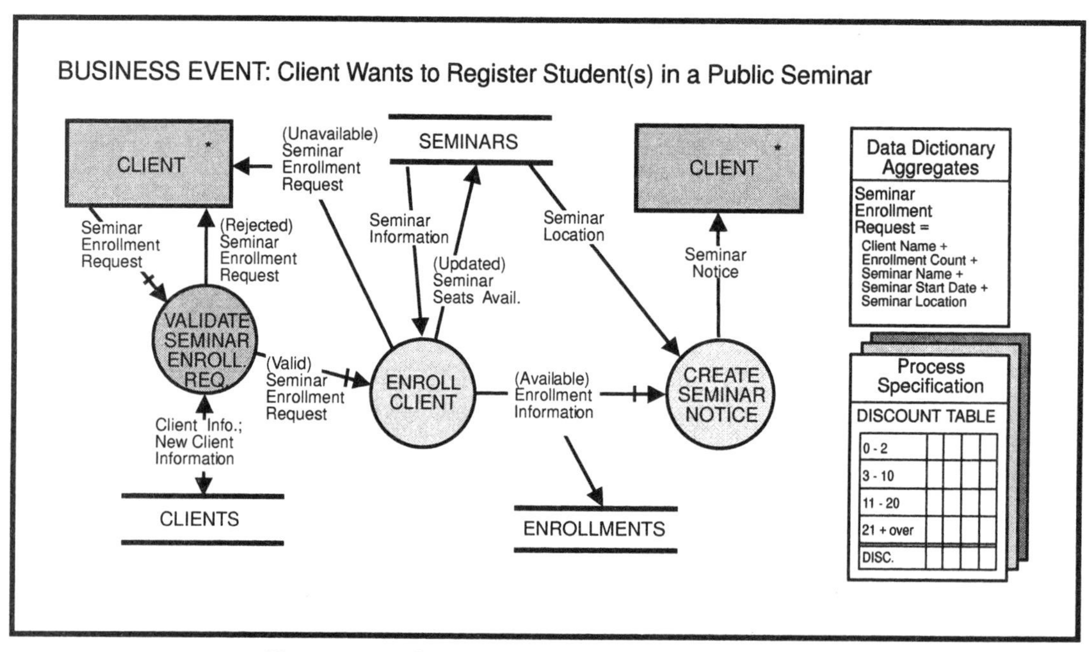

Fig. 12-11: LCI Example Business Process Model

Sample Business Event Analysis Specification: Data at Rest

For the static data side details, we can produce a Business Information/Data Model such as an Entity Relationship Diagram indicating that this Business Event associates three entities together in an **Enrolling** Relationship. LCI's policy is to form one enrollment per **Seminar Enrollment Request**. We don't create separate student entries in this Business Event, so we have a 1:1 cardinality between the Entities in this model. (Note that this is for one Business Event, and that after many occurrences of this Business Event, we may have many **Enrollments** associated with one **Seminar**, and even many **Enrollments** for the same **Client** on that **Seminar**.) The "Enrolling" Relationship Specification can specify this information.

Again, we would support the Information/Data Model in Figure 12-12 with Entity, Relationship, and Data Element Specifications.

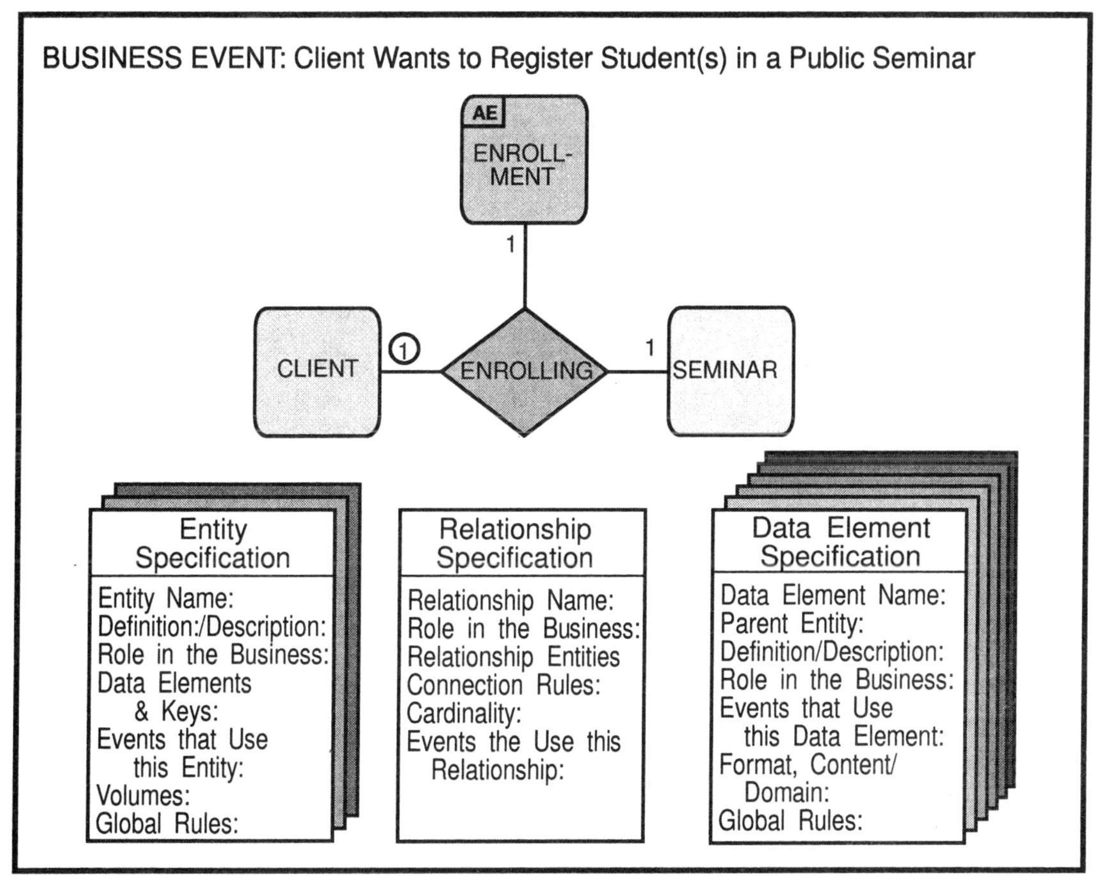

Fig. 12-12: LCI Example Business Information/Data Model

The Beginnings of Design: Data on the Move

We can now use the Business Process Model to show characteristics of our new system design.

> We can overlay an automation boundary on the Business Model. We may also decide to package our Business Event Partition further into two subsystems: an on-line Registration system and a batch Seminar Notification system. (You already know that I do not recommend fragmenting a Business Event in this way, but I want to show you one example of the repercussions of this fragmentation.) This will result in one Business Event Partition spanning two systems. If we do this, we must create a design store between the two subsystems with a second system stimulus for **'Time'** to trigger the batch processing system.

We would support the bounded and packaged Design Process Model of Figure 12–13 with specifications such as screen and record layouts for the flows that cross any design manual-to-automated boundary.

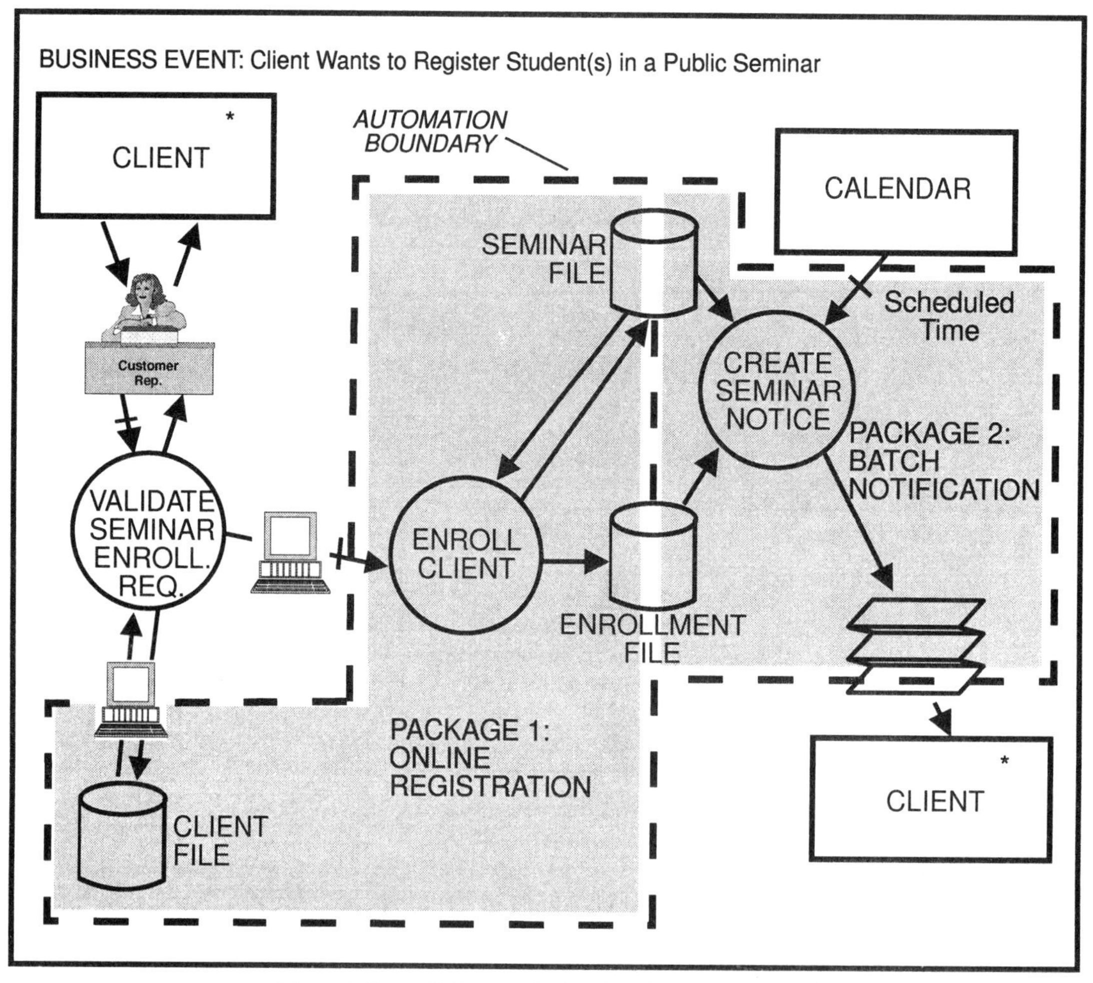

Fig. 12-13: LCI Example Design Process Model

The Beginnings of Design: Data at Rest

We can use the Business Information/Data Model to show characteristics of our new data design. We can form a relational computer database and/or separate manual lists or tables.

To implement our data we need to identify the structure of our resulting files and introduce (if necessary) any access keys. Figure 12-14 shows these relational tables with their access keys shaded and the data associated with each table. We would support these tables with design specifications of database definitions and/or record formats/layouts.

In the Process Model I have shown the three data stores as automated databases, but we could have developed a hybrid of manual folders with automated databases.

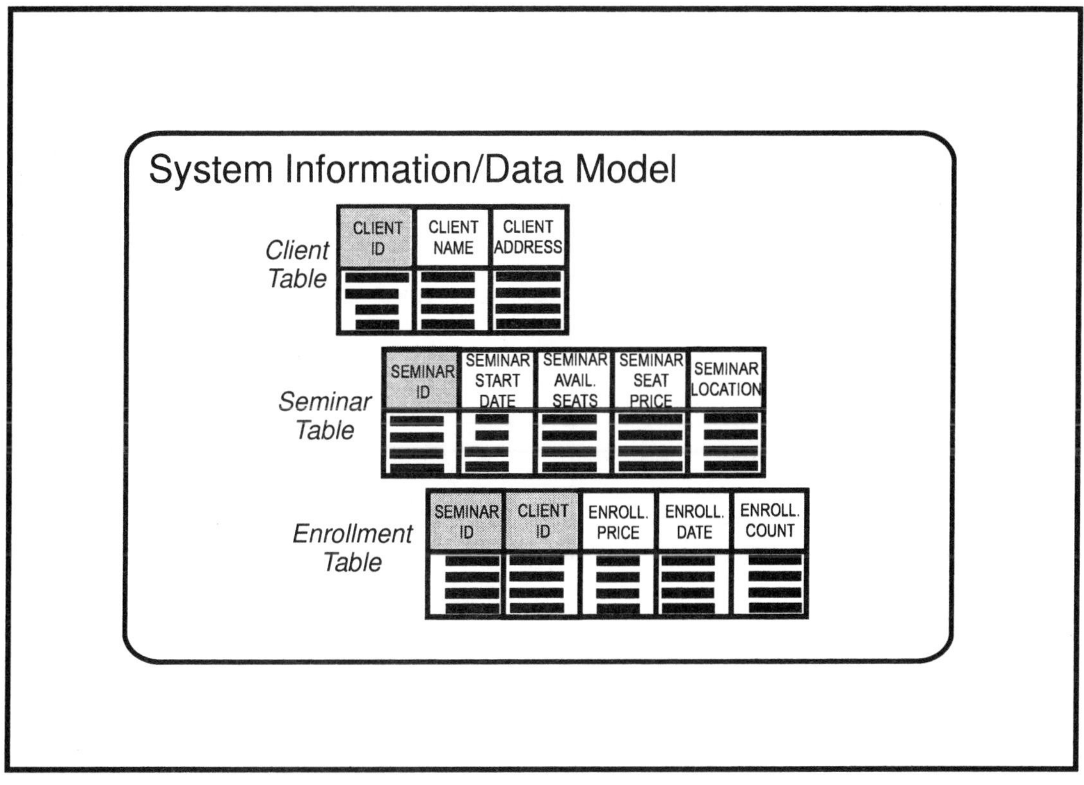

Fig. 12-14: LCI Example Design Data Model

Figure 12-15 shows a sample set of models and supporting specifications brought together to form one Business Event Specification entry in the Organizational Repository (which is the Business Library plus Design and Implementation Specifications and all required System Events needed to make the business run).

This Business Event Specification consists of:

- the detailed Business Process Model
- the detailed Business Information/Data Model
- the initial System Process Model
- the initial System Information/Data Model
- All supporting specifications for the above models.

The comprehensive diagram of Figure 12–15 is the final result of using the Business Event Methodology through the stages of analysis and design. (Note that in real-time systems, we may also add models such as Control Flow Diagrams or State Transition Diagrams.)

The complete set of Business Event Specifications (the Business Library) is the business at a detailed level.

We must take other aggregate models into consideration, such as system- and organization-level models. You should already know that we should bring together only collections of whole Business Event Partitions into system boundaries.

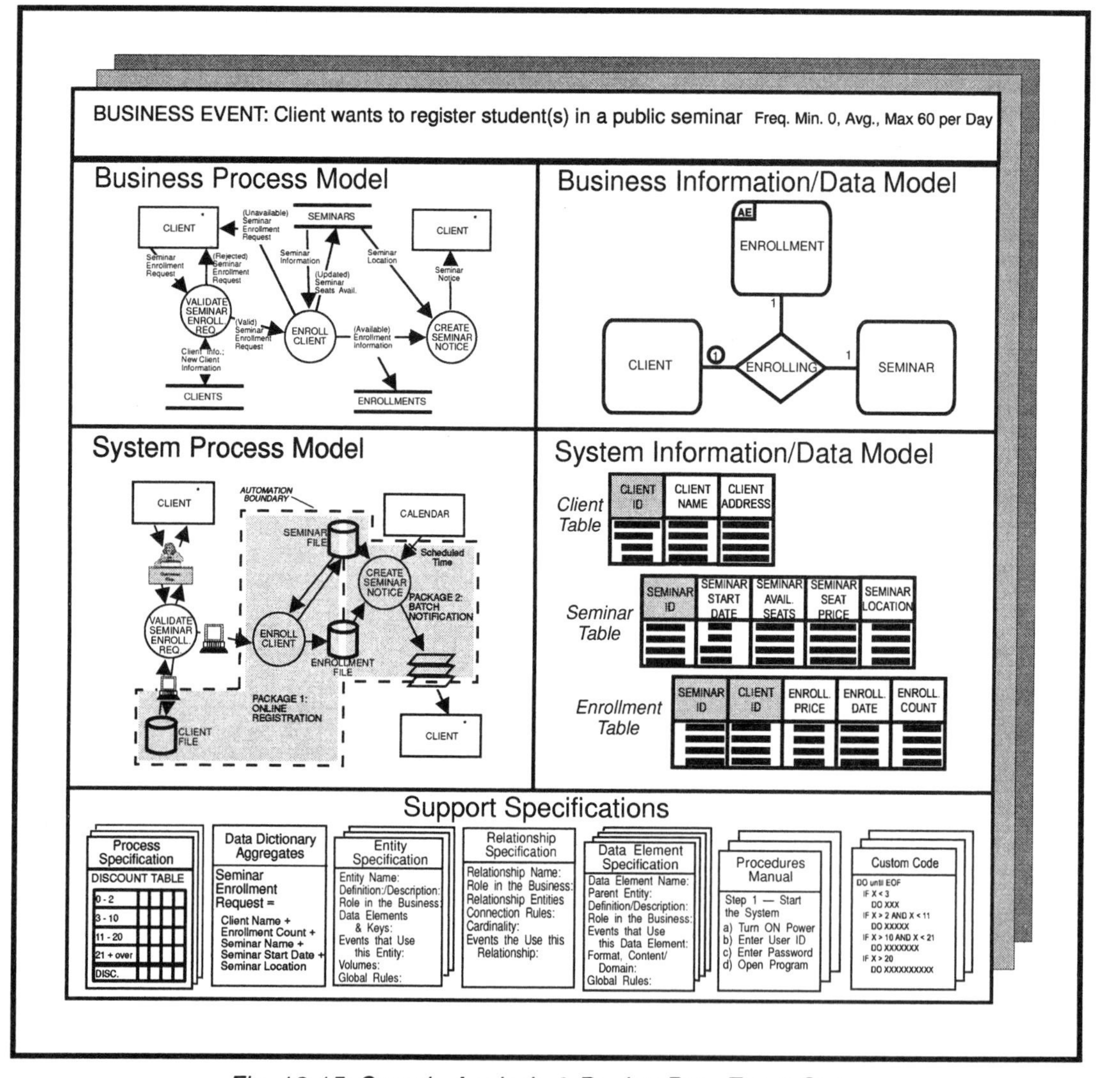

Fig. 12-15: Sample Analysis & Design Bus. Event Spec.

The methodology in Book II addresses these aggregate models for systems and the whole organization.

At the time of writing this book there are a number of powerful work flow and Computer Aided Systems Engineering (CASE) tools available on the market that can take these Business Event Models and generate robust application systems ready for the production environment. These tools point out the need for good "front end" analysis specifications; otherwise we will end up with faster bad systems based on old designs rather than quality, customer oriented business systems.

In my judgment, this complete set of Business Event Specifications *is* the Organizational Repository, the total description of the business — what it does and how it is currently implemented.

Summary

Given the Organizational Repository which contains the complete set of Business Model Event entries in the Business Library along with all of the Design and Implementation Specifications, the Business Library Conservator can fully assist with any future changes of the business and its implementation by having a link from the Business Models to the Design Models as they are implemented today. With the Business Library and the Business Model Event Matrices we can see that any change to the business will be an engineered change (i.e., no dead data, dead processing, or crossing our fingers to hope that nothing is going to get "messed up" by a simple change).

With a Customer Focused Organization and its engineered models we can focus on Strategically Planning where our business should be going in the future rather than having our time monopolized by legacy system issues.

Chapter 13

Strategic Planning via Business Event Partitioning

We build up whole cultural intellectual patterns based on past "facts" which are extremely selective. When a new fact comes in that does not fit the pattern we don't throw out the pattern. We throw out the fact. A contradictory fact has to keep hammering and hammering, sometimes for centuries, before maybe one or two people will see it. And then these one or two have to start hammering on others for a long time before they see it too.

Robert Pirsig
Lila

At the beginning of this book we talked about ***Customer Focused Engineering and Strategic Planning***. I had to bite my tongue (actually, my typing fingers) in that chapter because I needed you to understand the concepts of Business Model Events and Business Event Partitioning to truly put over my view of Strategic Planning. In that chapter I used terms such as "line of business" and "customer need" when I was really talking about what we now know as Business Model issues and Business Events. My concern in that chapter was that the critical issues of Strategic Planning would have been corrupted by the traditional "design-oriented" viewpoint. So let me rectify the wording in part of that chapter and use the correct terms to convey Strategic Planning issues.

The Strategic Business Planning I advocate here differs from classical Strategic Planning in that our first major task is to bring our organization in line with our customer's needs via Business Event Partitioning.

The Strategic Plan is presented in terms of:

- an agreed upon Organizational Goal and set of Objectives
- an organizational Business Model Event List
- a Business Model (void of any implementation characteristics)
- a set of Business Model Event Matrices

- lists of obstacles to achieving the organization's objectives
- a transition plan for Customer Focused Engineering the organization

Note that you would not need a transition plan when creating a Pre-Engineering Strategic Plan. Even if you're a management major straight out of traditional college programs and you're starting off fresh, you must be careful to avoid the trap of creating an Accounting Dept., a Stock Control Dept., a Shipping Dept., etc. with a traditional hierarchy of control, and creating a Strategic Plan for each department or bureau.

The focal point of any Strategic Planning approach is the set of Organizational Objectives that defines where the organization is today and where it expects to be in the future. Without this set the organization cannot define the processing and information it needs to support this movement.

The Key Organizational Questions

The key questions that an organization must answer in my view of Strategic Planning are repeated and further clarified in Figure 13–1. The details of "how to" accomplish them are discussed in ***Book Two: The Strategic Business Engineering Methodology***.

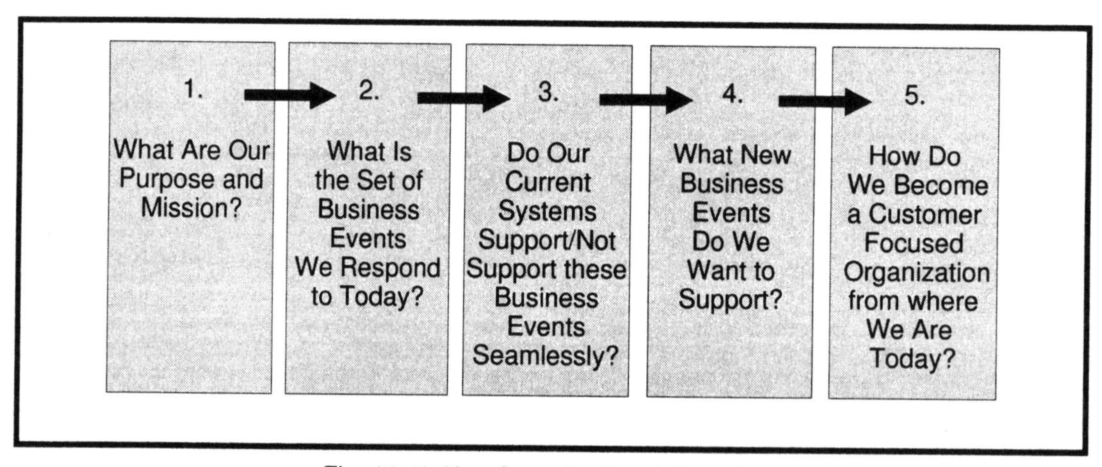

Fig. 13-1: Key Organizational Questions

1. What Are Our Purpose and Mission?

This is a description of the business our organization is in and what we want to be, or do, in current terms. This may be an overall purpose or it can be broken down into a set of quantifiable and measurable Organizational Objectives that contain the "vision" of the Strategic Planners. All projects use them to develop a further breakdown into individual objectives. The objectives in the Mission Statement should be displayed throughout the organization. It is this that each and that every employee and computer system are working to support

2. What Is the Set of Business Events We Respond to Today?

We have already seen in previous chapters that the managers in each department of an organization will typically have their own view on what the actual business is. The heads of an organization and the Customer Focused Engineers cannot take any particular department's view, so to answer this question we need:

- A definition of the existing business boundary between the organization and the external world and all outside customers with whom we interface.
- An identification of organizational Business Events at the boundaries of the organization (i.e., a Business Model Event List) with identification of the critical Business Events that are our reason for being in business or our main revenue generators. (Identify Regulatory and Dependent Events, but don't include them in our critical Business Events List.)
- A high-level map of our organization's existing implementation such as departments, divisions, computer systems, and computer platforms.

An Organizational Business Event Objective or set of Objectives can be developed for each Business Model Event using the Business Model Event List. These Organizational Business Event Objectives are quantifiable, measurable statements of a Business Event Partition's required performance. The Objectives must be stated such that it will be quite clear when they are, or are not, being met.

3. Do Our Current Systems Support/Not Support these Business Events Seamlessly?

We must evaluate how well our organization accomplishes the needs of our customers today. We do so by analyzing how the currently implemented systems respond and contribute to the organization's purpose and mission via the identified Business Model Events.

One way of assessing how well we support any specific Business Model Event is to observe how much fragmentation occurs across our existing manual and automated systems. We should be able to assess how we are meeting each Organizational Business Objective by projecting the optimum response/performance statistics onto each Business Event Partition and compare them with the current statistics.[1]

The organization's purpose and mission objectives decompose into specific Business Model Event Objectives, which in turn decompose into specific Business Policy for each Business Model Event. I believe that the only way the organization can truly track its performance is by gathering honest statistics (metrics) for each Business Event Partition as it is implemented in the current system(s) and then compare these with the new Customer Focused implementation advocated in this book. We should at least gather the statistics for our critical Business Events.

---

1 Organizations frequently have reasonable statistics or metrics of their operating costs for their manual systems, but not for their computer systems. This step may be difficult, but nonetheless essential, for the computer environment.

One of the main benefits of gathering statistics for a Business Model Event through existing systems is that it provides us with some very important metrics for both devising a Customer Focused Engineering Plan and monitoring its effectiveness over time. These metrics should reveal such things as:

- The long term savings in the cost of operations gained by Customer Focused Engineering.
- How best to use limited staff resources and avoid the common situation where a department is too busy keeping existing systems alive to give Customer Focused Engineering its necessary commitment.
- Whether some of an existing system can be salvaged.

It's during this step of Strategic Planning that we analyze the internal configuration of the organization by:

- Modeling current system boundaries. This task examines *how* we do business today. It involves gathering information on existing manual and automated systems, and identifying boundaries and interfaces between systems. Knowing the current system boundaries allows us to identify the systems (manual and automated) making up the organization, and determine the cost of operating each system.
- Mapping Business Events to current business design. Given current design boundaries we map Business Events to the current systems and see how the responses to these Events flow through our systems. We are also interested in how the systems talk to each other when responding to the same Business Event. We can use a matrix to map the Events. This yields a Business Model Event/Current System Matrix as shown in Figure 13-2.

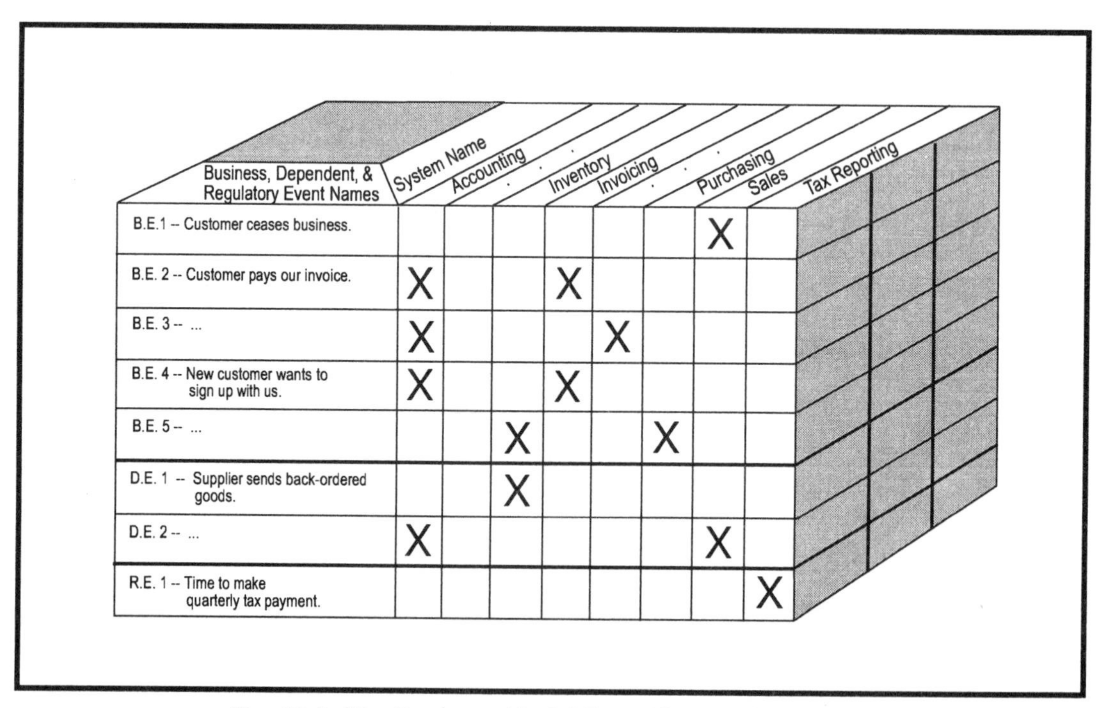

Fig. 13-2: The Business Model Event/Current Sys. Matrix

- Identifying manual and automated system cost and deterioration. This task examines the history of each system (manual and automated) to determine the rate of modification requests and the ease of satisfying them. Ideally, we would plan first to engineer the mission-critical Business Events, although we may be forced to start with the systems in most need of replacement.

4. What New Business Events Do We Want to Support?

In addition to evaluating the organization's current position, we must also identify what changes, additions, or deletions are required for future business opportunities. This can also be stated in high level terms — if we want to be more specific in our future Organizational Objectives we can:

- Identify future Business Events — By reviewing any new market business opportunities or the operations of competitive organizations, Strategic Planners can identify new Business Model Events to which the organization should respond in the future. Innovations in technology may also lead to new avenues of business and hence new Business Model Events.
- Assess feasibility and risk — We should analyze the feasibility and risk of responding to (or not responding to) the new Business Model Events, examine the technological and resource implications, and look for any existing processes and data that can be re-used.
- Produce the future organizational high-level Business Model.

One of the major advantages of having a Business Event Partitioned Business Model is it allows us to see the true details of the organization and therefore realize where we can take advantage of strategic alliances. These will be easily identified where a Business Event Response causes a Dependent Event to come back to our organization. In this regard we should especially evaluate Dependent Events that are timing-oriented and stimulated by a Control-triggered Stimulus from the clock or calendar. These types of Dependent Events are prime targets for one aspect of Strategic Planning. We can better serve our customers by being in control of as many aspects of filling a Business Event; especially those currently beyond our organization that show themselves as Dependent Events.

When forming strategic alliances, we need to think "outside the box," look at all External Interfaces, and ask what we could do for our Customers before they even stimulate our organization with a currently supported request.

Using the Business Model to Look to the Future

Some examples of this important part of Strategic Planning that apply to creating a Customer Focused Organization are:

> If we find that part of filling a **Customer Order** involves extracting **Materials** from our inventory and we get these **Materials** from a **Supplier**, then we will see these **Materials** trigger our organization via a Dependent Event on the Business Model. This Dependent Event is where we respond to a **Supplier** based on an outgoing requisition for more **Materials**.

Using this example we can see an opportunity to improve the business by bringing the Dependent Event under the wing of the Business Event. We can accomplish this by seeing on the Business Model that the **Materials** are extracted based on a **Customer Order**. It's at the point of extraction that items in our inventory store can fall below a re-order level and we can immediately issue a requisition for more supplies. In other words, the Business Event Stimulus, **Customer Order**, triggers the requisition and not a disjointed control-triggered (false) Event stimulus.

> We can go one step further and form a strategic alliance with that **Supplier** where they keep track of our reorder levels and our current inventory levels. (In this case we may not even have to keep track of our own inventory.) We send them a record of every item purchased as part of a Customer Order Business Event. Then they know before we do that it's time for them to fabricate and/or ship us the **Materials** (still on a just-in-time basis if that's part of the alliance agreement). After all, the **Supplier** is the one who has the most incentive to keep our **Materials** in stock to meet our **Customers'** needs because we pay them for what they supply us.

Dependent Events may also be service related and we may find services previously outsourced that we want to bring within our organization in order to have more direct control over satisfying our own customers' Business Events. I mentioned at the front of the book that I expect Customer Focused Engineering to make an organization grow.

In this way we should expect vendors with whom we outsource some of our tasks to prove that they produce efficient and effective products and services just as our people inside our organization must prove they are the most efficient and effective providers of products and services. This is just another example of the healthy nature of competition except that it is between an organization's staff and that of a vendor (as opposed to their organization's competition).

If we do bring in the Dependent Event and its associated processing and data that is beyond our organization, we will need to extend our Business Model. It would be very helpful if our supplier had a Business Event Partitioned model to use to easily extend ours.

An organization's Strategic Planners need to allow and even expect the Customer Focused Engineer to create radical new designs for implementing Business Events (e.g., no new Accounting, Order Entry, or Invoicing Systems). Along with radical new designs for Business Event Partitions we should also consider radical new designs for any System Event Partitions that are introduced to support our Business Event Partitions.

> For example, it's highly likely we will, and I hope we do, use human beings to implement some or all of a Business Event Partition. This will introduce System Events such as one for Pay Employees.
>
> A radical implementation might be a "real-time" payroll system. In this case payroll is no longer performed on a periodic batch basis. In the new implementation, payroll machines could be installed at the building exits. Employees could use a personal Personal Identification Number (PIN) and/or have a face scanning device that will allow them to access their earned payroll on demand. The employees could get paid in any increment they wish (e.g., they could withdraw their last hour's earnings or get all their earnings since the last time they made a withdrawal — which may be hours, weeks, or months.

Benefits from this design include keeping the payroll funds in the organization's internal bank and the company earning the interest on the balances. This may make the organization assume the task of managing the investments that earn the interest on the payroll dollars. The proceeds from the interest earned could be split between the organization and its employees. Part of this interest money could help fund things such as employee benefits programs providing more incentives for employees and employers. This way the money doesn't leave the organization to pay a bank that in turn pays the employees. In addition, the organization gets the benefits of having the funds available to it and the employees take on the responsibility of managing their own payroll disbursements.

As a third example, let's use one based on the government because, as I say in my seminars, the older and larger an organization, the harder it will be to restructure into a Customer Focused Organization. In this light, the U.S. government has two strikes against it, because in the U.S., organizations don't get much older or much larger. However, please keep in mind that it is exactly this type of institution that stands to gain the most potential benefits from Customer Focused Engineering provided that assigned project members are not aligned with any particular Department, Bureau, State, etc. *I may be cutting my own business throat here in that my business also survives on an internal corporate sponsor. I can only hope that an individual high enough in an organization is willing to stretch out his or her neck to implement the ideas in this book.*

So that I don't get associated with any political alliance, let me use the U.S. Post Office as an example. This scenario assumes I have a Business Model as input for my new designs. When I look at the operations of the Post Office, I see very little logical difference between its business requirements and those of any other shipping, package delivery, or courier service.

We could apply Customer Focused Engineering concepts in such an environment by forming an alliance with a phone or cable TV service company to provide a new type of service. This service would be where correspondence (letters) are formed on an interactive, voice recognition-based TV system. The system would of course provide options for video-, audio-, or text-based correspondence. In addition, for packages the system would replace the current addressing scheme with an absolute longitude/latitude based addressing system using the global positioning satellites (GPS) already available with no need for Zip Codes, Mail Stops, or other addressing schemes. Since the new addressing system would be entirely logical, the "mail" would be delivered to a person no matter where they were, instead of ending up in a mail box waiting for them to retrieve it. This new system would position the Post Office to become the natural "on-ramps" and "off-ramps" to the Information Super Highway.

Is the payroll scheme in example 2 above too far off the wall? Or, can we acknowledge that batch payroll is an artifact of the Payroll Department's operations?

Radical solutions should win praise from the Strategic Planners, even though the budgeting of ambitious designs may be prohibitive. After all, it's up to the Customer Focused Engineer to come up with the new designs and, it's the Strategic Planner's job to come up with the money.[2] As Strategic Planners, we should create an award for the most radical designs that our people propose, even if we don't implement them.

5. How Do We Become a Customer Focused Organization from where We Are Today?

Unfortunately, it's not practical for an existing organization to shut down its business and conduct a "big Bang" repartitioning of all its departments, divisions, etc. Therefore, we have to incrementally repartition organizational activities *in situ* via a Strategic Plan.

Obviously the plan will identify how we can create a functional repartitioning of our organization based on Business Model Events. To create this plan we need to know:

- the existing partitioning of the organization
- the Business Model Event/Current System Matrix developed in Step 3 (Figure 13-2)
- the resources available, and
- the Business Event Methodology

This plan shows the Business Model Event implementation sequence and the phasing in of Business Event-partitioned systems, and the phasing out of classically partitioned systems or their subsets. The methodology and project issues such as planning, budgeting, scheduling, etc. are the subject of Book II of the series.

Because we can isolate Business Model Events that are connected only by files, we can treat each Business Model Event (or collection of Business Model Events) as mini-projects (see Figure 13-3). This becomes very helpful when developing a Strategic Plan for phasing out old systems and phasing in Business Event Partitioned systems. When we look at how we might phase out old systems and engineer the entire organization, the problem is choosing which strategy to use. We could use a number of strategies for this.

- Identifying the most critical Business Events in our organization (e.g., those that generate revenue) and implementing them in total, one at a time.
- Identifying which existing systems are "on their last legs" and how many Business Events traverse through them.
- Identifying the Business Events that occur most frequently and implementing them first.

2 Of all of the technical people I've trained in Customer Focused Engineering, what comes in second to their primary obstacle of being warped by existing designs, is their self-imposed restraint and second guessing that management will automatically reject their designs based on budget restrictions. This may lead to a brilliant new design not even being presented to the Strategic Planners.

The strategy I recommend most is one where the Customer Focused Project members study and "turn off" (comment out code and procedure manuals) all of a Business Event's fragments in the existing systems one Business Event at a time. The initial stimulus for each Business Event is "trapped" at its entry into the organization and is then routed to a new Business Event Partitioned system (see Figure 13-4).

Another strategy would be to prioritize the sequence of system replacement so that the system with the most critical Business Event fragments is engineered first. Figure 13–3 shows that the Stock Control System is the first one to be replaced. Within the new system we will identify each Business Event fragment (see Scenario B in Figure 13-5). When Project #2 starts (Accounting), we will append the fragments of the Business Events together and so on until all systems have been replaced. The result will be a set of implemented systems as shown in Figure 13-4.

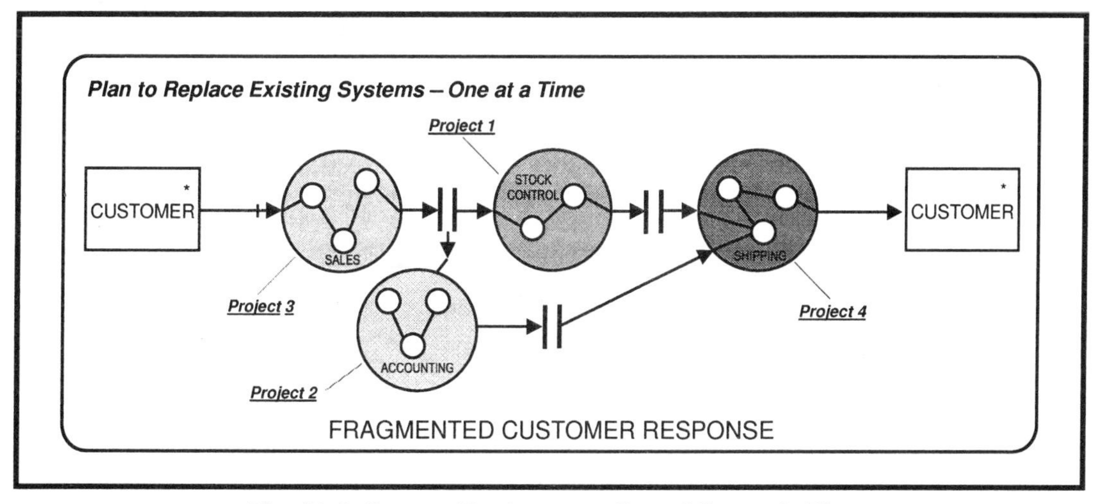

Fig. 13-3: System Replacement Based Strategic Plan

I find this "turning off systems one-at-a-time" strategy more awkward as we will probably need to provide the most temporary interface stores and procedures.

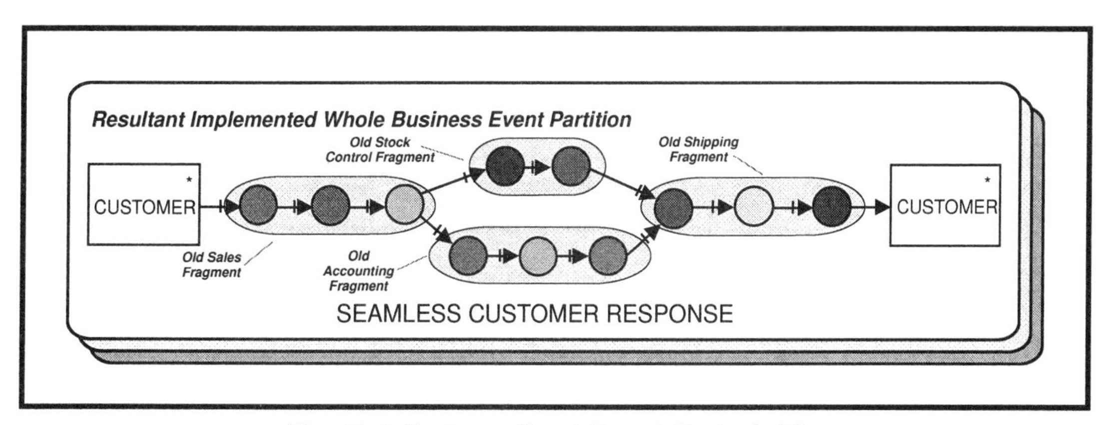

Fig. 13-4: Business Event Based Strategic Plan

Yet another strategy would be to identify the collection of Business Events that create the most important set of Data Elements for running the organization within the Business Model Event/Data Element Matrix. Once these are identified, the strategy would be to combine the Business Events that create this data with their appropriate Retrieve, Update, and Delete Events as the first phase of the Customer Focused Engineering project. With this strategy of implementing whole Business Events at a time we would immediately arrive at Scenario C in Figure 13–5.

The objective of all these strategies is to move from a fragmented Business Event structure as shown in Scenario A of Figure 13-5 to a seamless, Customer Focused Organization as depicted in Scenario C.

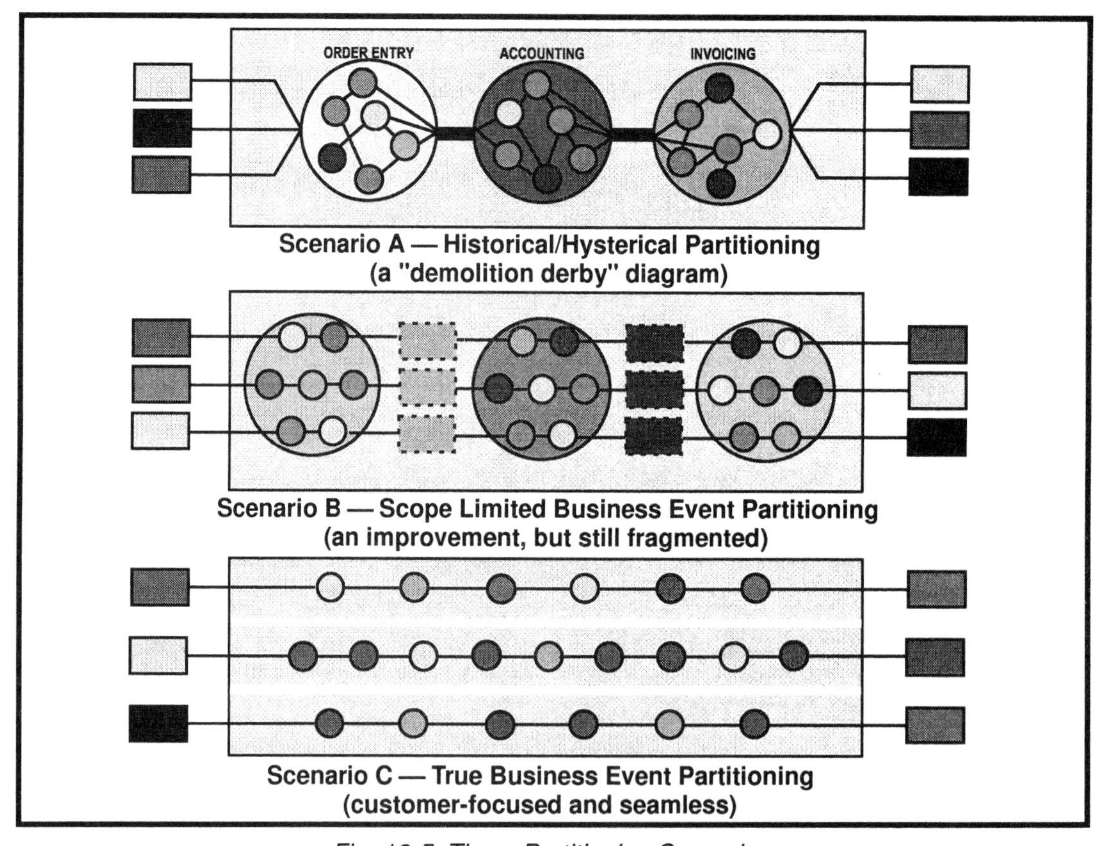

Fig. 13-5: Three Partitioning Scenarios

Note that for the above strategies, we will need to create some intermediate databases and files to support the migration of data during the conversion efforts.

Summary

Now that you understand the Business Event Methodology, I can give you my **real** succinct definition of Customer Focused Engineering.

Customer Focused Engineering is the re-alignment of an organization and its resources along Customer Business Event lines.

I believe that the Business Event Methodology is a significant new direction for the entire business community. Applying this approach to an organization will produce a business structure that I believe no other competing organization can beat.

Chapter

14

A Logical Conclusion — The Final Analysis

The greatest achievement was at first and for a time a dream. The oak sleeps in the acorn; the bird waits in the egg. Dreams are the seedlings of realities.

James Allen
As a Man Thinketh

In this book, we have examined how organizations took some archaeological wrong turns by implementing new systems based on their old ones. I hope that I have shown you the benefits of using the Business Event-based, Customer Focused Engineering Methodology to eliminate these archaeological wrong turns in future systems. I also hope you can now see what your business really does and make the fundamental structural changes needed to let your customers drive your business processes.

Customer Focused Engineering's Objectives and their Conclusions

Now we can look back at the objectives for creating a Customer Focused Organization as stated in the first chapter and see if we have addressed and satisfied them:

- To put the customer first — satisfy customers' needs and expectations by structuring our organization to seamlessly respond to these needs.

  The logical Business Event Partition is completely customer focused. It is not based on any existing implementations or old designs within an organization. Instead, it is driven entirely by the customer's need — the Business Event.

- To get back to business basics — focus on **what** we do rather than **how** we do it by knowing and following the organization's Mission Statement.

  Given the Mission Statement and by applying the concepts of business analysis modeling allows us to focus on the **what** we are in business for (Business Events) and to not be warped by the old **how** we accomplish it (the System Events).

- To cut red tape — achieve dramatic and measurable improvements in the performance of the organization's implemented systems by creating the most effective processes for delivering products and services.

  The majority of the red tape will be due to the old design. The most dramatic improvements will come from unfragmenting our Business Model Events, especially at the existing organization's seams (departments, bureaus, etc.). Therefore, our new Business Event-based system will remove a significant amount of processing and storage of data required by the old design.

- To replace old systems that may be hurting the organization today with quality engineered systems — the new engineered systems (be they manual or automated) will be faster and easier to install and maintain.

  Engineered systems based on Business Model Events should never hinder the organization; they are business based. When the new implementation follows a Business Event Partitioning, systems will be easier to install and maintain because they are based on business issues (as opposed to technical or internal political issues). As Business Event Partitions can be looked upon as mini systems in their own right, changes should be easily isolated, reducing if not eliminating (via the Business Model Event Matrices) maintenance "ripple effects."

- To create systems that promote data conservation and take advantage of the opportunities for re-using data and processing whenever possible.

  The Business Library Conservator, using the collection of Business Model Event Matrices and the Business Library can easily ensure the reusability and conservation of both data and processes.

- To attain the highest quality in the development and delivery of our products and services — use an engineering development discipline, empower employees, and use the best available technology in implementing our systems.

  We will only realize the full benefits of using the best technology and of having empowered employees after we have a Business Event Partitioned Business Model that allows us to take advantage of these designs and their implementations.

- To satisfy the organization's strategic mission — allow for constant improvement in the organization by producing a flexible environment for future change. This means that we create an environment in which we can make business changes that will not affect technology issues and where we can make technology changes without affecting business issues.

  The Business Event Methodology works in perfect unison with the principles of Total Quality Management. Constant system improvement becomes a far more powerful technique when applied at the analysis and design phases in addition to during the implementation phase of system development. The Customer Focused Engineering Methodology deliberately separates technology issues from business issues as shown in the Business Library entries. It is this separation that makes it possible to change technology issues independently of business issues and vice versa.

Now that we know the Business Event Methodology, we can add the goal of dividing up the budget along lines of Business Event Partitions rather than by traditional old design structures.

The problem we have in today's business world is structural and can't be solved with a quick techno-fix. What we have is 21st Century technology grafted onto a 19th Century business structure. As long as we keep focusing on new accounting systems, implementing document imaging systems on Intranets, and perpetuating fragmented responses to customers, we'll be mired in the past with the rest of the world.

I believe that we need to shift away from the typical, outdated, departmentalization based on similarity of task and its associated organizational structures of boss-subordinate relationships. This shift must be to an organizational partitioning based not on outdated human departmental or computer system boundaries, but on a Business Event Partitioning paradigm that is more fundamental to a true business view.

The Industrial Revolution paradigms have served us well to get to where we are today, but now they actually hinder us. The need to get beyond these old paradigms is the basis for Customer Focused Engineering.

I believe the use of the Business Event-based, Customer Focused Engineering Methodology is a significant new direction for the entire business community and for the systems industry. Applying the approach to an organization as a whole yields significant clarity regarding how the organization's old systems can get in the way of its business. I am convinced that using the Business Event Methodology, as described in this book, reveals a true business view, and as such, it is the most logical basis for a new structure of an organization.

If we use this business structure and apply engineering principles to the processes of creating manual and automated systems, we can gain the benefits of building safe, high-quality, cost-effective systems that are seamless and customer oriented.

When we have developed our Business Specification based on Business Event Partitioning we never need to conduct this analysis activity again — it's only maintenance from here on in.

When we've created a Customer Focused Organization, we will have reached the ultimate goal and broken the chains that bind us to 19th Century structures; the structures that will probably keep our competition in the thrall of every expensive new technological fad to come along.

I believe that we now have proven methods and models to engineer an organization successfully. The main ingredient I've found missing is a business-oriented methodology that encompasses these proven methods and models.

The Customer Focused Organization should grow and last as long as the business need it satisfies is present.

The methodology described in this book (and the management perspective in Book II of the series) allows us to bring the entire business view together into one unified model. By strategically applying the principles of Customer Focused Business Event Partitions, we're in a position to restructure our entire organization.

I hope that it won't be too long before we find retail outlets where we can purchase engineered Business Models that our organization can adapt to our business needs before using them as the basis for creating new manual and automated systems. In the near future it may be possible that any new systems needed to satisfy a customer's Business Event may be generated automatically (e.g., procedure manuals and computer code).

Championing Customer Focused Organizations

The full evolution to a Customer Focused Organization depends, to a great extent, on a group of people I call the "Internal Champions." These are typically the folks who have made my own business successful. These champions in each organization are the ones who want to "do it right" with the best techniques available (usually in the midst of chaos) and who often start the internal "revolution." These people can be at any level in the organization and their dedication to developing quality, Customer Focused Systems is, at times, even pursued at their own expense.

You have the knowledge now. You can revert back to the old way of doing things or go away and change the way you do business and make your organization the leader in its field of business.

Appendix A

Glossary

A

Analysis — The methodical study of a system or systems to understand, define, and document their business requirements independent of any existing design or implementation considerations.

Attribute — An indivisible item of data that takes on a value (also called a Data Element).

Automated — A qualifier for a portion of a system that is implemented with a computer.

B

Batch Processing — A processing approach in which transactions are accumulated over time for processing in a single execution of a system.

Bundled Store — A type of data store that contains multiple unrelated items of data.

Business — The essence of **what** an organization does. Those operations that pertain directly to the act of satisfying the organization's customers' needs.

Business Customer — Individuals, clients, agencies, systems, and other forms of impetus that are external to an organization and over which it has no control.

Business Event — An external incident originating at our organization's customer that places a demand on us to which we respond in order to accomplish our strategic mission.

Business Event Memory — The collection of all the stored (i.e., non-transient) Data Elements necessary to support the processing for one Business Event.

Business Event Methodology — The collection of techniques used for analysis, design, and implementation when engineering an organization's systems along business lines to achieve ultimate customer satisfaction.

Business Event Partition — The most natural business structure for satisfying one specific need of a customer. A Business Event Partition consists of a Business Model Event Stimulus plus all associated Processing, stored Memory, outgoing Responses, and Recipients that constitute the organization's complete reaction to an Event from an organization's external customer.

Business Event Partitioning The process of identifying and documenting the processing and data used by an organization to satisfy a single Event from an organization's external customer

Business Event Partition Elements The constituent parts of a Business Event Partition including the Business Event Source (the need), the Business Event Stimulus, the Business Event Processing, the Business Event Memory, the Business Event Response, and the Business Event Recipient.

Business Event Processing All the business logic and its associated transient data required to produce the total response to an Event from an organization's external customer.

Business Event Recipient An external individual, agency, organization, system, or other entity that receives the Response of an Event from an organization's external customer.

Business Event Response All output, such as data, control, products, and/or services resulting from one Event from an organization's external customer.

Business Event Source A customer — an external individual, agency, organization, system, or other entity that creates a stimulus to our organization.

Business Event Specification The complete set of models and supporting specifications and textual documentation needed to describe a Business Event Partition.

Business Event Stimulus A demand (input data, control trigger, or incoming material) resulting from a Business Event that activates part of our organization.

Business Issues Those things pertaining to **what** an organization does regardless of **how** it's implemented.

Business Library The documentation (the set of models and supporting specifications) for all Business Model Events in the organization.

Business Library Conservator The person or persons in an organization who are chartered to maintain the integrity of the organization's Business Library and its associated matrices.

Business Logic The rules governing the processing of data and/or control within an organization.

Business Model A non-redundant set of facts that describe the Customer Focused business of an organization. This is a business tool for declaring an implementation-independent view of what has to happen to satisfy the needs of an organization's customers.

Business Model Event Any of the three Event types (Business, Regulatory, and Dependent) that belong on the Business Model.

Business Model Event List A list of all the Business, Regulatory, and Dependent Events to which an organization responds.

Business Model Event Name A unique and descriptive name for a Business Model Event, preferably consisting of a whole sentence.

Business Objective A description of a business need that should be satisfied.

Business Policy The essential data and processing, as defined by the Business Policy Creator, that must be put in place to satisfy the organization's mission.

Business Policy Creator The organization's decision maker or makers who are authorized to dictate which data, processing, and control are used by the organization to respond to its customers' needs.

Business View An implementation-independent view. A qualifier for data or processing indicating that all design characteristics of the item have been removed.

C

Cardinality The numerical relationship between any two or more Entities on a data-oriented model (i.e., one-to-one, one-to-many, and many-to-one).

Cohesive Entity A grouping of all strongly related Data Elements. This collection of Data Elements is typically associated with one identifying key.

Control Flow Diagram A control oriented model that can also show data movement or be tied to an associated Data Flow Diagram. This model is helpful for modeling the complex control issues of a system.

Control Oriented Model A type of model suited for modeling the flow of control between different states in a system.

Control Triggered Stimulus A type of Event Stimulus that contains no data. One typical source of this stimulus is the calendar or a clock reaching a predetermined point in time.

Convenience Store A data store that is created to satisfy design or implementation reasons. A convenience store is not an essential business store.

Critical Business Event A Business Event that is essential for carrying out an organization's mission. In an organization that exists to make profits, this event is their main revenue generator.

Customer The external individuals, clients, agencies, systems, and other sources of impetus that are external to an organization and over which it has no control.

Customer Focused Engineer The person chartered to structure an organization along Business Event lines using an engineering discipline and a Customer Focused Engineering Methodology.

Customer Focused Engineering The re-alignment of an organization and its resources along Customer Oriented, Business Model Event lines.

Customer Focused Engineering Plan A well defined series of steps to engineer an organization's activities *in situ* with minimum impacts on the organization's day-to-day operations.

D

Data — The factual information used by an organization as the basis for calculations or processing.

Data Aggregate — A cohesive collection of Data Elements that may be a subset of a stored Entity or transient data flow.

Data Conservation — **(1) For Processing:** The practice of ensuring that data entering and leaving a process, procedure, or system within the organization is conserved (i.e., all data are used to derive the outputs and all outputs could be derived from their inputs).

(2) For Memory: The practice of ensuring that any data that is Created (captured) by a process, procedure, or system must be Retrieved (and optionally Deleted) by at least one other process, procedure, or system. Conversely, any data that is Retrieved, Updated, or Deleted must have been Created by at least one other process, procedure, or system.

Data Definition — A description of a Data Element or a Data Aggregate.

Data Dictionary — A repository or encyclopedia that contains supporting specifications for a Business Model and its implementation.

Data Element — An indivisible item of data that takes on a value (also known as an Attribute or Data Element Primitive).

Data Element Primitive — A data item that does not require any further breakdown in order to be defined.

Data Flow — A conduit for moving information between processes, data stores, or external interfaces.

Data Flow Diagram (DFD) — A model that is both process and data-oriented. This particular model allows the analyst to look at the data and the processing acting on that data as it travels through a system, business, or organization.

Data Integrity — A feature of a Data Element indicating that it is either dependent or independent of another Data Element.

Data Transformation — A process that changes the status or content of an item of data.

Data Triggered Stimulus — An Event Stimulus consisting of a cohesive set of data necessary for processing the event.

Deliverable — A defined result — usually the output of a process or procedure.

Dependent Event — An external incident originating at our our organization's suppliers/vendors that is in response to an outgoing request made by us in order to satisfy one or more Business Events.

Design **(1) as a verb** — The process of inventing a solution to a problem.

(2) as a noun — The solution to a problem or requirement.

(3) as an adjective for data and processes — It indicates that implementation features are specified such as the media and format or the people or program I.D.s.

Dynamically Defined Event An Event that is put together to completely satisfy the specific need of a customer at the time they interface with an organization.

E

Engineering The application of science and mathematics by which the properties of matter and the sources of energy in nature are made useful to man in structures, machines, products, systems, and processes.

Engineering Discipline Applying the concepts of engineering in the production of a product or service.

Entity A cohesive set of related Data Elements captured by an organization.

Entity Relationship Diagram (ERD) A model declaring Entities (cohesive sets of data) and Relationships (the associations between those Entities) that are needed to support an organization's Business Policy. An ERD can show "n-ary" (multiple) relationships as opposed to binary relationships.

Essential Boundary Those boundaries over which an organization has no control (i.e., governed by external Events).

Essential Business Those things that directly satisfy an organization's mission.

Entity Specifications The definitions of cohesive sets of stored data.

External Interface A point beyond the organization where a need occurs to which the organization responds.

F

Flow Chart A model that depicts processing in a linear (flow-of-control) manner.

Fragmented Event A portion of a Business Model Event that has been broken apart or batched to accommodate an existing design.

Fragmented Store A data store that does not contain a complete, cohesive set of data.

Functional Decomposition Diagram A process oriented model that provides a process-only view. These models are a good way to decompose a process or a problem and to break down that process or system from a high-level to a low-level view.

Functional Primitive A detailed-level, specific process that does not require further decomposition to be understood.

H

Hierarchical Relationship Diagram — A data-oriented model that shows the design structure of a manual or automated file system.

I

Implementation — The process of building and testing a design prior to installation.

Information/ C.R.U.D. Event — A Business Event that requires a simple Create, Retrieve, Update, or Delete.

Information/Data Oriented Model — A graphical representation of the static data of a system or an organization.

Internal Interface — A point where a Business Model Event stopped in relation to our area of study (i.e., it's a point where a Business Model Event got fragmented).

L

Leveling — The process of decomposing a single-function process into discrete subprocesses.

M

Maintenance — The clean-up process performed on a production system to remove features that were inserted by error during its development.

Manual — A qualifier for a portion of a system that is implemented with human beings.

Material Triggered Stimulus — The actual products or items that arrive at and stimulate an organization.

Memory — A record of relevant stored data that an organization needs to respond to a stimulus.

Message — The communications between Objects on an Object Oriented Model.

Method — A reusable process acting on shared variables — Data or States— within an Object.

Metrics — Detailed historical records of deliverable related usage of project resources (also known as statistics).

Mission Statement — A brief, clearly stated expression of an organization's reason for being in existence.

Model — A representation to be used as a pattern or guide for conceptualizing, specifying, planning, or executing a system or deliverable.

Modification — The necessary change to a product or system initiated by the Strategic Planners/Business Policy Creators when business needs change.

N

Network Structure Diagram — A data-oriented model used to show the access of one Entity (cohesive set of data) from any other.

Non-essential Boundary — Those boundaries based on some design or implementation or other historical reasons.

Normalization — The process of systematically eliminating data redundancy and dependence within the Data Elements of an Entity.

O

Object — An encapsulation of data and processes.

Object Oriented Model (OO) — A model that declares Object classes and the interactions between those Objects via Messages.

Oragnizational Events — Any of the five Event types (Strategic Events, System Events, Business Events, Regulatory Events, and Dependent Events) to which an organization typically responds.

Organizational Repository — The complete documentation (the set of models and supporting specifications) for all Business Model Events in the organization. When complete, each Business Model Event's entry in the repository will show its essential business view, design view, and implementation view along with the necessary System Events to support the implementation of the Business Model Events.

P

Pre-Engineering — The application of the discipline and the methodology of Customer Focused Engineering to a new organization.

Process — The set of forces, actions, laws, rules, and operations that act on a stimulus, generate the response, and usually alter the state, memory, or material within a system.

Process Hierarchy Diagram — A model used to depict top-down control issues such as showing who is in charge of whom in an organization or the control structure of computer systems to show "boss" modules and "subordinate" modules.

Process Integrity — A feature of a single-function process indicating that all data input into it is utilized and that all data output from it is derived only from its input.

Process Oriented Model — A type of model that focuses exclusively on the processing aspects of a system.

Process Specification The specification of a primitive process that describes the policy or procedure necessary to transform incoming data into outgoing data.

Project Charter A public document that declares the scope and objectives of a specific project.

Project Model A model depicting project control issues such as systems development standards and plans.

Project Objective The directions as to the manner in which a project should be conducted.

Q

Quality Quality is recognized in a product or service when it satisfies both the ethical and measurable requirements of the requester. It is accomplished with pride of ownership on the part of everybody involved in satisfying those requirements. Succinctly, "quality is conformance to requirements."

R

Regulatory Customer This is typically an external governing organization that regulates an organization and with which it has to interface because of the business it is in and/or its location.

Regulatory Event An external incident originating at a governing body that places a demand on our organization to which we respond in order to comply with legal requirements.

Relationship An association between two or more Entities that is important to an organization.

Relationship Specification The optional definitions of the rules of association between Entities.

Reusability A characteristic of data or a process that determines its suitability for use across Events or applications.

Reusable Library A well documented and cataloged collection of quality-engineered processes, objects, or Entities that are available for re-use.

Reusable Process An engineered, single-function process available for use across Events or applications.

S

Seamless Business System A system in which a Business Event Partition is implemented as a unit with no artificial design boundaries or stores.

State Transition Diagram (STD) A control-oriented model used to model the flow of control between the different states of a system.

Strategic Event/ Meta Event An incident, typically originating at our organization's leaders, that makes us change the way we do business and hence change our Business Model in some way.

Strategic Objective A quantifiable and measurable statement of a specific line of business' required performance. An objective contains the "vision" of the Strategic Planners.

Strategic Planning A planning approach aimed at organizational issues that address where the organization is today, where it expects to be in the future and how to get there.

System A collection of Event Partitions (or fragmented partitions) usually based on project resource availability or historical issues.

Systems Analyst A person who studies, models, and documents an organization's Business Policy.

System Builder The persons (Technical Writers and Programmers) who implement an organization's Business Policy.

System Customer These are internal persons, departments, or systems that exist because of the existing structure of an organization.

System Designer The person who designs the implementation of the organization's Business Policy.

System Development A sequence of steps that involves a progression of specifications from gathering objectives in a Project Charter, to identifying requirements in an Analysis Specification, to creating solutions in a Design Specification, and on into building the systems in an Implementation Specification.

System Event An internal incident created during or caused by the design of an organization's structure and systems in order to satisfy some aspect of technology (human or computer).

System Issue A technology aspect that pertains to **how** an organization is designed to run its day-to-day operations.

System Model An implementation view of the Business Model. This is a design tool for declaring how the business runs in the real world.

System Store A memory store that is not shared across Business Model Events. These are invented to satisfy a design (see Convenience Store).

T

Transient Data Flow The intermediate data between two processes within a Business Model Event. The data in a transient data flow are not stored and "burn up" after the processing has terminated.

Appendix B

Bibliography

Chen, Peter. 1976. The Entity Relationship Diagram — Towards a Unified View of Data. The Transactions of the IEEE.

Constantine, Larry, and Yourdon, Ed. 1979. Structured Design. New Jersey. Prentice-Hall.

Crosby, Philip. 1979. Quality Is Free. New York, New York. Mentor.

Ram Dass. 1975. Be Here Now. New York, New York. Crown Publishing.

DeMarco, Thomas. 1978. Structured Analysis and System Specification. New York, New York. Yourdon, Inc.

Dickinson, Brian. 1981. Developing Structured Systems. New York, New York. Yourdon Press.

Dickinson, Brian. 1989. Developing Quality Systems: A Methodology Using Structured Techniques. New York, New York. McGraw-Hill.

Dickinson, Brian. 1991. Strategic Business Engineering: A Synergy of Software Engineering and Information Engineering. Brisbane, California. LCI Press.

Editors, Webster's Encyclopedic Unabridged Dictionary of the English Language. Springfield, MA. G. & G. Merriam Company.

Editors, 1975. A Course on Miracles. Glen Ellen, California. Foundation for Inner Peace.

Gane, C., and Sarson, T. 1977. Structured Systems Analysis: Tools and Techniques. New York. Improved System Technologies.

Hatley, D., and Pirbhai, I. 1987. Strategies for Real-time System Specification. New York, New York. Dorset House.

Humphrey, Watts. 1989. Managing the Software Process. Reading, MA. Addison–Wesley.

McMenamin, Steve, and Palmer, John. 1984. Essential Systems Analysis. New York, New York. Yourdon Press.

Pirsig, Robert. 1991. Lila — an Inquiry into Morals. New York, New York. Bantam Books.

Pirsig, Robert. 1974. Zen and the Art of Motorcycle Maintenance — an Inquiry into Values. New York, New York. Bantam Books.

Santayana, George. 1905. The Life of Reason — Volume 1 – Reason in Common Sense. New York, New York. Charles Scribner & Sons.

Smith, Adam. 1776. The Wealth of Nations.

Taylor, Frederick Winslow. 1911. Scientific Management.

Appendix C

Index

A

A basis for Strategic Planning, 27
Archaeological wrong turns, 41
Asterisks on Data Flow Diagrams, 76

B

Bachman/Martin-style Entity Relationship Diagram, 74
Basis for a competitive edge, 27
Beginnings of the Business Library, 137
Beware of Bundled Events, 144
Beware of Fragmented & Bundled Data Stores, 147
Beware of Historical Events, 146
Bundled and fragmented stores, 130
Bundled data stores, 148
Bundled Events, 144
Bundled stores, 147
Business Customer
  and Business Events, 105
  defined, 105
Business Engineering
  hallmarks, 24
  needs of the customer, 24
Business Event Memory, 127, 131 - 132
  repartitioning, 167
  sources for, 168
Business Event Methodology
  a synthesis, 178
  effect on development schedules, 195
  foundation, 88
Business Event Partition, 117 - 118, 132, 156, 164, 187
  defined, 116
  design options, 200
  policy responses, 163
Business Event Partitioning, 90, 204
  via Strategic Planning, 221 - 232
Business Event Processing, 126
Business Event Recipient, 135
Business Event Response, 135
Business Event Specification, 155 - 176, 213, 217

Business Event Stimulus, 163
  defined, 120
Business Event-driven systems design, 201
Business Events, 94, 105
  as the basis for Customer Focused Engineering, 231
  basis, 116
  composition, 118
  control-oriented stimuli, 124
  data-oriented stimuli, 121
  defined, 103
  example of partition packaging, 204
  external stimulus, 89
  listing, 111
  material-oriented stimuli, 123
  modeling, 90
  naming, 105
  Partitioning, 115 - 116
  recognizing, 103 - 104
  reusable data and processes, 181
  separating from Strategic Events, 98
  Source, 119
  Stimulus, 120
  Strategic Planning, 27
  user view partitioning, 201
Business Information/Data Model, 217
Business Library, 174
  contents, 175
  defined, 174
Business Library Conservator, 181, 185, 187, 191
Business Model, 6, 56, 97
  defined, 6
Business Model Event List, 110
Business Model Event Matrices, 178
Business Model Event Partition, 196
Business Model Event/Data Element Matrix, 187
Business Model Event/Data Entity Matrix, 191
Business Model Event/Engineered System Matrix, 193 - 194
Business Model Event/Relationship Matrix, 192
Business Model Event/Reusable Process Matrix, 181
Business Model Events, 87
  defined, 87
Business Policy Creator, 92, 161, 190
  and the Meta Model, 190
Business Process Model, 217
Business view, iv

C

C.R.U.D. Events, 104, 133, 185, 187
Cardinality, 73
Characteristics of an effective model, 67
Chen-style Entity Relationship Diagram, 74
Choosing the Right Model, 80
Cohesion
    data, 163
    processes, 37
Cohesive sets of stored data, 50, 166
    modeling, 73
Constantine, Larry, 37
Control Flow Diagram
    defined, 80
    Hatley/Pirbhai-style, 80
Control Oriented Models
    defined, 79
Control-oriented Business Event stimuli, 124
Convenience data stores, 129
Convenience Stores and fragmented Events, 129
Creating a Customer Focused Organization via Strategic Planning, 25 - 32
Crosby, Philip, 4
Customer Focused Engineering, 224
    defined, 231
    minimizing disruption, 24, 56
Customer Focused Organizations
    Objectives, 8
    Organizational Goal, 8
    Requirements, 8

D

Data cohesion vs. data conservation, 163
Data conservation, 178, 183, 187
Data Dictionary, 133, 166 - 167, 181, 185, 190, 195
Data Element Specification, 171
Data Elements
    cohesive sets, 50
    conservation, 185
    pools, 132
Data Flow Diagram, 38, 175
    defined, 75
    DeMarco-style, 76
    Gane/Sarson-style, 76
Data integrity, 178, 183
Data normalization, 168
Data ownership, 186
Data reusability, 187
Data-oriented Business Event stimuli, 121
DeMarco, Tom, 38
Dependent Events, 155
    defined, 108

Designing and implementing Business Event Systems, 197 - 220
Discovering the business purpose, 36
Dynamically defined Business Events, 150

E

Elements of Business Event Partitions, 118
Empowerment, 205
Entity Relationship Diagram, 168, 171, 175
    Bachman/Martin-style, 74
    Chen-style, 74
Entity Relationship Diagrams, 73
Entity Specification, 171
Essential boundaries
    defined, 28
Essential data stores, 131, 166
Events, 89
    Business Model, 87
    categorizing, 92, 95
    Organizational, 87
    types, 92, 95

F

False Events, 146 - 147
Flow Charts, 70
Fragmented and Bundled Business Events, 146
Fragmented and bundled Events, 198
Fragmented Business Events
    defined, 143
Fragmented Events, 109, 141
Fragmented stores, 33, 50, 147
Functional Decomposition Diagrams, 71
Functional partitioning, 46, 48, 90, 115, 156, 204

G

Goals for Creating a Customer Focused Organization, 12

H

Hatley/Pirbhai Control Flow Diagram, 80
Hierarchical Data-oriented Models, 74

I

Identifying the customer and their Events, 93
Implementation issues
    removing, 35
Information Events, 104
Information/data-oriented models, 73

K

Kennedy, John F., 31
Key questions in Customer Focused Engineering, 29, 222

L

Logical data stores, 132

M

Manual system design, 201
Martin/Bachman-style data models, 73
Material-oriented Business Event stimuli, 123
Memory, 57, 60
    defined, 59
Meta Model, 185, 188
Mission Statement, 22, 31
    defined, 29 - 30
Modeling
    cohesive sets of data, 73
    data, 68
    one-to-many relationships, 73
    processing, 68
    redundancy, 138, 171
Models
    analysis, 157
    Bachman/Martin-style Entity Relationship Diagram, 74
    Business, 6, 27, 56, 63, 89, 95, 97, 123, 164, 185, 196
    Business Event Level, 157
    business/analysis, 81 - 82
    characteristics of effective, 67
    Chen-style Entity Relationship Diagram, 74
    Control Flow Diagram, 64, 80
    control oriented, 79
    Data Flow Diagram, 64, 75
    Entity Relationship Diagram, 73
    Flow Charts, 70
    Functional Decomposition Diagrams, 71
    graphical vs. text, 63
    Hatley/Pirbhai Control Flow Diagram, 80
    hierarchical data-oriented, 74
    important process model levels, 166
    Information/Data, 64
    information/data-oriented, 73
    level of detail, 66
    Meta, 185, 188
    Object Oriented, 64, 78
    process and data-oriented, 75
    Process Hierarchy Diagrams, 72
    process oriented, 70
    project, 6
    selecting, 63, 80
    source of data, 130
    State Transition Diagram, 64, 79
    System, 6
    system/design, 81 - 82

N

Naming Business Event Memory, 134
Naming Business Event Processing, 127
Naming Business Event Recipients, 136
Naming Business Event Responses, 135
Naming Business Event Sources, 119
Naming Business Events, 105
Naming control-oriented stimuli, 125
Naming data-oriented stimuli, 123
Naming material-oriented stimuli, 124
Naming standard, 181
Non-essential boundaries
    defined, 28

O

Object Oriented Models
    defined, 78
Organizational Business Repository, 175
Organizational Events
    defined, 87
Organizational Repository
    definition, 217
Overcoming the obstacles, 22

P

Partitioning
    defined, 203
    functional, 37
    historical reasons, 42 - 43, 90
Partitioning by Business Events, 115
Problems with existing systems, 33
Process, 57, 60
    defined, 59
Process and data-oriented models, 75
Process Hierarchy Diagrams, 72
Process integrity, 178, 181
Process Specification, 171
Process-memory view, 57, 59, 81
Process-oriented models, 70
Processing reusability, 178
Project Charter, 4, 6, 16, 81
Project Model
    defined, 6

Q

Quality, 4
    after-the-fact inspections, 52
    building in, 23
    chain of dependence, 7, 53
    in reusable components, 179

R

Ram Dass, 34
Recognizing Business Events, 103
Recognizing Dependent Events, 108
Recognizing Regulatory Events, 107
Recognizing Strategic Events, 98
Recognizing System Events, 101
Regulatory Events, 94, 155
    defined, 106
Relationship Specification, 171
Repartitioning Business Event Memory, 167
Requirements, 57
    customer, 145
Reusability, 163
    data, 166
    data naming standard, 186
    incentives, 180
    potential for, 178
    process naming standard, 183
    processing, 161, 178
    processing and data, 181
Reusability Pyramid, 180
Reusable Library, 182, 195

S

Significant vs. trivial Business Events, 104
Specification, 57
State Transition Diagrams
    defined, 79
Stimulus-response, 57 - 58, 60, 81
Stored data between Business Event Partitions, 131
Strategic Events
    defined, 97
    separating from customer issues, 98
Strategic Planning
    defined, 25
Strategic Planning via Business Event Partitioning, 221 - 232
Strategic Planning via Business Events, 27
Structured Design, 37
Subordinate processes, 182
Subpartitioning business processes, 163
Subpartitioning for reusability, 181
Subpartitioning memory, 166
    reasons for, 166
Subpartitioning processing, 156
    reasons for, 160
Superordinate processes, 182
Symbology for stimulating Control Flows, 158
Symbology for stimulating Data Flows, 158

System
  analysis, 55
  determining which to engineer first, 229
  development project characteristics, 6
  interfaces, 37
  stimulus-response, 57
  stores, 128
  understanding, 55
System Customers
  and System Events, 100
System Events, 94, 200
  defined, 101
System Model
  defined, 6
System stores
  required, 200
Systems archaeology, 33

T

Transaction Types and Fragmented Events, 142
Transient data flows, 130 - 131, 164, 190

U

Understanding the nature of systems, 55

W

What obscures the essential business, 35

Y

Yourdon, Ed, 37

The Author as a Stimulus-Response System

For over 30 years I've seen a whole bunch of methodologies and business fads come and go. In the last four years I've had the luxury of contemplating my naval at my home at Lake Tahoe, California and I've looked back on the methodologies that have come and gone along with all of the different flavors of restructuring (such as restructuring organizations by product or region, "just-in-time" inventory, flattening of the hierarchy, etc.). I've seen many organizations spend incredible amounts of money cascading through these various business fads with little (if any) payback. The point is, I truly believe structuring the organization based on Business Event Partitions is the ultimate structure that cannot be beat. It doesn't matter which methodologies (and their models) come along in the future, they all can and should use Business Event Partitioning as their guiding structure for analyzing, designing, and implementing the organization's systems.

Just as the Industrial Revolution paradigm was refined incrementally over the years to arrive at a system that got us to where we are today, I fully expect (and hope) that the Customer Focused Engineering Methodology will be refined constantly. I believe that all work is, in a sense, iterative and that we should constantly be refining our ideas. This book has gone through many iterations and reflects my latest thinking on the *need for* and the *how to* of Customer Focused Engineering. Of course, this refinement can't occur in a vacuum. It depends on the critical thinking and input of my audience including System Engineers, Business Policy Creators, Business Library Conservators, and other champions of change for the better. I refine my thinking based on this input, so I welcome both your questions and comments about what you encounter when implementing the ideas in this book. Please contact me via email at LCI's Web Site.

Brian Dickinson
Logical Conclusions, Inc.
Web: http://www.logical-inc.com